Smart Nanomaterials Technology

Nanotechnology is a rapidly growing scientific field and has attracted a great interest over the last few years because of its abundant applications in different fields like biology, physics and chemistry. This science deals with the production of minute particles called nanomaterials having dimensions between 1 and 100 nm which may serve as building blocks for various physical and biological systems. On the other hand, there is the class of smart materials where the material that can stimuli by external factors and results a new kind of functional properties. The combination of these two classes forms a new class of smart nanomaterials, which produces unique functional material properties and a great opportunity to larger span of application. Smart nanomaterials have been employed by researchers to use it effectively in agricultural production, soil improvement, disease management, energy and environment, medical science, pharmaceuticals, engineering, food, animal husbandry and forestry sectors.

This book series in Smart Nanomaterials Technology aims to comprehensively cover topics in the fabrication, synthesis and application of these materials for applications in the following fields:

- Energy Systems—Renewable energy, energy storage (supercapacitors and electrochemical cells), hydrogen storage, photocatalytic water splitting for hydrogen production
- Biomedical—controlled release of drugs, treatment of various diseases, biosensors,
- Agricultural—agricultural production, soil improvement, disease management, animal feed, egg, milk and meat production/processing,
- Forestry—wood preservation, protection, disease management
- Environment—wastewater treatment, separation of hazardous contaminants from wastewater, indoor air filters.

Azamal Husen
Editor

Nanobiosensors for Agricultural and Other Related Sectors

Springer

Editor
Azamal Husen
Wolaita Sodo University
Wolaita, Ethiopia

Graphic Era (Deemed to be University)
Dehradun, India

ISSN 3004-8273 ISSN 3004-8281 (electronic)
Smart Nanomaterials Technology
ISBN 978-981-96-6498-6 ISBN 978-981-96-6499-3 (eBook)
https://doi.org/10.1007/978-981-96-6499-3

This Springer imprint is published by the registered company Springer Nature Singapore Pte Ltd.
The registered company address is: 152 Beach Road, #21-01/04 Gateway East, Singapore 189721, Singapore

Preface

The agricultural sector is a promising area that plays a key role in determining a country's sustainability and economic growth. The main principle of sustainable agriculture practice, which aims to meet present needs without compromising resources for future generations, relies on two key aspects. First, agricultural practices should be self-sustaining by conserving protective resources, for instance maintaining soil fertility, protection of groundwater, developing renewable energy sources, and finding alternatives to cope with the adverse effects of climate change. Second, sustainability should extend to managing surrounding urban areas through proper sewage waste recycling, fostering rural employment, and contributing to the development of a sustainable rural landscape. Positive agricultural productivity is undermined by issues such as soil erosion, groundwater pollution, river eutrophication, weed proliferation, and resistance to chemical controls, all of which have a negative impact on the environment. Thus, for sustainable agricultural practices, it is crucial to further develop emerging technologies that can boost plant yields without depleting resources, causing pollution and so on.

In this connection, nanotechnology-based sensing has gained significant momentum, offering a wide range of applications in the agricultural sector. While both nanosensors and nanobiosensors fall under the broader category of nanotechnology, they differ in their functionality, applications, and the types of materials they interact with. Nanosensors typically consist of nanomaterials (e.g., carbon nanotubes, quantum dots, or nanowires) and a transducer to convert the signal into a readable form. They do not necessarily require biological components. Nanosensors are used in a wide variety of fields, including environmental monitoring, industrial applications, and security. They can monitor pollutants in the air or water, detect explosives or toxic chemicals, and measure physical parameters like pressure and temperature. Some of the example of nanosensors are sensors that used in environmental monitoring to detect gas concentrations such as carbon dioxide, methane, etc. These types of nanosensors are also used a pressure sensors used in mechanical applications and also sensors for detecting water contamination from chemicals or heavy metals. However, nanobiosensors consist of three components. The first component is the biologically sensitized element (probe), which includes biologically derived substances, namely receptors, enzymes, antibodies, nucleic acids, molecular imprints, lectins, tissues, microorganisms, organelles, and others. These elements, or bio-mimetic constituents, capture signals from the target sample and transmit them to the transducer. The second component is the transducer, which serves as an interface. It detects the physical changes that occur during the reaction at the bioreceptor or sensitive biological element and then converts this energy into a measurable electrical output. The third component is the detector, which captures the signals from the transducer. These signals are then sent to a microprocessor, where they are amplified, analyzed, and converted into a user-friendly output. The data can be displayed or stored for further use. Nanobiosensors can be used to detect a comprehensive range of fertilizers, herbicides, pesticides, and insecticides as well as pathogens, fungicides, moisture, soil pH, soil contaminants, etc. They are also useful in bioremediation, recycling of agricultural wastes, wastewater treatment and disinfection.

Nanobiosensors, when integrated with satellite farming systems, can provide localized, real-time information at the molecular level. For instance, nanosensors deployed in the soil or on the plants can monitor specific variables like nutrient levels, pest activity, or the presence of pathogens. These sensors could be linked to the satellite system, providing a comprehensive overview of farm health and enabling more accurate decision-making. The combination of satellite farming, site-specific crop management, and nanobiosensors in precision agriculture can offers significant potential to improve efficiency, reduce costs, and promote sustainability. Nanobiosensors can provide real-time, localized data on plant health, soil conditions, and environmental factors, allowing farmers to make more informed decisions and tailor their practices to specific needs. This advanced approach not only boosts plant yield and quality but also ensures that farming practices are more environmentally responsible and economically viable. However, there are several challenges to overcome such as regulatory issues, environmental, and technical obstacles, the future of nanosensors in the agricultural sector holds immense potential for transforming farming practices, making them more efficient, sustainable, and environmentally friendly.

Taken together, the aim of the book is to provide an overview on the most important and selected topics related to nanosensors in agricultural system and they are going to be explored. This book will provide valuable information to scientists, researchers, and upper level of students, working specially in agricultural sectors, crop science, plant nanobiotechnology, plant biology, plant biochemistry, plant microbiology, and other allied subjects and or science. I am grateful to all contributors for readily accepting my invitation, sharing their knowledge in specialized areas of research, and readily adjusting the suggestions for improving the shape of their contributions. Finally, my special thanks go to Shagufta, Zaara, Mehwish, and Huzaifa for providing their time and overall extended support to put everything together.

Wolaita, Ethiopia Azamal Husen

Contents

Nanosensors in the Agricultural System—the Current Understanding, Role and Prospects 1
Patriciani Estela Cipriano, Matias Siueia Júnior, Deivisson Ferreira da Silva, Geslin Mars, Rodrigo Fonseca da Silva, and Flávio Henrique Silveira Rabêlo

Precision Agriculture and Nanosensor 33
Vikas Sharma, Animesh Gosh Bag, Nitin M. Changade, and Neeraj Singh Parihar

Nanosensors, Climate Change and Agricultural Sectors 49
Maryam Zakavi and Shahnoush Nayeri

Role of Chemical Nano Sensors for Alleviation of Sustainable Agricultural and Biomass Production 65
Walaa Yasir, Nadia Sharif, Mona F. A. Dawood, Irfan Aziz, and Zainul Abideen

Nanosensors and Their Role in Soil Quality, Moisture, pH, Contaminants, and Other Molecules Detection 101
Patriciani Estela Cipriano, Matias Siueia Júnior, Deivisson Ferreira da Silva, Geslin Mars, Rodrigo Fonseca da Silva, and Flávio Henrique Silveira Rabêlo

Nanosensors for Detection of Agricultural Soil Microbes 137
Ishaan Lakhani, Proxima Dhiman, Vikas Dutt, Hina Alim, Ahmad Ali, and Nimisha Patel

Nanosensors and Their Role in Bioremediation 155
Shrejal Tiwari and Anant

Nanosensors for Heavy Metal Pollution Detection in Agricultural System 175
Shefali Singh, Smita Rai, Hiba Khan, Mariyam Shakeel, Archana Vimal, Reena Vishvakarma, Poonam Sharma, and Swati Sharma

Gold Nanomaterials and Composites Based Nanosensors for Agricultural Sectors 193
Aisha Kamal, Ayeesha Khatoon, Nida Sultan, and Sazia Siddiqui

Silver Nanomaterial and Composite-Based Nanosensors for the Agriculture Sector 221
Sneh Lata and Nasreen Ghazi Ansari

Zinc Oxide Nanosensors in Precision Agriculture: A Green Approach to Improve Global Food Security 247
Swati Sachdev and Mohammad Israil Ansari

Various Metal-Oxide Based Nanosensors for Agricultural Practices 261
Himshikha Malviya, Diksha Singh, and Jay Singh

Nanosensors in Plant Disease Diagnosis and Management 287
Sadia Haider, Pakeeza Afzal, Sabir Hussain, Faisal Mahmood, Tanvir Shahzad, Muhammad Sadiq, Hassan Nawaz, Awais Maqsood, and Muhammad Shahid

Nanosensors for Wastewater Treatment and Disinfection 307
R. Ridhi

Nanosensors and Their Role in Recycling of Agricultural Wastes 331
Mohammad Areeb Siddiqui, Reena Vishvakarma, and Archana Vimal

About the Editor

Azamal Husen served as a Full Professor and Head of the Department of Biology at the University of Gondar, Ethiopia, and currently working as a Foreign Delegate at Wolaita Sodo University, Ethiopia. He is also an Adjunct Professor at Graphic Era University, India. Previously, he was a Full Professor at Sankalchand Patel University in India and a Visiting Faculty at the Forest Research Institute and Doon College of Agriculture and Forest in Dehra Dun, India. With 25 years of expertise in Nanobiotechnology, Plant Physiology, Environmental Botany, Medicinal Plants, and Clonal Propagation. He has led research funded by organizations such as the World Bank, the National Agricultural Technology Project, the Indian Council of Agricultural Research, the Indian Council of Forest Research Education, and the Japan Bank for International Cooperation. He has delivered numerous talks at conferences and workshops, led academic staff orientation courses, and developed B.Sc., M.Sc., and Ph.D. curricula. He was honored with an award from the University of Gondar for his decade of service in teaching, research, and community work. Husen has published extensively and served on editorial boards of prestigious journals. Acknowledged as one of the "World's Top 2% Scientists" (2022–2024) by Stanford University, he has also contributed as a Book Series Editor and Journal Guest Editor for several scientific publications in Plant Biology, Medicinal Plants, and Nanotechnology.

Nanosensors in the Agricultural System—the Current Understanding, Role and Prospects

Patriciani Estela Cipriano, Matias Siueia Júnior, Deivisson Ferreira da Silva, Geslin Mars, Rodrigo Fonseca da Silva, and Flávio Henrique Silveira Rabêlo

Abstract Agriculture is undergoing a revolution thanks to the integration of nanotechnology, an innovative field that uses materials and devices at the nanoscale. Nanotechnology transforms agricultural practices and enhances precision agriculture by addressing food security, environmental sustainability, and effective resource management. Nanosensors are analytical devices designed to monitor parameters related to soil and plants. These nanoscale sensors provide real-time information on moisture, nutritional status, pH, and microbial activity. Their high sensitivity and specificity enable early detection of plant diseases and environmental stresses, enabling rapid interventions that minimize crop losses and maximize productivity. In addition, they simplify soil quality assessment and pollutant monitoring, encouraging sustainable and environmentally responsible practices. Nanofertilizers and nanopesticides improve the absorption of nutrients and agrochemicals, reducing losses and environmental impact. Nanoclays and nanozeolites improve soil water retention, which is essential in areas with water scarcity. These advances aim to

P. E. Cipriano (✉) · D. F. da Silva
Department of Soil Science, Lavras School of Agriculture, Federal University of Lavras, Lavras, Minas Gerais, Brazil
e-mail: patricianiestela@gmail.com

P. E. Cipriano · F. H. S. Rabêlo
Luiz de Queiroz College of Agriculture, University of São Paulo, Piracicaba 13418-900, Brazil

M. Siueia Júnior
Laboratory of Plant Nutritional Physiology, Graduate School of Integrated Sciences for Life, Hiroshima University, Higashi-Hiroshima, Japan

College of Agricultural Sciences, Aquila University, Maputo, Mozambique

D. F. da Silva
Instituto Federal Catarinense—Campus Araquari, Santa Catarina, Brazil

G. Mars
Department of Soil Science and Agricultural Engineering, Ponta Grossa State University, Ponta Grossa, Brazil

R. F. da Silva
Department of Agricultural Sciences, Federal University of Piauí, Piauí, Brazil

A. Husen (ed.), *Nanobiosensors for Agricultural and Other Related Sectors*, Smart Nanomaterials Technology, https://doi.org/10.1007/978-981-96-6499-3_1

increase agricultural production without expanding land or harming ecosystems. Nanosensors improve food safety and quality by identifying pollutants, toxins, and pathogens in processed products. Rapid analysis capabilities facilitate food safety monitoring, ensuring regulatory compliance and increasing consumer confidence. Adopting nanotechnology faces high costs, regulatory challenges, and a need for more awareness. Research aims to make these tools more accessible, especially in food-insecure developing regions. Nanotechnology is a game-changing innovation for sustainable agriculture. The approach optimizes resources, improves crop resilience, and ensures food quality, meeting the global population's needs while preserving the environment. In this chapter, we explore nanosensor applications in the agricultural system, highlighting the current understanding, significant role, and prospects of nanotechnology in this field.

Keywords Food security · Nanofertilizers · Nanopesticides · Sustainable agriculture

1 Introduction

Agriculture faces many challenges worldwide, especially in achieving future food availability and sustainability goals that respond to the population growth estimated to be 9.9 billion by 2050 [181]. Based on these facts, a new agri-tech industry revolution called nanotechnology is emerging to sustain global food demand, and Richard Feynman proposed it in 1959 [19].

The word 'nano' derives from the Greek word 'dwarf' which means a one-billionth segment of a meter, whereas for this proposed Nanotechnology "smart agriculture" refers to using structures and devices with an average size between 1 and 100 nm [19], emerged as a promising approach to develop advanced sensors for various applications, including environmental monitoring, healthcare, and agriculture [58].

The advancement in nanotechnology led to a boom in nanomaterial applications, which depicts its potential to overcome these issues [218]. Nanomaterials are also utilized for various applications, ranging from introductory physics and materials science to medicine and information technology [76, 77, 85].

Nanotechnology, an innovative and interdisciplinary scientific approach, involves designing, formulating, and applying instruments, devices, and systems at the nanometer scale. This complexity and depth offer several advantages over conventional methods, such as real-time monitoring, high spatial resolution, low energy consumption, and integration with wireless communication technologies [156]. Leveraging unique phenomena, this approach enables novel applications [151] and holds significant potential for promoting greener, more sustainable agriculture [85].

According to reports, the earliest application of nanotechnology in agriculture was in 2004 for crop growth, animal breeding, and aquaculture. However, after that period, reports have been published on a wide range of potential applications in the agro-environmental sector, such as agrochemical delivery [79], nano fertilizers [55], stress tolerance [221], nano pesticides [217], soil protection [115], food packaging [5], and contaminant removal [93], amongst others.

The application of the nanotechnology approach in agriculture has demonstrated tremendous potential for crop management, one of the proposed tools for sustainable agricultural intensification and the production of superior products that are highly beneficial in agriculture, water, environment, drug development, and health [92]. Moreover, nanotechnology's potential in agriculture is being explored to reduce the use of chemicals for plant protection chemicals, minimize nutrient losses, enable rapid pest detection, and enhance yields through improved nutrient management [64]. This potential for sustainable agriculture is a beacon of hope for the future of farming.

Additionally, nanotechnology has demonstrated potential in addressing various food-related challenges, including nutrient delivery, protein bioseparation, rapid detection of chemical and biological contaminants, solubilization, and nutraceutical nanoencapsulation. Food nanotechnology leverages nanocarrier processes and technologies to enhance the bioactivity of ingredients, improve their biological availability, and provide protection against environmental or chemical changes [140].

Nanotechnology has numerous applications in agriculture, such as: (i) using nanoclays and nanozeolites to enhance soil's water retention [167], providing a slow-release mechanism for water to reduce drought periods during the growing season and improve herbicide delivery, (ii) deploying nanosensors for soil analysis, water management, and optimizing pesticide and nutrient use; (iii) utilizing nanomagnets to remove soil contaminants; (iv) applying nanoparticles (NPs) in the development of advanced insecticides, pesticides, and repellents; and (v) using nanomaterials to enable the controlled release of genetic material for crop improvement.

Consequently, it can positively affect the agricultural sector by improving crop production and food security while promoting social and economic equity [143], These technical innovations in agriculture are essential to combating the worldwide challenges of population growth, climate change, and restricted food availability by promoting the sustainable agricultural intensification that aims to increase food production from the same existing farmland without adverse environmental impacts [190]. In this chapter, we present an overview of different approaches, strategies, and applications of nanosensors in the agricultural environment, practically in any plant species, and we provide past and recently developed technologies that have been applied.

2 Nanotechnology and Nanosensores in Agriculture

Agriculture greatly benefits from practical nanotechnology, where it was possible to develop or design devices to help reduce biotic and abiotic stress in plants before they are severely affected and reduce plant productivity [123]. This component or technology, the nanosensor or biosensor, is defined in this chapter as a compact analytical device that consists of biological constitutes, viz, proteins, oligonucleotides, cells, and tissues, which are utilized for the production of output or signal that humans can use, and selective transducer with a characteristic dimension that is nanometre in scale [89].

Even though biosensors have existed since the early twentieth century, their applications were limited to labs only [193]. Currently, among all the approaches, a biosensor seems like a modest and compact investigative device that can produce definite systematic data either quantitatively or semi-quantitatively by employing a recognition component of biological origin joined to a signal transformation unit.

According to [37], recent advances in nanotechnology have opened various new ways to design biosensors, such as the hybridization of nano-materials with different biosensing devices (nano-bio sensors) offers many conjoining and multipurpose approaches for enhanced sensitivity for detection [198], and the nanomaterial employed in sensing is called a nanosensor constructed at the atomic scale for data collection.

Nanomaterials can be transformed into data that can be analyzed for various applications, such as monitoring physical and chemical indicators in hard-to-reach areas, detecting biological chemicals in different cellular organelles, and identifying nanoscale particles in both environmental and industrial settings [179]. Nanosensors are capable of detecting even a single virus particle or substances in extremely low concentrations. A nanosensor consists of a bio-sensitive layer covalently bonded to a transducer. When the target analyte interacts with the bioreceptor, the resulting physiochemical change is converted into an electrical signal [164].

Nanomaterial-based biosensors now have exciting opportunities compared to conventional biosensors. Nano-biosensors have distinct benefits, such as increased identification specificity, and have vast potential for application in various areas, such as bioprocess and environmental management, irrigation, food quality control, biodefense, and, most notably, medical applications. However, in this instance, the role of nanobiosensors in agriculture and agricultural products is of interest. The nanosensor has several advantages over conventional biosensors, as shown in Fig. 1.

3 Nanobiosensores and Agriculture

Based on the increasing population in the world, this results in a high demand for more food supplies, which leads to a continuous advance in agricultural production, prioritizing sustainable agriculture. Addressing these concerns seems a difficult challenge,

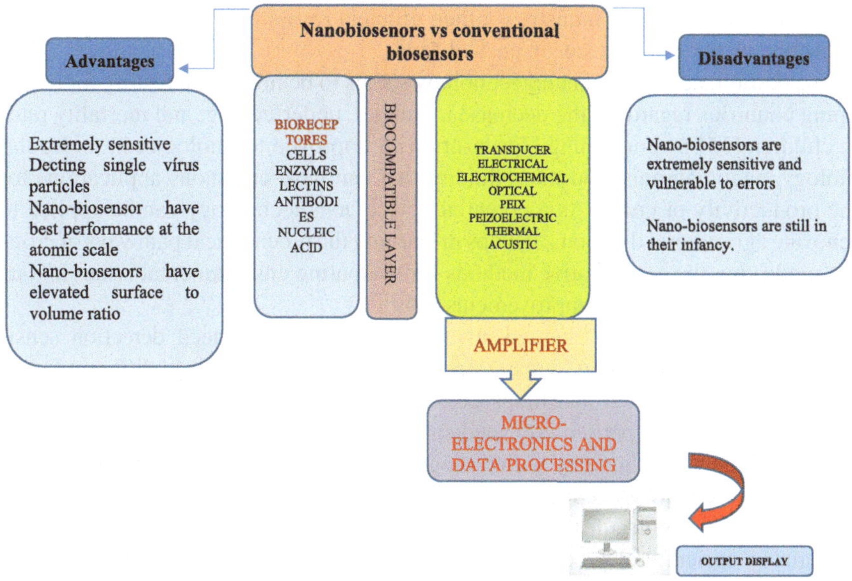

Fig. 1 Diagrammatic representation of major components of biosensor and Advantages and Disadvantages of nanobiosensor versus conventional biosensor

given the continuous population growth [197], profound changes in the demand for food and the threat of extensive migration of rural youth in search of a better life [124]. As a result, scientists and engineers are already practicing new methods to increase agricultural productivity [18]. However, agriculture nanotechnology has focused on research and application to resolve agriculture and environmental issues such as sustainability, crop improvement, and enhanced productivity.

Although different types of nanosensors have also been utilized in plants, including plasmonic nanosensors, Förster resonance energy transfer (FRET)-based nanosensors, carbon-based electrochemical nanosensors, nanowire nanosensors, and antibody nanosensors have allowed the study of cellular functions [99] and metabolic flux [15], the monitoring of spatiotemporal dynamics of analytes [191], and the detection of viral and fungal pathogens [45]. Several reviews have been published on applying nanosensors in agriculture and plant sciences. For instance, this review focused on the specific applications of nanotechnology, e.g., to agriculture [222].

Nanosensors are used to enhance the efficiency of natural agricultural practices by monitoring nutrients, water, and chemicals through precision farming techniques. This includes tools such as satellite monitoring, geographic systems, and remote sensing technologies that can detect pests on crops or signs of stress like drought [167]. The use of autonomous sensors linked to GPS for real-time monitoring is expected to play a crucial role in nanotechnology-driven agricultural tools [35]. Nanosensors can be arranged throughout the field to monitor crop growth and soil parameters. Nanosensors are getting attention and playing a significant role in the development of the food industry for their efficient responding tools to detect gases, microbes, or toxic substances in packed foods.

In addition, agricultural nanotechnology seems to be highly interesting for developing countries regarding the decrease in hunger, underfeeding, and mortality rates in children [65] by combining biosensors with improved technologies in molecular biology, nanomaterials, and microfluidics that can have enormous applications for the productivity of crops. As a potential device, nanotechnology can be applied to renovate agricultural divisions, it helps in learning the biochemical pathways of crops via modifying the conservative methods for evaluating environmental issues and its application to production improvements [149].

Nanobiossensors have marked advantages such as enhanced detection sensitivity/specificity and possess great potential for their applications in different fields, including environmental and bioprocess control, quality control of food, agriculture, biodefense, and, particularly, medical applications. They are innovative devices that combine nanotechnology with biological recognition elements, with the ability to detect and quantify chemical or biological substances [183]. These devices can measure specific molecules in biological samples, giving data that is necessary for optimizing agricultural production and ensuring food quality, from a measurable signal, typically electrical, optical or mechanical, that provides real-time data on the target analyte [130]. Due to their nanometric size, these sensors offer high sensitivity and specificity, which makes them ideal for applications in various areas such as agriculture, health, environmental monitoring and food safety. In this way, agriculture tends to improve every day, to meet the global demand for agricultural products.

In agriculture, nanobiosensors have shown their important role in improving precision in agriculture, as they can facilitate the monitoring of soil health by detecting pH, moisture, nutrients and microbial activity [216]. In addition, they can detect diseases in plants at early stages, allowing for rapid and more accurate actions to avoid significant loss of production [201]. Nanobiosensors have been testified for detecting pathogens in processing plants or alerting clients, protocols, and providers on the safety position of food [28]. It has also been employed for food impurities, mycotoxins, and microbes [183]. Allergens have also been detected via biosensor tools with the assistance of nanoparticles, and the report is about to be commercialized [209]. Considering their unique properties, derived from nanomaterials, they improve sensitivity and specificity, making them valuable tools for modern agriculture.

Another application of nanobiosensors in agriculture is soil quality monitoring. These sensors can accurately detect nutrient deficiencies in crops, environmental conditions such as soil moisture and water quality, and track the presence of pollutants or harmful chemicals in the environment [189]. Unlike traditional methods that usually require larger samples and longer laboratory processing times, nanobiosensors deliver fast results, allowing professionals and farmers to make informed decisions quickly. The use of these real-time nanobiosensors can allow monitoring of various properties of soil, such as temperature, pH, soil moisture content, soil microenvironment, and nutrient status [130]. Nanobiosensors can enhance productivity by improving fertilization management, reducing input costs, and protecting the environment when combined with precision farming.

Despite their promise, the adoption of nanobiosensors can face challenges such as the need for regular calibration, initial costs and their advancement in developing countries. However, ongoing research and technological advances should improve their effectiveness and accessibility in agricultural practices, as they have the potential to revolutionize sustainable agriculture and contribute to global food security in an ever-changing environment. There is a need to raise awareness of the potential benefits of nanobiosensors. Ultimately, nanobiosensors represent a significant advance in agriculture by providing accurate and timely data on nutrient levels and plant health, optimizing the use of resources and minimizing waste.

4 Nanotechnologies and Precision Agriculture

Employing nano-based technologies such as nanocarriers, nanocatalysis, nanobiosensors, nanopesticides, and various others can revolutionize the agricultural sector. These innovations enable precision agriculture, enhance nutritional quality, boost crop yields, and reduce toxicity levels [163]. Precision agriculture is revolutionizing the agricultural landscape worldwide, unlocking many benefits that enhance productivity, improve crop quality, boost profitability, promote sustainability, safeguard the environment, ensure food safety, elevate on-farm quality of life, and stimulate rural economic development. Precision agriculture is now an essential component of planning to achieve this goal [221].

According to [132], implementing nanoparticle-mediated materials and advanced biosensors in precision agriculture solely relies on nanochips or nanosensors. These advancements enhance data collection, enable precise decision-making, and optimize resource management in agricultural systems by effectively measuring various stress factors, soil quality parameters, and heavy metal detection.

Each type of nanomaterial exhibits distinct physicochemical properties that make it particularly well-suited for a wide range of sensing techniques [48]. The most frequently utilized nanoparticles in precision agriculture include silver nanoparticles (Ag NPs), zinc oxide nanoparticles (ZnO NPs), titanium dioxide nanoparticles (TiO_2 NPs), carbon nanotubes (CNTs), quantum dots (QDs), and nanorods [133]. Moreover, nanobased materials have significantly advanced the capabilities of biosensors and analytical instruments, enabling the precise identification of biological components through signals generated by transducers [89].

In agricultural research, nano-biosensors made from materials such as copper (Cu), silver (Ag), and gold (Au) have demonstrated precision and accuracy in monitoring pathogens, nutrient loss, moisture variations, fertilizer distribution, pH stability, and elemental concentrations. These advancements significantly enhance agricultural productivity [54]. Ragaei and Sabry [155] noted that nanoparticles affect plant growth and metabolism differently. Nanoencapsulation plays a critical role in environmental protection by preventing harmful compounds from evaporating and seeping into the environment. To address this challenge successfully, creating more effective, non-persistent insecticides, especially those featuring controlled-release formulations, is crucial. Researchers can leverage this data to predict disease symptoms and determine crops' nutritional and water requirements [121].

Nanotechnology is revolutionizing precision agriculture through various interventions, including nutrient enhancements like nano-fertilizers and nano-biofertilizers, improved seed germination and crop growth, NPs for effective pest and disease management, and innovative nanoherbicides. Additionally, it plays a significant role in post-harvest applications, plant genetic manipulation, and the development of advanced nanosensors such as eletronic noses (e-Nose), and eletronic tongues (e-Tongue), and smart dust (Table 1). Nano-barcodes are also utilized alongside bioremediation techniques to tackle persistent agrochemicals [133].

Nanosensors are revolutionizing precision agriculture through innovative applications. For instance, near-infrared fluorescent H_2O_2 nanosensors can detect early signs of plant stress [210]. Red selenium (Se) NPs produced via bioreduction also serve as effective optical biosensors for testing heavy metal toxicity [6]. Furthermore, nanosensor-based devices, such as e-Nose and e-Tongue, are employed to identify a variety of gases, flavors, and aromas, along with their fluctuating concentrations. Wireless nanosensors are deployed in agricultural fields to collect real-time data on crop health, soil quality, and weather conditions [170]. These advanced technologies enable smart irrigation, intelligent fertilization, effective pest control, and comprehensive greenhouse monitoring [132].

Electrochemical sensors play a crucial role in gathering soil chemical information by identifying specific ions within the soil [132]. These sensors effectively measure parameters such as soil pH and nutrient concentrations [153], as well as detect elements like nitrogen, phosphorus, and potassium (NPK) [26]. Additionally, they can identify pesticide residues [192], pathogenic *Staphylococcus aureus* [22], chloramphenicol [213], and various N species [162]. In the future, deploying nanostructured catalysts will revolutionize fertilizer applications, enhancing their efficacy in soil treatments and crop spraying while allowing for reduced dosages [163]. Moreover, integrating nanotools with biosensors will significantly amplify the sensitivity of advanced agricultural practices, leading to improved outcomes in these applications [58].

5 Nanotechnology in Crop Protection

According to Hajji-Hedfi and Chhipa [69], the conventional use of pesticides and chemicals, known for their detrimental impact on the environment, remains common in managing plant pests and diseases. To mitigate the risks of toxicity to both humans and the environment, there has been a growing focus on developing pesticide formulations that are effective, safe, and environmentally sustainable. Recent technological advancements have led to the development of nanopesticides, which offer targeted pesticide delivery and present a potentially environmentally friendly solution for sustainable agriculture. High adsorption, decreased volatilization, enhanced

Table 1 Nanobiosensor application in agriculture

Nano biosensor	Application	References
Disease assessment and soil quality	Weak microbes and the other impregnate the elements by strong microbes with an oxygen consumption of two microns in the suspension of a soil sample in a buffer solution. Comparing them makes it easy to determine which microbe promotes the soil	
Promoting sustainable agriculture	Nano-fertilizers contain nutrients stored and distributed to crops in a nanoparticle to minimize nanofertilizers' leaching, emission, and long-term absorption of nitrogen: Improve the growth of plants, Improve the quality and value of fertilizers, improve the penetration and retention of water	[38, 94, 111]
Detection of contaminants	Developed for identifying dryness, temperature or stress contaminants, pests, nutrient levels, and plant stresses. They can also enable agriculturists to develop their skills only if necessary by using inputs	[169, 196, 223]
Detection of protein and DNA	Ultra-sensitive identification of special protein molecules by the use of protein linkage-interaction propensities. These biosensors can be useful for the detection of plant parasites, plant defects, biomarkers and for distinguishing between one plant species and another	[60, 224]
Analyzing the food products	Analysis of Vitamin Detection of Antibiotics Microbial Infection Detection	

tissue penetration, and controlled release are some of the additional benefits that nanopesticides have demonstrated above traditional pesticides [84].

The nanopesticide aims to release the ideal quantity of its active component formulations in a precise or regulated manner in response to environmental changes [226]. Numerous nanopesticide formulations are currently in development, aimed at achieving the controlled release of various organic and inorganic components for effective pest control strategies [59, 176]. Several key factors propel the advancement of nanotechnology in pesticides, including decreased pesticide usage, enhanced solubility, controlled release mechanisms, targeted delivery systems, and greater stability of active ingredients [82].

As an example of studies with nanopesticides, the work of Zhao et al. [226, 227] with several plants, including *Zea mays*, *Cucumiss nativus*, and *Ocimum basilicum*, demonstrated that they exhibited metabolic responses to nanopesticides. [187] observed the effects of $Cu(OH)_2$ nanowires on Cu transport and the accumulation of several biochemical substances, including pigments, essential oils, and fatty acids of two types of *O. basilicum*. Kah et al. [82] found that clothianidin encapsulated using polymeric nanoparticles had a better pesticide half-life, photodegradation, and sorption rate than pure clothianidin.

Ojha et al. [137] highlight a range of nanobased systems designed for crop protection. These include encapsulated nanosystems, such as biopolymers, synthetic polymers, polymeric-based nanomaterials, lipid-based and inorganic porous nanomaterials, and clay-based options. Additionally, the study discusses nonencapsulated nanopesticides featuring silver, ZnO, silica, Cu, TiO_2, and MgO nanoparticles. The diversity extends to nano-based systems like nanocomposites, all aimed at enhancing crop protection. Encapsulation protects light-sensitive pesticides from degradation, limiting evaporation [1]. Nanocapsule formulations enhance pesticide droplets, improving resistance to environmental factors and increasing wettability and targeted adsorption [136].

According to Ojha et al. [137], nanomaterials can penetrate plants and facilitate the controlled release of agrochemicals in several modes. These modes include an array of advanced polymers responsive to various stimuli, such as ultrasound, light, redox/thiol conditions, magnetic fields, enzymatic activity, antigen–antibody interactions, electric fields, and variations in pH and temperature.

Research has demonstrated the potential of nanosensors as practical diagnostic tools for identifying pesticides, pests, and pathogens in crops and plants. For example: 2,4,5-Trichlorophenoxy acetic acid (systemic herbicide) [91], fenitrothion in water [103], melamine in milk [101], aflatoxin (AFM1) in milk [141], organochlorine and organophosphorus pesticides in cabbage [228], dimethoate [61], methyl parathion in water [142], *Bacillus cereus* in diverse food samples [139], methyl parathion in vegetables (cabbage, spinach, lettuce) [83], methyl parathion, chlorpyrifos [202], fenamithion and acetamiprid in water [154], herbicide metsulfuron-methyl, an acetolactate synthase (ALS) inhibitor [31], parathion in water [203], dichlorvos and paraoxon in drinking water [196], parathion in vegetables [109], pirimicarb in vegetables [184], imazaquin, metsulfuron-methyl and atrazine (herbicide detection) [36], 2,4-Dichlorophenoxyacetic acid (2,4-d) (herbicide) [21, 200], atrazine (herbicide) [114, 188], dichlorodiphenyltrichloroethane (pesticide) [11], insecticide cartap (nereistoxin derivative), OP neurotoxin [177], paraoxon [78, 112] and OP compounds [113].

Numerous NPs exhibit promising antimicrobial properties against various phytopathogens. Silver NPs have demonstrated efficacy against pathogens such as *Bipolaris sorokiniana* and *Magnaporthe grisea* [81], *Raffaelea sp.*, the oak wilt pathogen [97], and *Colletotrichum gloeosporioides* [4]. They are also effective against other fungal threats like *Colletotrichum sp.*, which causes pepper anthracnose [106], *Powdery mildews* [107], and a range of pathogens including *Alternaria alternata*, *Sclerotinia sclerotiorum*, *Macrophomina phaseolina*, *Rhizoctonia solani*,

Botrytis cinerea, and *Curvularia luneta* [100]. Furthermore, they target *Alternaria* sp., including *A. alternata, Alternaria brassicicola, and Alternaria solani*, as well as *Cladosporium cucumerinum, Corynespora cassiicola, Cylindrocarpon destructans, Didymella bryoniae*, various *Fusarium sp., Glomerella alternata, Monosporascus cannonballus*, and *Pythium* sp. [96]. Additional pathogens such as *Xanthomonas campestris* pv. malvacearum [158], *Colletotrichum coccodes, Monilinia* sp., and *Pyricularia* sp. [108], along with *Citrobacter freundii* and *Erwinia cacticida* [146], and *Bipolaris sorokiniana* [127] are also impacted by these NPs.

The Ag-silica hybrid complex enhances the expression of defense-related genes, specifically markers for systemic acquired resistance [30]. Additionally, nanosized silver and silica effectively combat pathogens such as *Pythium ultimum, Magnaporthe grisea, Botrytis cinerea, Rhizoctonia solani*, and *Colletotrichum gloeosporioides* [144]. The silver-chitosan nanocomposite demonstrates efficacy against *B. cinerea* Pers [131]. Colloidal Ag NPs are effective against *R. solani, S. sclerotiorum, and Sarracenia minor* [126].

Nanoparticles of ZnO are effective against *Botrytis cinerea* and *Penicillium expansum* [70], as well as *Fusarium graminearum* [42]. Sulfur NPs demonstrate efficacy against *F. solani* and *Venturia inaequalis* [160], as well as *Drosophila melanogaster* [12]. Surface-modified sulfur nanoparticles have been shown to impact *Aspergillus niger*and *Fusarium oxysporum* [29]. Silver and Cu NPs are effective against the pathogens *A. alternata and B. cinerea* [138].

6 Nanotechnologies and Soil Remediation

The persistence of potentially toxic elements (PTEs) in agroecosystems is challenging for current and future generations. This problem is caused mainly by the inadequate disposal of industrial and urban waste and the use of low-quality agricultural inputs. These PTEs cause significant and long-lasting damage, interfering with the cellular activities of living organisms [178]. Nanosensors, have emerged as an important tool for detecting and quantifying the presence of these PTEs in samples, as well as determining their activity, concentration and other relevant parameters [171].

Nanotechnology sensors have been developed to monitor PTEs in the soil in real time, allowing early identification of polluted areas and remediation of the situation. Carbon nanotube sensors have been developed to detect lead (Pb) ions in soil, with a detection accuracy of 0.1 nM and a fast response in less than 10 s [51].

The synthesis of multimodal nanosensors using nanochemistry concepts is highly attractive, as it not only efficiently detects but also removes PTEs ions in aqueous solutions. Satapathi et al. [166] developed a multimodal nanosensor composed of a thin silica layer surrounding magnetic Fe_2O_3 NPS, an immobile spacer arm, and a fluorescent quantum dot for identification and removal of mercury (Hg) ions, with a detection limit at nanomolar levels.

An optical nanosensor made with CdSe quantum dots could detect cadmium (Cd), restoring its green luminescence in the presence of the metal [116]. Graphene sensors were used to identify Cd ions, with a sensitivity of 0.01 nM and a high selectivity about other metals [215]. Vahab and Keshispour [194] developed a nanosensor with a graphene quantum dot modified with Eriochrome Black T to detect cobalt (Co) and Cu. Both cations were successfully detected, and the linear ranges obtained were 0.02–0.08 M and 0.015–0.2 M, with the detection limit of 0.014 and 0.018 M, respectively.

Another interesting approach involves using metal oxide nanosensors, which have proven effective in environmental remediation. Metal oxide nanosensors, in various forms, including chemical, gaseous, biological, optical, and humidity sensors, have been widely studied to identify environmental pollutants and volatile organic compounds in the environment [62]. In order to improve the stability, sensitivity, and response time of metal oxide nanosensors, one alternative is to combine them with other organic counterparts, thus creating a nanocomposite that not only improves the stability of the sensors but also results in a synergistic effect that is essential for effective detection [63]. Bao et al. [17] constructed a nanosensor modified with carbon nanotubes with chitosan and graphene oxide multilayers to rapidly detect Cu and Hg ions, with detection limits of 1.57×10^{-7} M and 4.98×10^{-6} M, respectively.

7 Nanotechnologies and Water Quality Assessment

Water quality is a critical factor in environmental preservation, public health, and the sustainable management of water resources. With the growing challenges of drinking water scarcity and increasing water pollution, nanotechnology has emerged as a promising and innovative tool for assessing and monitoring water quality. Unlike traditional methods, which are often time-consuming and lack sensitivity to detect low concentrations of pollutants, nanosensors provide precise and efficient solutions for real-time water quality monitoring [199].

One of the most significant advantages of nanotechnology is its ability to enable real-time analysis. Nanosensors integrated with electronic devices provide continuous monitoring of water quality in rivers, lakes, and supply systems. This technology facilitates the early detection of contaminants or issues, enabling prompt and effective interventions to prevent environmental harm and protect human health (Fig. 2).

Nanotechnology sensors utilize materials at the nanoscale to detect chemical and biological contaminants in water. These sensors can be engineered to target specific pollutants, including potentially toxic elements (PTEs), pesticides, and pathogens. Selecting the appropriate nanomaterial is critical to ensuring optimal sensor performance, with key factors such as sensitivity, selectivity, and stability carefully considered based on the contaminants to be detected. Sensor design also plays a vital role and involves configuring the sensing platform, which may operate on electrochemical, optical, or mass-sensing principles. The choice of design depends on the intended application and the required level of sensitivity (Fig. 3).

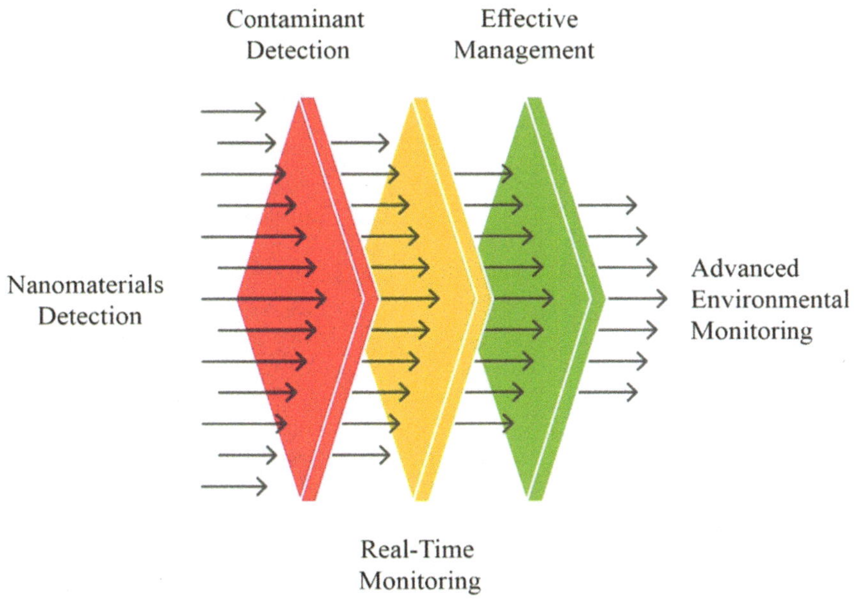

Fig. 2 Water quality monitoring method using nanotechnology sensors in real time

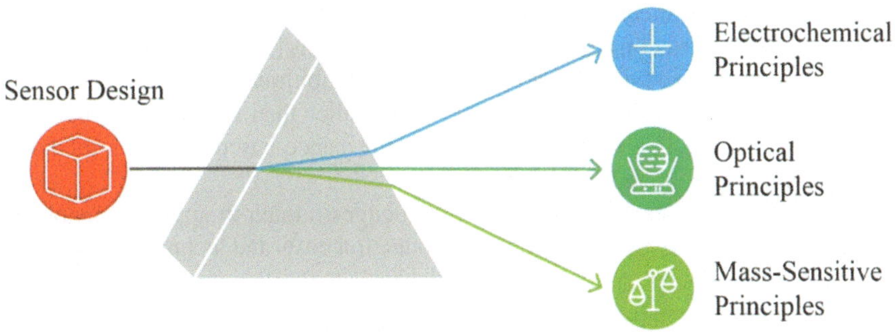

Fig. 3 Different designs of nanotechnology sensors based on specific operating mechanisms

The use of nanotechnology-based sensors for pollutant detection represents one of the most advanced areas of environmental monitoring. These nanosensors can detect trace amounts of heavy metals, pesticides, volatile organic compounds, and pathogens with remarkable precision, often at concentrations that conventional methods cannot reliably measure [7].

In addition to their use in sensors, nanomaterials have been widely applied in water separation and purification processes (Fig. 4). Metallic NPs, such as Au and Ag, are commonly utilized for their unique optical and electronic properties, which make them effective for contaminant detection [14]. Carbon nanotubes, known for their

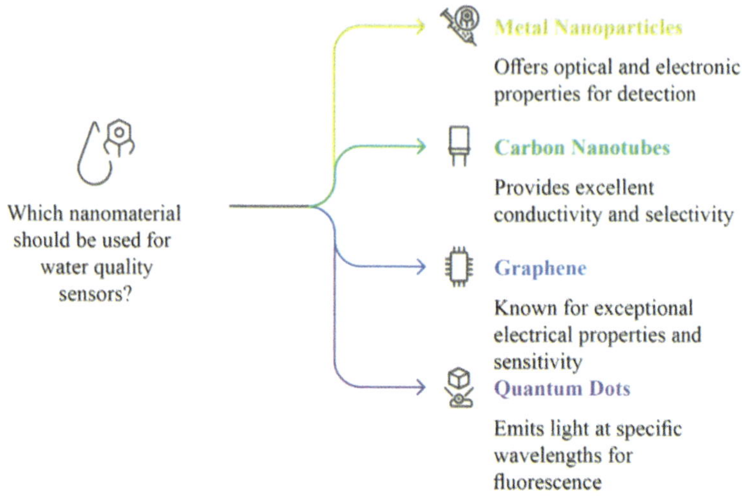

Fig. 4 Different materials used in the construction of nanotechnological sensors

exceptional conductivity, can be functionalized to enhance selectivity for specific analyses. They are also highly efficient in filtering contaminant particles and offer superior durability compared to conventional materials [128]. Similarly, graphene, with its exceptional electrical properties, is employed to create highly sensitive sensors capable of detecting a wide range of contaminants [7]. Additionally, quantum dots, which are nanocrystalline semiconductors, exhibit quantum confinement due to their small physical dimensions (1–10 nm, smaller than the Bohr radius of an exciton). This property enables their use in fluorescence-based detection methods, offering high precision and sensitivity [73].

Despite its immense potential, several challenges remain in applying nanotechnology to water quality assessment. High production costs and technical complexity continue to hinder large-scale adoption, particularly in developing countries. Furthermore, the potential environmental and human health impacts of nanomaterials in aquatic systems require thorough investigation.

Establishing clear regulations and standardized protocols is essential to ensure the safe and effective use of nanotechnology in water management. Greater investment in research and development, coupled with partnerships among academic institutions, government agencies, and private enterprises, can accelerate the adoption of these innovative solutions.

Looking ahead, integrating nanotechnology with artificial intelligence and sensor networks holds the promise of revolutionizing water monitoring. Intelligent systems powered by nanosensors could predict water quality trends, facilitate real-time decision-making, and enable more proactive and sustainable water management practices.

8 Nanotechnologies and Food Security

Food contamination can be classified into three main groups: toxins, infectious agents (such as bacteria and viruses), and chemical contaminants (including drugs, antibiotics, and food additives). Food can be contaminated at various stages, including cultivation, handling, storage, and food preparation, and this highlights food safety as a fundamental aspect to consider. Excessive consumption of antibiotics in foods of animal origin and contamination by chemical residues in agricultural products can seriously harm human health [27]. According to [188], pesticides can be classified into four main categories: carbamates, pyrethroids, organophosphates, and organochlorines. Hu et al. [75] developed a colorimetric sensor capable of detecting dimethoate in tomatoes, cucumbers, and cabbage, simulating Au NPs activity through peroxide inhibition.

Toxins, which include marine biotoxins, mycotoxins, and cyanogenic glycosides, occur naturally in certain foods, such as poisonous mushrooms [27]. Mycotoxins are widely recognized as ubiquitous contaminants in cereals [8]. According to [110], fungi such as *Fusarium*, *Aspergillus*, and *Penicillium* produce mycotoxins, secondary compounds resulting from their metabolism. Due to their carcinogenic nature, these agents can contaminate a wide range of foods. Microorganisms, such as bacteria and viruses, are important sources of food contamination. By contaminating food and water, these microorganisms have the potential to cause disease in humans [8].

Ensuring food safety is essential today, especially in a world where the population is growing. Daily monitoring of food is widely recognized as essential to ensure its safety and quality, allowing the identification of toxins, bacteria, and other hazardous substances [161]. Nanosensors have generated great enthusiasm compared to traditional techniques, as they allow rapid and local analysis, eliminating the need for high voltages or inert gases. Bulbul et al. [23] explored the use of portable biosensors incorporating Au, Ag, and magnetic NPs, as well as TiO_2 nanocrystals, to detect food contaminants. These pollutants include microorganisms, pesticides, heavy metals, and mycotoxins, achieving accuracy even at extremely low concentrations.

Traditional methodologies have several limitations, including gas chromatography, liquid chromatography, mass spectrometry, and high-performance liquid chromatography (HPLC). The main challenges include long processing times, high costs, complexity in sample preparation, operational difficulties, the need for specialized professionals, and high reagent consumption. Nanobiosensors have emerged as a revolutionary solution to overcome these limitations [46], and this has resulted in a significant transformation in the food industry, promoting the adoption of agile, accurate, and financially sustainable practices [3].

Nanobiosensors, recognized for their remarkable stability, selectivity, sensitivity, and portability, have been significantly improved by integrating nanomaterials with biosensors and developing analytical measurement methods [220]. According to [159], nanobiosensors demonstrate remarkable efficacy in detecting various analytes, such as pesticides and antibiotics, and ensure food safety and quality. Nanomaterials have improved chemical, optical, and electrical characteristics, making them extremely attractive for the development of nanobiosensors (Fig. 5). Thus, the solution for several advantages, such as rapid detection, ease of use, portability, and reliability, lies in the use of sensors developed with nanotechnology [161].

According to [195], certain nanosensors detect the gases released by food and accelerate its deterioration. Carbon dioxide (CO_2) sensors can detect the onset of deterioration by measuring the amount of CO_2 in a package. They are also essential for storage and packaging in controlled and modified atmospheres. Therefore, these devices have immense market potential for the food and agricultural industries [134].

Milk is a product that requires special attention. Jin et al. [80] developed an innovative piezoelectric immunosensor using Au NPs to detect aflatoxin B1 in milk.

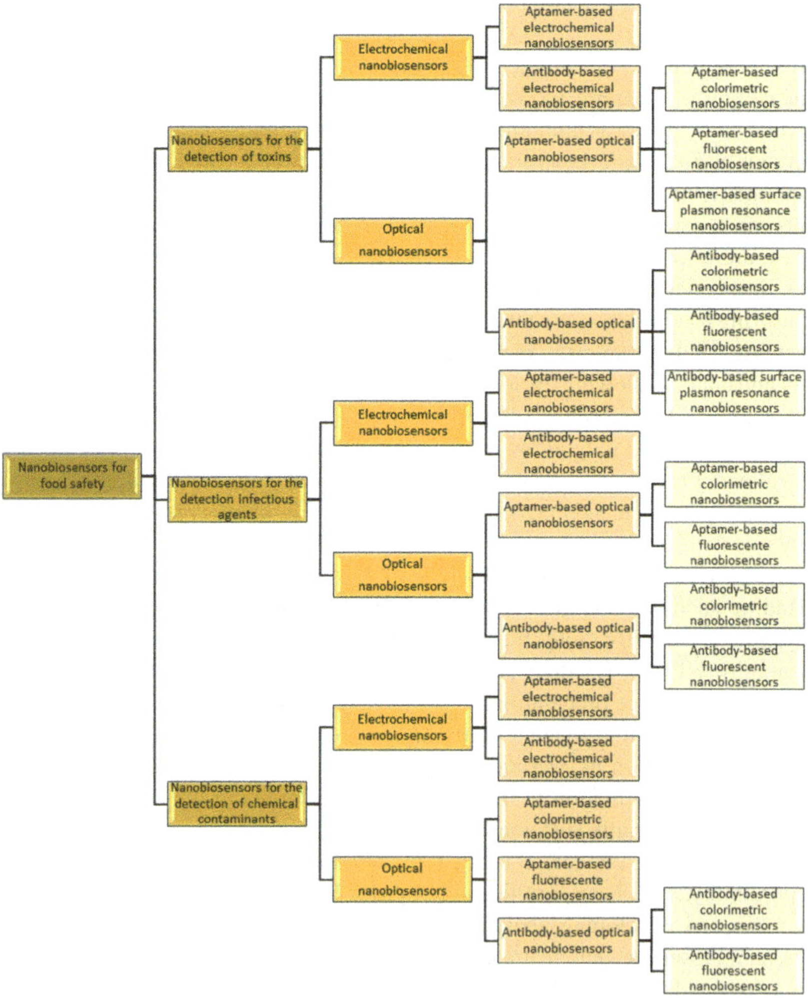

Fig. 5 Types of nanosensors for food safety

In addition, Dinckaya et al. [43] created an impedimetric biosensor based on Au NPs specifically to identify aflatoxin M1 in milk and dairy products. In addition, Kasoju et al. [88] also introduced a paper-based colorimetric microfluidic technique, employing unmodified Au NPs to monitor the presence of aflatoxin M1 in milk samples.

Global industries are turning to the implementation of these advanced detection techniques as they offer faster, more effective, and more reliable detection [72]. Carbon nanotubes, graphene, magnetic nanoparticles NPs, QDs, and AuNPs are examples of nanomaterials used in various biosensing and sensor applications. Focusing on AuNPs, a nanomaterial that has garnered great interest due to its unique properties, researchers have exploited these characteristics in various sensing applications, as evidenced in Table 2 [34].

9 Nanotechnologies and Biofortification

Biofortification increases the concentration of micronutrients in staple foods and can be implemented through various approaches such as agronomic biofortification, genetic biofortification, and genetic modification [95]. Agronomic biofortification enriches food crops with micronutrients using foliar spraying, seed treatment, soil fertilizers and nutrient solution immersion of seedlings [40].

Nanoparticles exhibit properties, such as a high surface area to volume ratio and dissolution profiles that can be adjusted as needed [90]. Foliar spraying or soil NPs application can enhance growth, increase crop productivity, and increase plant micronutrient levels [174]. Nutrients can be managed with a controlled release via NPs, reducing leaching and increasing accumulation in edible tissues [87].

Nanoparticles with specific structures and properties can be a safe alternative for food biofortification [47]. Nanoparticles can be synthesized by chemical and biological methods or physical processes [13]. Nanomaterial biosynthesis uses microbial or plant strategies to transform elements and oxides into nanoscale materials [122]. Microorganism-based methods use algae, fungi, actinomycetes, and bacteria, while plant-based methods use tissues such as fruits, sap, stems, bark, roots, leaves, and agricultural residues [119]. Plant-assisted methods use phytochemicals in the preparation of NPs and as stabilizers. These approaches are sustainable, cost-effective, efficient, rapid, and less toxic [87]. Biosynthetic approaches have limitations, with batch variability being the main shortcoming [16].

Nanoparticles offer innovative solutions for the controlled release of micronutrients for plant growth [145]. Nanofertilizers are an effective alternative to combat micronutrient deficiency [41]. Nanofertilizers can be enhanced with nanoclays, zeolites, and encapsulation, improving soil fertility, plant growth, and micronutrient levels [148]. The most frequently studied nanofertilizers for the biofortification of food crops include zinc (Zn), iron (Fe), Cu, and Se [150].

Table 2 Gold nanoparticle (AuNP) sensors are used in the food industry

Class	Application	Detection mechanism	References
Food additive	Hydrazine	UV– visible, colorimetry, HPLC, GC, LCMS and Electrophoresis	[44]
	Hydrazine, sulfite, nitrite		[118]
	Nitrite		[129]
	Metanil yellow		[172]
	Fasta green		
	Tartrazine		[214]
Pathogens	*Escherichia coli*	Electrochemical, Lateral flow immune assay aptamer-nanoZyme and electrical assay	[9, 33, 159, 209]
	Salmonella strains		[204]
	Shigella flexneri		[56]
	Escherichia coli DH5		[117]
Antibiotics	(fluoro)quinolones	Surface plasmon resonance, Immunoassay	[147]
	β-agonist		[71],
	Kanamycin		[2, 183]
	Diclazuril		[208]
	Aminoglycoside antibiotics		[212]
Food spoilage	Xanthine	Amperometric and fluorescence assay	[39]
	Histamine		[20]
	Biogenic amines in fish		[49]
Toxins	Deoxynivalenols (DONs), fumonisins (FBs), T-2 toxins (t-2 s), zearalenones (ZEAs) and aflatoxins (Afs)	Rabbit pyrogen test, chromogenic assay, turbidity method, solid phase extraction	[98]
	Aflatoxin B1		[173]
	Staphylococcus enterotoxin B (SEB)		[214]
	Afloxin B1, aflatoxin G1, aflatoxin M1, ochratoxin A and zearalenone		[175]
Heavy metals	Mercury	Fluorescence quenching assay	[32]
Adulterant	Ferulic acid	Colorimetric and aptamer based assay	
	Melamine		[25]
	Urea		[102]
Pesticide	Chlorpyrifos (CP), azinphos-methyl (AM), pirimiphos-methyl (PM), fenamiphos (FP), and phosalone (PS)	Colorimetric and fluorescence quenching	[52]
	Dithiocarbamates		[168]

Adapted of Rastogi et al [161]

Nanoparticles of ZnO stand out for their remarkable plant absorption, accumulation, and metabolization capacity, which is attributed to their high surface area-to-volume ratio and makes them essential in combating Zn deficiency [125]. However, γ-Fe_2O_3 NPs are more efficient in transferring nutrients from roots to other parts of plants due to their reduced volatility and smaller nanometric size than granulated iron oxide (FeO) [10]. According to [67], the use of FeO NPs in seed treatment emerges as an innovative and sustainable approach. It offers an efficient Fe solution supply and promotes the biofortification of food crops.

According to [211], copper oxide (CuO) NPs are used as fertilizers, plant growth regulators, and soil remediation additives. Kapoor et al. [87] highlighted the importance of adequate nutrition and nanoscale platforms to effectively deliver micronutrients such as Cu in the early stages of plant growth and disease response. However, high levels of CuO can result in phytotoxicity, compromising plant growth and reducing water availability [211], and it may occur due to the release of Cu ions, which can disrupt redox reactions in cells [207].

When used as nano fertilizers, Se NPs demonstrate significantly higher availability due to their high surface area-to-volume ratio. This characteristic results in remarkable improvements in antioxidant, antimicrobial, and anticancer properties, surpassing the Se effects in its raw form [74]. It rapidly binds to ligands and acts as a powerful heavy metal detoxification agent. Selenium NPs have proven to be an economically viable alternative for crop biofortification. They also contribute to optimizing the production of biomolecules such as proteins, amino acids, phenolic compounds, and glucosinolates [24].

Innovative approaches, such as nanotechnology, including the use of chelators and zeolites, can achieve biofortification. Chelators are small molecules that establish strong bonds with metal ions, becoming effective soil amendments. They improve the availability of metal ions through two or more coordinated bonds, which results in greater availability and absorption by plants [57]. Some chelators, such as ethylenediaminetetra acetic acid (EDTA), are synthetic, while others can be produced biochemically, such as transferrins and phytochelators [135]. Nanoparticles combined with nanochelators play a significant role in micronutrient biofortification strategies [53]. The main characteristics of nanochelators include a low affinity for harmful metals, a high affinity for essential metals, and remarkable lipid solubility [86].

Zeolites show significant potential as fertilizers for food crops [86]. Zeolites have a three-dimensional crystalline structure with remarkable porosity and a large surface area, ranging from 500 to 800 m^2g^{-1}. This characteristic allows for highly efficient cation exchange and preserving negatively and positively charged nutrients for an extended period [68]. Zeolite nanoparticles can release elements in a controlled manner, promoting more efficient absorption by plants and preventing loss in the system [50]. Zeolites have great potential to carry essential nutrients, such as N and K [50]. Zeolite nanocomposites enriched with NPK, when associated with other micronutrients, appear to be promising innovative mineral fertilizers. Control of the release profile can be optimized through ball milling, facilitating the availability of desired nutrients for plants [219]. Nanozeolites emerge as an innovative and promising fertilizer capable of stimulating plant growth and promoting the biofortification of food crops [120].

10 Conclusions and Future Perspectives

Nanoparticles facilitate the production of precise nanosensors for soil condition assessment, identifying diseases, and gauging factors like acidity, wetness, and sustenance. These sensors improve how we use farm supplies and lessen the damage to our planet by keeping soil good and clean water. In biofortification, nanotechnology has escalated the accessibility of trace nutrients in principal sustenances, employing nanofertilizers and nanoparticles possessing distinct traits to enhance plant absorption.

Additionally, it must integrate at least two statistics related to malnutrition. Tiny particles called nanomaterials help clean up dirty earth, making bad land better and less destructive for nature. 'A potential use lies in agriculture, involving the genesis of micronized insecticides and micronized compounds that provide regulated discharge and increased effectiveness against the harmful agents.' Cutting down on harmful substances and lessening the danger to our surroundings. "Supplement, real-time surveillance facilitated by micro-devices incorporated into smart systems has transformed watering and nutrient application, enhancing sustainability and yield."

Nanotechnology Sentence: In the future, a mix of tiny tech, innovative computer thinking, and smart-sensing gadgets will make farming even better. Let us use modern tech to handle natural stuff better, make farming better, and handle weather changes easier. Continuous research and development in this field is essential to ensure these innovations are widely accessible and environmentally safe. Nanotechnology, therefore, represents a strategic pillar for the future of global agriculture, reconciling productivity, sustainability, and food security.

Competing Interests The authors declare that they have no competing interest.

Declaration of Generative AI and AI-Assisted Technologies in the Writing Process During the preparation of this work the book chapter author(s) have not taken any assistance from AI, and the book chapter author(s) reviewed and edited the content as needed and take(s) full responsibility for the content of the publication. All authors read and approved the final version of the manuscript.

References

1. Abas Shah M, Wani SH, Akhtar AK (2016) Nanotechnology and insecticidal formulations. J Food Bioeng Nanoproc 1(3):285–310
2. Abedalwafa MA, Tang Z, Qiao Y, Mei Q, Yang G, Li Y, Wang L (2020) An aptasensor strip-based colorimetric determination method for kanamycin using cellulose acetate nanofibers decorated DNA–gold nanoparticle bioconjugates. Microchim Acta 187(6)

3. Adley C (2014) Past, present and future of sensors in food production. Foods 3(3):491–510
4. Aguilar-Méndez MA, San Martín-Martínez E, Ortega-Arroyo L, Cobián-Portillo G, Sánchez-Espíndola E (2011) Synthesis and characterization of silver nanoparticles: effect on phytopathogen *Colletotrichum gloesporioides*. J Nanopart Res 13(6):2525–2532
5. Ahmad A, Qurashi A, Sheehan D (2023) Nano packaging—progress and future perspectives for food safety, and sustainability. Food Packag Shelf Life 35:100997
6. Ahmed F, Dwivedi S, Shaalan NM, Kumar S, Arshi N, Alshoaibi A, Husain FM (2020) Development of selenium nanoparticle based agriculture sensor for heavy metal toxicity detection. Agriculture 10:610
7. Ahmed YM, Eldin MA, Galal A, Atta NF (2024) Electrochemical sensor based on PEDOT/CNTs-graphene oxide for simultaneous determination of hazardous hydroquinone, catechol, and nitrite in real water samples. Sci Rep 14:5654
8. Alamer S, Eissa S, Chinnappan R, Herron P, Zourob M (2018) Rapid colorimetric lactoferrin-based sandwich immunoassay on cotton swabs for the detection of foodborne pathogenic bacteria. Talanta 185:275–280
9. Ali MA, Eldin TAS, Moghazy GMEI, Tork IM, Omara II (2014) Original research article detection of *E. coli* O157: H7 in feed samples using gold nanoparticles sensor. Int J Curr Microbiol Appl Sci 3(6):697–708
10. Alidoust D, Isoda A (2013) Effect of γFe_2O_3 nanoparticles on photosynthetic characteristic of soybean (*Glycine max* (L.) Merr.): foliar spray versus soil amendment. Acta Physiol Plant 35:3365–3375
11. Alvarez M, Calle A, Tamayo J, Lechuga LM, Abad A, Montoya A (2003) Development of nanomechanical biosensors for detection of the pesticide DDT. Biosens Bioelectron 18(5):649–653
12. Araj SEA, Salem NM, Ghabeish IH, Awwad AM (2015) Toxicity of nanoparticles against *Drosophila melanogaster* (Diptera: Drosophilidae). J Nanomater 5
13. Arole VM, Munde S (2014) Fabrication of nanomaterials top-down and bottom-up approaches—an overview. JAAST Mater Sci (Special Issue) 1:2–89
14. Azeez L, Lateef A, Olabode O (2023) An overview of biogenic metallic nanoparticles for water treatment and purification: the state of the art. Water Sci Technol 88(4):851–873
15. Bagal-Kestwal D, Kestwal RM, Chiang BH (2015) Invertase-nanogold clusters decorated plant membranes for fluorescence-based sucrose sensor. J Nanobiotechnol 13:30
16. Baig N, Kammakakam I, Falath W, Kammakakam I (2021) Nanomaterials: a review of synthesis methods, properties, recent progress, and challenges. Mater Adv 2:1821–1871
17. Bao Q, Li G, Yang Z, Pan P, Liu J, Li R, Wei J, Hu W, Cheng W, Lin L (2021) In situ detection of heavy metal ions in sewage with screen-printed electrode-based portable electrochemical sensors. Analyst 146(18):5610–5618
18. Baruah S, Dutta J (2009) Nanotechnology applications in pollution sensing and degradation in agriculture: a review. Environ Chem Lett 7:191–204
19. Bayda S, Adeel M, Tuccinardi T, Cordani M, Rizzolio F (2020) The history of nanoscience and nanotechnology: from chemical-physical applications to nanomedicine. Molecules 25(1):1–15
20. Bi J, Tian C, Zhang GL, Hao H, Hou HM (2020) Detection of histamine based on gold nanoparticles with dual sensor system of colorimetric and fluorescence. Foods 9(3)
21. Boro RC, Kaushal J, Nangia Y, Wangoo N, Bhasin A, Suri CR (2011) Gold nanoparticles catalyzed chemilumi-nescence immunoassay for detection of herbicide 2, 4-dichlorophenoxyacetic acid. Analyst 136(10):2125–2130
22. Braiek M, Rokbani KB, Chrouda A, Mrabet B, Bakhrouf A, Maaref A, Jaffrezic- Renault NJB (2012) An electrochemical immunosensor for detection of Staphylococcus aureus bacteria based on immobilization of antibodies on self-assembled monolayers-functionalized gold electrode. Biosensors 2:417–426
23. Bulbul G, Hayat A, Andreescu S (2015) Portable nanoparticle-based sensors for food safety assessment. Sensors (Switzerland) 15(12):30736–30758

24. Carvalho KM, Gallardo-Williams MT, Benson RF, Martin DF (2003) Effects of selenium supplementation on four agricultural crops. J Agric Food Chem 51:704–709
25. Chang K, Wang S, Zhang H, Guo Q, Hu X, Lin Z, Sun H, Jiang M, Hu J (2017) Colorimetric detection of melamine in milk by using gold nanoparticles-based LSPR via optical fibers. PLoS ONE 12(5):1–12
26. Chauhan N, Urooj S (2019). The effect of Azadirachta Indica extract on the soil nutrients and the NPK value determination by electrochemical sensor. Int J Biosens Bioelectron 5
27. Chen H, Zhou K, Zhao G (2018) Gold nanoparticles: from synthesis, properties to their potential application as colorimetric sensors in food safety screening. Trends Food Sci Technol 78:83–94
28. Cheng MM-C, Cuda G, Bunimovich YL, Gaspari M, Heath JR, Hill HD, Mirkin CA, Nijdam AJ, Terracciano R, Thundat T (2006) Nanotechnologies for biomolecular detection and medical diagnostics. Curr Opin Chem Biol 10:11–19
29. Choudhury SR, Ghosh M, Mandal A, Chakravorty D, Pal M, Pradhan S, Goswami A (2011) Surface-modified sulfur nanoparticles: an effective antifungal agent against *Aspergillus niger* and *Fusarium oxysporum*. Appl Microbiol Biotechnol 90(2):733–743
30. Chu H, Kim HJ, Kim JS, Kim MS, Yoon BD, Park HJ, Kim CY (2012) A nanosized Ag–silica hybrid complex prepared by γ-irradiation activates the defense response in Arabidopsis. Radiat Phys Chem 81(2):180–184
31. Da Silva AC, Deda DK, da Róz AL, Prado RA, Carvalho CC, Viviani V, Leite FL (2013) Nanobiosensors based on chemically modified AFM probes: a useful tool for metsulfuron-methyl detection. Sensors 13(2):1477–1489
32. Darbha GK, Ray A, Ray PC (2007) Gold nanoparticle-based miniaturized detection of mercury in soil, water, and. ACS Nano 1(3):208–214
33. Das A, Kamle M, Bharti A, Kumar P (2020) Nanotechnology and it's applications in environmental remediation: an overview. Vegetos 32(3):227–237
34. Das M, Shim KH, An SSA, Yi DK (2011) Review on gold nanoparticles and their applications. J Toxicol Environ 3(4):193–205
35. Davari MR, Kazazi SB, Pivehzhani OA (2017) Nanomaterials: implications on agroecosystem. Nanotechnology. In: Prasad R, Kumar M, Kumar V (eds) Nanotechnology. Springer, Singapore, pp 59–71
36. Deda DK, Pereira B, Bueno CC, Silva AND, Ribeiro GA, Amarante AM, Franca EF, Leite FL (2013) The use of functionalized AFM tips as molecular sensors in the detection of pesticides. Mater Res 16(3):683–687
37. Dehghani S, Nosrati R, Yousefi M, Nezami A, Soltani E, Taghdisi SM, Abnous K, Alibolandi M, Ramezani M (2018) Aptamer-based biosensors and nanosensors for the detection of vascular endothelial growth factor (VEGF): a review. Biosens Bioelectron 110:23–37
38. DeRosa MC, Monreal C, Schnitzer M, Walsh R, Sultan Y (2010) Nanotechnology in fertilizers. Nat Nanotechnol 5:91
39. Dervisevic M, Dervisevic E, Çevik E, Şenel M (2017) Novel electrochemical xanthine biosensor based on chitosan–polypyrrole–gold nanoparticles hybrid bio-nanocomposite platform. J Food Drug Anal 25(3):510–519
40. Dimkpa CO, Bindraban PS (2016) Fortification of micronutrients for efficient agronomic production: a review. Agron Sustain Dev 36:1–26
41. Dimkpa CO, Bindraban PS (2018) Nanofertilizers: new products for the industry? J Agric Food Chem 66:6462–6473
42. Dimkpa CO, McLean JE, Britt DW, Anderson AJ (2013) Antifungal activity of ZnO nanoparticles and their interactive effect with a biocontrol bacterium on growth antagonism of the plant pathogen *Fusarium graminearum*. Biometals 26(6):913–924
43. Dinckaya E, Kınık O, Sezginturk MK, Altu C, Akkoca A (2011) Development of an impedimetric aflatoxin M1 biosensor based on a DNA probe and gold nanoparticles. Biosens Bioelectron 26:3806–3811
44. Dorostkar S, Hemmateenejad B (2013) Label-free colorimetric detection of picomolar amounts of hydrazine using a gold nanoparticle based assay. Jiran Chem Soc 10(3):513–519

45. Duhan S, Kumar R, Kumar N, Kaur P, Nehra K, Duhan S (2017) Nanotechnology: the new perspective in precision agriculture. Biotechnol Rep 15:11–23
46. Eivazzadeh-Keihan RP, Pashazadeh R, Hejazi M, de la Guardia M, Mokhtarzadeh A (2017) Recent advances in nanomaterial-mediated bio and immune sensors for detection of aflatoxin in food products. CTrAC Trends Anal Chem 87:112–128
47. Elemike EE, Uzoh IM, Onwudiwe DC, Babalola OO (2019) The role of nanotechnology in the fortification of plant nutrients and improvement of crop production. Appl Sci 9:1–32
48. Elizabath A, Babychan M, Mathew AM, Syriac GM (2019) Application of nanotechnology in agriculture. Indian J Pure Appl Biosci 7:131–139
49. El-Nour KMA, Salam ETA, Soliman HM, Orabi AS (2017) Gold nanoparticles as a direct and rapid sensor for sensitive analytical detection of biogenic amines. Nanoscale Res Lett 12(1)
50. Eroglu N, Emekci M, Athanassiou CG (2017) Applications of natural zeolites on agriculture and food production. J Sci Food Agric 97:3487–3499
51. Eustis S, Krylova G, Eremenko A, Smirnova N, Schill AW, El-Sayed MA (2005) Growth and fragmentation of silver nanoparticles in their synthesis with a FS laser and CW light by photo-sensitization with benzophenone. Photochem Photobiol Sci 4(1):154–159
52. Fahimi-Kashani N, Hormozi-Nezhad MR (2016) Gold-nanoparticlebased colorimetric sensor array for discrimination of organophosphate pesticides. Anal Chem 88(16):8099–8106
53. Fakharzadeh S, Hafizi M, Baghaei MA, Etesami M, Khayamzadeh M, Kalanaky S, Akbari ME, Nazaran MH (2020) Using nanochelating technology for biofortification and yield increase in rice. Sci Rep 10:1–9
54. Fatima F, Pathak N, Verma SR, Bajpai P (2018) Toxicity and immunomodulatory efficacy of biosynthesized silver myconanosomes on pathogenic microbes and macrophage cells. Artif Cells Nanomed Biotechnol Rep 46:1637–1645
55. Fellet G, Pilotto L, Marchiol L, Braidot E (2021) Tools for nano-enabled agriculture: fertilizers based on calcium phosphate, silicon, and chitosan nanostructures. Agronomy 11:1239
56. Feng J, Shen Q, Wu J, Dai Z, Wang Y (2019) Naked-eyes detection of *Shigella flexneri* in food samples based on a novel gold nanoparticle-based colorimetric aptasensor. Food Control 98:333–341
57. Flora SJS (2015) Arsenic: chemistry, occurrence, and exposure. Handbook of arsenic toxicology. Elsevier, pp 1–49
58. Fraceto LF, Grillo R, De Medeiros GA, Scognamiglio V, Rea G, Bartolucci C (2016) Nanotechnology in agriculture: which innovation potential does it have? Front Environ Sci 4:20
59. Gahukar RT, Das RK (2020) Plant-derived nanopesticides for agricultural pest control: challenges and prospects. Nanotechnol Environ Eng 5(1):3
60. Galandova J, Ziyatdinova G, Labuda J (2008) Disposable electrochemical biosensor with multiwalled carbon nanotubes-chitosan composite layer for the detection of deep DNA damage. Anal Sci 24(6):711–716
61. Gan N, Yang X, Xie D, Wu Y, Wen W (2010) A disposable organophosphorus pesticides enzyme biosensor based on magnetic composite nano-particles modified screen printed carbon electrode. Sensors 10(1):625–638
62. Ganie AS, Bano S, Khan N, Sultana S, Rehman Z, Rahman MM, Sabir S, Coulon F, Khan MZ (2021) Nanoremediation technologies for sustainable remediation of contaminated environments: recent advances and challenges. Chemosphere 275:130065
63. Ganie AS, Bano S, Sultana S, Sabir S, Khan MZ (2021) Ferrite nanocomposite based electrochemical sensor: characterization, voltammetric and amperometric studies for electrocatalytic detection of formaldehyde in aqueous media. Electroanalysis 33:233–248
64. Ghormade V, Deshpande MV, Paknikar KM (2011) Perspectives for nano-biotechnology enabled protection and nutrition of plants. Biotechnol Adv 29(6):792–803
65. Gogos A, Knauer K, Bucheli TD (2012) Nanomaterials in plant protection and fertilization: current state, foreseen applications, and research priorities. J Agric Food Chem 60:9781–9792

66. Guan H, Du S, Han B, Zhang Q, Wang D (2023) A rapid and sensitive smartphone colorimetric sensor for detection of ascorbic acid in food using the nanozyme paper-based microfluidic chip. LWT 184:115043
67. Guha T, Mukherjee A, Kundu R. (2021) Nano-scale zero valent iron (nZVI) priming enhances yield, alters mineral distribution and grain nutrient content of *Oryza sativa* L. cv. Gobindobhog: a field study. J Plant Growth Regul 1–24
68. Guo J (2004) Synchrotron radiation, soft-X-ray spectroscopy and nanomaterials. Int J Nanotechnol 1:193–225
69. Hajji-Hedfi L, Chhipa H (2021) Nano-based pesticides: challenges for pest and disease management. Euro-Mediterr J Environ Integr 6:69
70. He P, Shen L, Liu R, Luo Z, Li Z (2011) Direct detection of β-agonists by use of gold nanoparticle-based colorimetric assays. Anal Chem 83(18):6988–6995
71. He L, Liu Y, Mustapha A, Lin M (2011) Antifungal activity of zinc oxide nanoparticles against *Botrytis cinerea* and *Penicillium expansum*. Microbiol Res 166(3):207–215
72. Heo J, Hua SZ (2009) An overview of recent strategies in pathogen sensing. Sensors (Switzerland) 9(6):4483–4502
73. Hong Y, Wu Y, Wu S, Wang X, Zhang J (2019) Overview of computational simulations in quantum dots. Isr J Chem 59:661–672
74. Hosnedlova B, Kepinska M, Skalickova S, Fernandez C, Ruttkay-Nedecky B, Peng Q, Baron M, Melcova M, Opatrilova R, Zidkova J, Bjørklund G, Sochor J, Kizek R (2018) Nano-selenium and its nanomedicine applications: a critical review. Int J Nanomed 13:2107–2128
75. Hu Y, Wang J, Wu Y (2019) A simple and rapid chemosensor for colorimetric detection of dimethoate pesticide based on the peroxidase-mimicking catalytic activity of gold nanoparticles. Anal Methods 11:5337–5534
76. Husen A, Siddiqi KS (2014) Phytosynthesis of nanoparticles: concept, controversy and application. Nanoscale Res Lett 9(229):1–24. https://doi.org/10.1186/1556-276X-9-229
77. Husen A, Siddiqi KS (2023) Advances in smart nanomaterials and their applications. Elsevier Inc., 50 Hampshire St., 5th Floor, Cambridge, MA 02139, USA. https://doi.org/10.1016/C2021-0-02202-1
78. Ji X, Zheng J, Xu J, Rastogi VK, Cheng TC, DeFrank JJ, Leblanc RM (2005) (CdSe) ZnS quantum dots and organophosphorus hydrolase bioconjugate as biosensors for detection of paraoxon. J Phys Chem B 109(9):3793–3799
79. Ji Y, Ma S, Lv S, Wang Y, Lu S, Liu M (2021) Nanomaterials for targeted delivery of agrochemicals by an all-in-one combination strategy and deep learning. ACS Appl Mater Interfaces 13:43374–43386
80. Jin X, Jin X, Chen L, Jiang J, Shen G, Yu R (2009) Piezoelectric immunosensor with gold nanoparticles enhanced competitive immunoreaction technique for quantification of aflatoxin B1. Biosens Bioelectron 24:2580–2585
81. Jo YK, Kim BH, Jung G (2009) Antifungal activity of silver ions and nanoparticles on phytopathogenic fungi. Plant Dis 93(10):1037–1043
82. Kah M, Kookana RS, Gogos A, Bucheli TD (2018) A critical evaluation of nanopesticides and nanofertilizers against their conventional analogues. Nat Nanotechnol 13(8):677–684
83. Kang TF, Wang F, Lu LP, Zhang Y, Liu TS (2010) Methyl parathion sensors based on gold nanoparticles and Nafion film modified glassy carbon electrodes. Sens Actuators B Chem 145(1):104–109
84. Kannan P, Maduraiveeran G (2023) Metal oxides nanomaterials and nanocomposite-based electrochemical sensors for healthcare applications. Biosensors 13:542
85. Kanooujiya R, Saroj S, Rajput V, Alimuddin Srivastava S, Minkina T, Igwegbe C, Singh M, Kumar A (2023) Emerging application of nanotechnology for mankind. Emergent Mater 6:3
86. Kapoor P, Dhaka RK, Sihag P, Mehla S, Sagwal V, Singh Y, Langaya S, Balyan P, Singh KP, Xing B, White JC, Dhankher OP, Kumar U (2022) Nanotechnology-enabled biofortification strategies for micronutrients enrichment of food crops: current understanding and future scope. NanoImpact 26:100407. https://doi.org/10.1016/j.impact.2022.100407

87. Kapoor A, Raghunathan M, Lal B, Kumar P, Srivastava N, Devnani G, Pal D B (2024) Sustainable valorization of waste plastic into nanostructured materials for environmental, energy, catalytic and biomedical applications: a review. Chemosphere 364:143279. https://doi.org/10.1016/j.chemosphere.2024.143279
88. Kasoju A, Shahdeo D, Khan AA, Shrikrishna NS, Mahari S, Alanazi AM, Bhat MA, Jyotsnendu Giri J, Gnadhi S (2020) Fabrication of microfluidic device for aflatoxin M1 detection in milk samples with specific aptamers. Sci Rep 10:4627
89. Kaushal M, Wani SP (2017) Nanosensors: frontiers in precision agriculture. Nanotechnol Agric Paradigm 279–291
90. Khan A, Singh J, Upadhayay VK, Singh AV, Shah S (2019) Microbial biofortification: a green technology through plant growth promoting microorganisms. In: Sustainable green technologies for environmental management. Springer, pp 255–269
91. Khan AA, Akhtar T (2011) Adsorption thermodynamics studies of 2, 4, 5-trichlorophenoxy acetic acid on poly-o-toluidine Zr(IV) phosphate, a nano-composite used as pesticide sensitive membrane electrode. Desalination 272(1):259–264
92. Khan F, Pandey P, Upadhyay TK (2022) Applications of nanotechnology-based agrochemicals in food security and sustainable agriculture: an overview. Agriculture 12:1672
93. Khan S, Naushad M, Al-Gheethi A, Iqbal J (2021) Engineered nanoparticles for removal of pollutants from wastewater: current status and future prospects of nanotechnology for remediation strategies. J Environ Chem Eng 9:106160
94. Khodakovskaya M, Dervishi E, Mahmood M, Xu Y, Li Z, Watanabe F, Biris AS (2009) Carbon nanotubes are able to penetrate plant seed coat and dramatically affect seed germination and plant growth. ACS Nano 3(10):3221–3227
95. Khush GS, Lee S, Cho JI, Jeon JS (2012) Biofortification of crops for reducing malnutrition. Plant Biotechnol Rep 6:195–202
96. Kim SW, Jung JH, Lamsal K, Kim YS, Min JS, Lee YS (2012) Antifungal effects of silver nanoparticles (AgNPs) against various plant pathogenic fungi. Mycobiology 40(1):53–58
97. Kim SW, Kim KS, Lamsal K, Kim YJ, Kim SB, Jung M, Sim SJ, Kim HS, Chang SJ, Kim JK, Lee YS (2009) An in vitro study of the antifungal effect of silver nanoparticles on oak wilt pathogen Raffaelea sp. J Microbiol Biotechnol 19(8):760–764
98. Kong D, Liu L, Song S, Suryoprabowo S, Li A, Kuang H, Wang L, Xu C (2016) A gold nanoparticle-based semi-quantitative and quantitative ultrasensitive paper sensor for the detection of twenty mycotoxins. Nanoscale 8(9):5245–5253
99. Kosaka PM, Pini V, Ruz JJ, Da Silva RA, González MU, Ramos D, Calleja M, Tamayo J (2014) Detection of cancer biomarkersin serum using a hybrid mechanical and optoplasmonic nanosensor. Nat Nanotechnol 9:1047–1053
100. Krishnaraj C, Ramachandran R, Mohan K, Kalaichelvan PT (2012) Optimization for rapid synthesis of silver nanoparticles and its effect on phytopathogenic fungi. Spectrochim Acta A Mol Biomol Spectrosc 93:95–99
101. Kuang H, Chen W, Yan W, Xu L, Zhu Y, Liu L, Chu H, Peng C, Wang L, Kotov NA, Xu C (2011) Crown ether assembly of gold nanoparticles: melamine sensor. Biosens Bioelectron 26(5):2032–2037
102. Kumar P, Ramulu Lambadi P, Kumar Navani N (2015) Non-enzymatic detection of urea using unmodified gold nanoparticles based aptasensor. Biosens Bioelectron 72:340–347
103. Kumaravel A, Chandrasekaran M (2011) A biocompatible nano TiO_2/nafion composite modified glassy carbon electrode for the detection of fenitrothion. J Electroanal Chem 650(2):163–170
104. Kwak S-Y, Wong MH, Lew TTS, Bisker G, Lee MA, Kaplan A, Dong J, Liu AT, Koman VB,Sinclair R (2017) Nanosensor technology applied to living plant systems. Annu Rev AnalChem 10:113–140
105. Lamri M, Bhattacharya T, Boukid F, Chentir I, Dib AL, Das D, Djenane D, Gagaoua M (2021) Nanotechnology as a processing and packaging tool to improve meat quality and safety. Foods 10(11):2633

106. Lamsal K, Kim SW, Jung JH, Kim YS, Kim KS, Lee YS (2011) Application of silver nanoparticles for the control of Colletotrichum species in vitro and pepper anthracnose disease in field. Mycobiology 39(3):194–199
107. Lamsal K, Kim SW, Jung JH, Kim YS, Kim KS, Lee YS (2011) Inhibition effects of silver nanoparticles against powdery mildews on cucumber and pumpkin. Mycobiology 39(1):26–32
108. Lee KJ, Park SH, Govarthanan M, Hwang PH, Seo YS, Cho M, Lee WH, Lee JY, Kamala-Kannan S, Oh BT (2013) Synthesis of silver nanoparticles using cow milk and their antifungal activity against phytopathogens. Mater Lett 105:128–131
109. Li C, Wang C, Wang C, Hu S (2006) Development of a parathion sensor based on molecularly imprinted nano-TiO_2 self-assembled film electrode. Sens Actuators B Chem 117(1):166–171
110. Li Y, Liu X, Lin Z (2012) Recent developments and applications of surface plasmon resonance biosensors for the detection of mycotoxins in foodstuffs. Food Chem 132:1549–1554
111. Lin DH, Xing BS (2008) Root uptake and phytotoxicity of ZnO nanoparticles. Environ Sci Technol 42(15):5580–5585
112. Lin TJ, Huang KT, Liu CY (2006) Determination of organophosphorous pesticides by a novel biosensor based on localized surface plasmon resonance. Biosens Bioelectron 22(4):513–518
113. Lin Y, Lu F, Wang J (2004) Disposable carbon nanotube modified screen-printed biosensor for amperometric detec¬tion of organophosphorus pesticides and nerve agents. Electroanalysis 16(1–2):145–149
114. Liu R, Guan G, Wang S, Zhang Z (2011) Core-shell nanostructured molecular imprinting fluorescent chemosensor for selective detection of atrazine herbicide. Analyst 136(1):184–190
115. Liu W, Li Y, Feng Y, Qiao J, Zhao H, Xie J, Fang Y, Shen S, Liang S (2020) The effectiveness of nanobiochar for reducing phytotoxicity and improving soil remediation in cadmium-contaminated soil. Sci Rep 10:858
116. Liu Y, Wang Z, Zhao G (2020) Carbon nanotube-based sensor for chlorpyrifos detection in soil. J Hazard Mater 389(121845):118
117. Ma Q, Gong N, Li Y et al (2015) Gold nanoparticles as dual functional sensor to detect *E. coli* DH5α as a model for gram-negative bacteria. J Chin Chem Soc 62:521–527
118. Maduraiveeran G, Ramaraj R (2017) Gold nanoparticle-based sensing platform of hydrazine, sulfite, and nitrite for food safety and environmental monitoring. Anal Sci Technol 8(1):1–10
119. Malik P, Shankar R, Malik V, Sharma N, Mukherjee TK (2014) Green chemistry based benign routes for nanoparticle synthesis. J Nanoparticles 2014:1–14
120. Manikandan A, Subramanian K (2016) Evaluation of zeolite based nitrogen nanofertilizers on maize growth, yield and quality on inceptisols and alfisols. Int J Plant Soil Sci 9:1–9
121. Mariano M, El Kissi N, Dufresne A (2014) Cellulose nanocrystals and related nanocomposites: review of some properties and challenges. J Polym Sci B Polym Phys 52:791–806
122. Mateus MP, Tavanti RFR, Tavanti TR, Santos EF, Jalal A, Reis AR (2021) Selenium biofortification enhances ROS scavenge system increasing yield of coffee plants. Ecotoxicol Environ Saf 209:111772
123. Miguel-Rojas C, Pérez-de-Luque Alejandro A (2023) Nanobiosensors and nanoformulations in agriculture: new advances and challenges for sustainable agriculture. Emerg Top Life Sci 7:229–238
124. Miladinov G (2023) Impacts of population growth and economic development on food security in low-income and middle-income countries. Front Hum Dyn 5
125. Milani N, Hettiarachchi GM, Kirby JK, Beak DG, Stacey SP, McLaughlin MJ (2015) Fate of zinc oxide nanoparticles coated onto macronutrient fertilizers in an alkaline calcareous soil. PLoS ONE 10:1–16
126. Min JS, Kim KS, Kim SW, Jung JH, Lamsal K, Kim SB, Jung MY, Lee YS (2009) Effects of colloidal silver nanoparticles on sclerotium-forming phytopathogenic fungi. Plant Pathol J 25(4):376–380
127. Mishra S, Singh BR, Singh A, Keswani C, Naqvi AH, Singh HB (2014) Biofabricated silver nanoparticles act as a strong fungicide against *Bipolaris sorokiniana* causing spot blotch disease in wheat. PLoS ONE 9(5):e97881

128. Mishra S, Sundaram B (2023) Efficacy and challenges of carbon nanotube in wastewater and water treatment. Environ Nanotechnol Monit Manage 19:100764
129. Mo R, Wang X, Yuan Q, Yan X, Su T, Feng Y, Lv L, Zhou C, Hong P, Sun S, Wang Z, Li C (2018) Electrochemical determination of nitrite by au nanoparticle/graphene-chitosan modified electrode. Sensors (Switzerland) 18(7):1–12
130. Mondal R, Dam P, Chakraborty J, Paret ML, Katı A, Altuntas S, Sarkar R, Ghorai S, Gangopadhyay D, Mandal AK, Husen A (2022) Potential of nanobiosensor in sustainable agriculture: the state-of-art. Heliyon 8(12)
131. Moussa SH, Tayel AA, Alsohim AS, Abdallah RR (2013) Botryticidal activity of nanosized silver-chitosan composite and its application for the control of gray mold in strawberry. J Food Sci 78(10):M1589–M1594
132. Zain M, Ma H, Rahman SU, Nuruzzaman Md, Chaudhary S, Azeem I, Mehmood F, Duan A, Sun C (2024) Nanotechnology in precision agriculture: advancing towards sustainable crop production. Plant Physiol Biochem 206:108244
133. Chanu NB, Singh MCK (2022) Applications of nanotechnology in precision agriculture. In: Ghorbanpour M, Shahid MA (eds) Nano-enabled agrochemicals in agriculture. Academic Press, pp 175–187
134. Neethirajan S, Jayas DS, Sadistap S (2009) Carbon dioxide (CO_2) sensors for the agri-food industry—a review. Food Bioprocess Technol 2(2):115–121
135. Nurchi VM, Crisponi G, Lachowicz JI, Medici S, Peana M, Zoroddu MA (2016) Chemical features of in use and in progress chelators for iron overload. J Trace Elem Med Biol 38:10–18
136. Nuruzzaman M, Rahman MM, Liu Y, Naidu R (2016) Nanoencapsulation, nano-guard for pesticides: a new window for safe application. J Agric Food Chem 64:1447–1483
137. Ojha S, Kumar B (2018) A review on nanotechnology based innovations in diagnosis and treatment of multiple sclerosis. J Cellular Immun 4(2):56–64
138. Ouda SM (2014) Antifungal activity of silver and copper nanoparticles on two plant pathogens, *Alternaria alternata* and *Botrytis cinere*a. Res J Microbiol 9(1):34–42
139. Pal S, Ying W, Alocilja EC, Downes FP (2008) Sensitivity and specificity performance of a direct-charge transfer biosensor for detecting *Bacillus cereus* in selected food matrices. Biosyst Eng 99(4):461–468
140. Pandey G (2018) Challenges and future prospects of agri-nanotechnology for sustainable agriculture in India. Environ Technol Innov 11:299–307
141. Paniel N, Radoi A, Marty JL (2010) Development of an electrochemical biosensor for the detection of aflatoxin M1 in milk. Sensors 10(10):9439–9448
142. Parham H, Rahbar N (2010) Square wave voltammetric determination of methyl parathion using ZrO_2-nanoparticles modified carbon paste electrode. J Hazard Mater 177(1):1077–1084
143. Parisi C, Vigani M, Rodríguez-Cerezo E (2015) Agricultural nanotechnologies: what are the current possibilities. Nano Today 10(2):124–127
144. Park HJ, Kim SH, Kim HJ, Choi SH (2006) A new composition of nanosized silica-silver for control of various plant diseases. Plant Pathol J 22(3):295–302
145. Park S, Choi KS, Kim S, Gwon Y, Kim J (2020) Graphene oxide-assisted promotion of plant growth and stability. Nanomaterials 10:1–11
146. Paulkumar K, Gnanajobitha G, Vanaja M, Rajeshkumar S, Malarkodi C, Pandian K, Annadurai G (2014) Piper nigrum leaf and stem assisted green synthesis of silver nanoparticles and evaluation of its antibacterial activity against agricultural plant pathogens. Sci World J 2014:1–9
147. Peng J, Liu L, Xu L, Song S, Kuang H, Cui G, Xu C (2017) Gold nanoparticle-based paper sensor for ultrasensitive and multiple detection of 32 (fluoro)quinolones by one monoclonal antibody. Nano Res 10(1):108–120
148. Polat E, Karaca M, Demir H, Onus AN (2004) Use of natural zeolite (clinoptilolite) in agriculture. J Fruit Ornam Plant Res 12:183–189
149. Prasad R, Bhattacharyya A, Nguyen QD (2017) Nanotechnology in sustainable agriculture: recent developments, challenges, and perspectives. Front Microbiol 8:1–13

150. Prasad R, Kumar M, Kumar V (2017b) Nanotechnology: an agricultural paradigm. Springer Nature, Singapore, p 371
151. Prasad R, Pandey R, Barman I (2016) Engineering tailored nanoparticles with microbes: quo vadis? WIREs Nanomed Nanobiotechnol 8:316–330
152. Prexha K, Rahul KD, Pooja S, Sheetal M, Vijeta S, Yogita S, Sonu L, Priyanka B, Krishna PS, Baoshan X, Jason CW, Om PD, Upendra K (2022) Nanotechnology-enabled biofortification strategies for micronutrients enrichment of food crops: current understanding and future scope. Nano Impact 26:100407
153. Pudake RN, Jain U, Kole C (2021) Biosensors in agriculture: recent trends and future perspectives. Springer
154. Qu F, Zhou X, Xu J, Li H, Xie G (2009) Luminescence switching of CdTe quantum dots in presence of p-sulfo-natocalix[4]arene to detect pesticides in aqueous solution. Talanta 78(4):1359–1363
155. Ragaei M, Sabry AH (2014) Nanotechnology for insect pest control. Int J Sci Environ Technol 3:528–545
156. Rai V, Acharya S, Dey N (2012) Implications of nanobiosensors in agriculture. J Biomater Nanobiotechnol 3(02):315
157. Raj V, Vijayan AN, Joseph K (2015) Cysteine capped gold nanoparticles for naked eye detection of *E. coli* bacteria in UTI patients. Sens Bio-sens Res 5:33–36
158. Rajesh S, Raja DP, Rathi JM, Sahayaraj K (2012) Biosynthesis of silver nanoparticles using Ulva fasciata (Delile) ethyl acetate extract and its activity against *Xanthomonas campestris* pv. malvacearum. J Biopestic 5(119):119–128
159. Ramezani M, Esmaelpourfarkhani M, Taghdisi SM, Abnous K, Ali-bolandi M (2020) Application of nanosensors for food safety. In: Nanosensors for smart cities. INC
160. Rao KJ, Paria S (2013) Use of sulfur nanoparticles as a green pesticide on *Fusarium solani* and *Venturia inaequalis* phytopathogens. RSC Adv 3(26):10471–10478
161. Rastogi S, Kumari V, Sharma V, Ahmad FJ (2022) Gold nanoparticle-based sensors in food safety applications. Food Anal Methods 15:468–484
162. Ryu H, Thompson D, Huang Y, Li B, Lei YJS (2020) Electrochemical sensors for nitrogen species: a review. Sens Actuators Rep 2:100022
163. Sahoo A, Sethi J, Satapathy KB, Sahoo SK, Panigrahi GK (2023) Nanotechnology for precision and sustainable agriculture: recent advances, challenges and future implications. Nanotechnol Environ Eng 8:775–787
164. Saini RK, Bragri LP, Bajapai AK (2017) Smart nanosensors for pesticide detection. In: New pesticides and soil sensors. Elsevier, pp 519–559
165. Sangeetha J, Thangadurai D, Hospet R, Harish ER, Purushotham P, Mujeeb MA, Shrinivas J, David M, Mundaragi AC, Thimmappa AC, Arakera SB, Prasad R (2017) Nanoagrotechnology for soil quality, crop performance and environmental management. In: Prasad R, Kumar M, Kumar V (eds) Nanotechnology. Springer Nature Singapore Pte Ltd, pp 73–97
166. Satapathi S, Kumar V, Chini MK, Bera R, Halder KK, Patra A (2018) Highly sensitive detection and removal of mercury ion using a multimodal nanosensor. Nano-Struct Nano-Obj 16:120–126
167. Sekhon BS (2014) Nanotechnology in agri-food production: an overview. Nanotechnol Sci Appl 7:31–53
168. Senkbeil S, Lafleur JP, Jensen TG, Kutter JP (2012) Gold nanoparticle-based fluorescent sensor for the analysis of dithiocarbamate pesticides in water. In: Proceedings of the 16th international conference on miniaturized systems for chemistry and life sciences, MicroTAS, pp 1423–1425
169. Seo S, Dobozi-King MF, Young R, Kish LB, Cheng M (2008) Patterning a nanowell sensor biochip for specific and rapid detection of bacteria. Microelectron Eng 85(7):1484–1489
170. Shafi A, Ahmad N, Sultana S, Sabir S, Khan MZ (2019) Ag2S-sensitized NiO-ZnO heterostructures with enhanced visible light photocatalytic activity and acetone sensing property. ACS Omega 4(7):12905–12918

171. Shafi U, Mumtaz R, García-Nieto J, Hassan SA, Zaidi SAR, Iqbal N (2019) Precision agriculture techniques and practices: from considerations to applications. Sensors 19:3796
172. Shah A (2020) A novel electrochemical nanosensor for the simultaneous sensing of two toxic food dyes. ACS Omega 5(11):6187–6193
173. Sharma A, Matharu Z, Sumana G, Solanki PR, Kim CG, Malhotra BD (2010) Antibody immobilized cysteamine functionalized gold nanoparticles for aflatoxin detection. Thin Solid Films 519(3):1213–1218
174. Sharma G, Sharma AR, Bhavesh R, Park J, Ganbold B, Nam JS, Lee SS (2014) Biomolecule-mediated synthesis of selenium nanoparticles using dried vitis vinifera (raisin) extract. Molecules 19:2761–2770
175. Sheini A (2020) Colorimetric aggregation assay based on array of gold and silver nanoparticles for simultaneous analysis of aflatoxins, ochratoxin and zearalenone by using chemometric analysis and paper based analytical devices. Mikrochim Acta 187(3):167
176. Shoaib A, Waqas M, Elabasy A, Cheng X, Zhang Q, Shi Z (2018) Preparation and characterization of emamectin benzoate nanoformulations based on colloidal delivery systems and use in controlling *Plutella xylostella* (L.) (Lepidoptera: Plutellidae). RSC Adv 8(28):15687–15697
177. Simonian AL, Good TA, Wang SS, Wild JR (2005) Nanoparticle-based optical biosensors for the direct detection of organophosphate chemical warfare agents and pesticides. Anal Chim Acta 534(1):69–77
178. Singh S, Sangwan S, Sharma P, Devi P, Moond M (2021) Nanotechnology for sustainable agriculture: an emerging perspective. J Nanosci Nanotechnol 21:3453–3465
179. Singh V, Joung D, Zhai L, Das S, Khondaker SI, Seal S (2011) Graphene based materials: past, present and future. Prog Mater Sci 56:1178–1271
180. Song KM, Cho M, Jo H, Min K, Jeon SH, Kim T, Han MS, Ku JK, Ban C (2011) Gold nanoparticle-based colorimetric detection of kanamycin using a DNA aptamer. Anal Biochem 415(2):175–181
181. Sosunova I, Porras J (2022) IoT-enabled intelligent waste management systems for smart cities: a systematic review. IEEE Access 10(4):1–1
182. Spagnolo S, De La Franier B, Davoudian K, Hianik T, Thompson M (2022) Detection of E. coli bacteria in milk by an acoustic wave aptasensor with an anti-fouling coating. Sensors 22(5):1853
183. Srivastava AK, Dev A, Karmakar S (2018) Nanosensors and nanobiosensors in food and agriculture. Environ Chem Lett 16(1):161–182
184. Sun H, Fung Y (2006) Piezoelectric quartz crystal sensor for rapid analysis of pirimicarb residues using molecularly imprinted polymers as recognition elements. Anal Chim Acta 576(1):67–76
185. Sunita O, Deepika S, Arghya S, Hasnahana C, Debajyoti K, Utpal B (2018) Nanotechnology in crop protection. Nanomater Plants, Algae, Microorganisms Concepts Controversies 1:345–391
186. Suri CR, Kaur J, Gandhi S, Shekhawat GS (2008) Label-free ultra-sensitive detection of atrazine based on nanomechanics. Nanotechnology 19(23):235502
187. Tan W, Gao Q, Deng C, Wang Y, Lee WY, Hernandez-Viezcas JA et al (2018) Foliar exposure of $Cu(OH)_2$ nanopesticide to basil (*Ocimum basilicum*): variety-dependent copper translocation and biochemical responses. J Agric Food Chem 66(13):3358–3366
188. Tehri N, Kumar N, Vashishth A (2020) Current trends in enzymatic biosensors for pesticides determination. Int Res J Environ Sci 9:87–107
189. Thakur M, Wang B, Verma ML (2022) Development and applications of nanobiosensors for sustainable agricultural and food industries: recent developments, challenges and perspectives. Environ Technol Innov 26
190. The Royal Society (2009) Reaping the benefits: science and the sustainable intensification of global agriculture. The Royal Society, London
191. Tsuchiya Y, Yoshimura M, Sato Y, Kuwata K, Toh S, Holbrook-Smith D, Zhang H, McCourt P, Itami K, Kinoshita T et al (2015) Probing strigolactone receptors in striga hermonthica with fluorescence. Science 349:864–868

192. Umapathi R, Ghoreishian SM, Sonwal S, Rani GM, Huh YS (2022) Portable electrochemical sensing methodologies for on-site detection of pesticide residues in fruits and vegetables. Coord Chem Rev 453:214305
193. Updike SJ, Hicks GP (1967) The enzyme electrode. Nature 214:986–988
194. Vahab L, Keshipour S (2022) Novel nanosensor of cobalt (II) and copper (II) constructed from graphene quantum dots modified with Eriochrome black T. Sci Rep 12:13179
195. Valdes M, Valdes Gonzalez A, Garcia Calzon J, Diaz-Garcia M (2009) Analytical nanotechnology for food analysis. Microchim Acta 166:1–19
196. Vamvakaki V, Chaniotakis NA (2007) Pesticide detection with a liposome-based nano-biosensor. Biosens Bioelectron 22(12):2848–2853
197. van Dijk M, Morley T, Rau ML, Saghai Y (2021) A meta-analysis of projected global food demand and population at risk of hunger for the period 2010–2050. Nat Food 2(7):494–501
198. Vigneshvar S, Sudhakumari CC, Senthilkumaran B, Prakash H (2016) Recent advances in biosensor technology for potential applications-an overview. Front Bioeng Biotechnol 4:11
199. Vikesland PJ (2018) Nanosensors for water quality monitoring. Nat Nanotech 13:651–660
200. Vinayaka AC, Basheer S, Thakur MS (2009) Bioconjugation of CdTe quantum dot for the detection of 2, 4-dichlo-rophenoxyacetic acid by competitive fluoroimmunoassay based biosensor. Biosens Bioelectron 24(6):1615–1620
201. Virk, V, Deepak H, Taneja K, Srivastava R, Giri S (2024) Amelioration in nanobiosensors for the control of plant diseases: current status and future challenges. Front Nanotechnol (6)
202. Viswanathan S, Radecka H, Radecki J (2009) Electrochemical biosensor for pesticides based on acetylcholinesterase immobilized on polyaniline deposited on vertically assembled carbon nanotubes wrapped with ssDNA. Biosens Bioelectron 24(9):2772–2777
203. Wang M, Li Z (2008) Nano-composite ZrO_2/Au film electrode for voltammetric detection of parathion. Sens Actuators B Chem 133(2):607–612
204. Wang P, Lombi E, Zhao FJ, Kopittke PM (2016) Nanotechnology: a new opportunity in plant sciences. Trends Plant Sci 21(8):699–712
205. Wang W, Liu L, Song S, Xu L, Kuang H, Zhu J, Xu C (2016) Gold nanoparticle-based strip sensor for multiple detection of twelve Salmonella strains with a genus-specific lipopolysaccharide antibody 基于菌属特异性抗体的金标试纸传感器同时检测12种沙 门氏菌. Sci China Mater 59(8):665–674
206. Wang Y, Alocilja EC (2015) Gold nanoparticle-labeled biosensor for rapid and sensitive detection of bacterial pathogens. J Biol Eng 9(1):1–7
207. Wang Y, Jiang F, Ma C, Rui Y, Tsang DCW, Xing B (2019) Effect of metal oxide nanoparticles on amino acids in wheat grains (*Triticum aestivum*) in a life cycle study. J Environ Manage 241:319–327
208. Wang Z, Wu X, Liu L et al (2020) Rapid and sensitive detection of diclazuril in chicken samples using a gold nanoparticle-based lateral-flow strip. Food Chem 312:126116
209. Warriner K, Reddy SM, Namvar A, Neethirajan S (2014) Developments in nanoparticles for use in biosensors to assess food safety and quality. Trends Food Sci Technol 40:183–199
210. Wu H, Nißler R, Morris V, Herrmann N, Hu P, Jeon S-J, Kruss S, Giraldo JP (2020) Monitoring plant health with near-infrared fluorescent H_2O_2 nanosensors. Nano Lett 20:2432–2442
211. Xiong T, Dumat C, Dappe V, Vezin H, Schreck E, Shahid M, Pierart A, Sobanska S (2017) Copper oxide nanoparticle foliar uptake, phytotoxicity, and consequences for sustainable urban agriculture. Environ Sci Technol 51:5242–5251
212. Yan S, Lai X, Du G, Xiang Y (2018) Identification of aminoglycoside antibiotics in milk matrix with a colorimetric sensor array and pattern recognition methods. Anal Chim Acta 1034:153–160
213. Yang G, Zhao FJB (2015) Electrochemical sensor for chloramphenicol based on novel multi-walled carbon nanotubes@molecularly imprinted polymer. Biosens Bioelectron 64:416–422
214. Yang X, Luo N, Tan Z, Jia Z, Liao X (2017) A fluorescence probe for tartrazine detection in foodstuff samples based on fluorescence resonance energy transfer. Food Anal Methods 10(5):1308–1316

215. Yılmaz M, Erkartal M, Ozdemir M, Sen U (2017) Preparation of ZnO nanoparticles using PS-b-PAA reverse micelle cores for UV protective, self-cleaning and antibacterial textile applications. Carbohyd Polym 159:1–10
216. Yin H, Cao Y, Marelli B, Zeng X, Mason AJ, Cao C (2021) Soil sensors and plant wearables for smart and precision agriculture. Adv Mater 33(20):2007764
217. Yin J, Su X, Yan S, Shen J (2023) Multifunctional nanoparticles and nanopesticides in agricultural application. Nanomaterials 13:1255
218. Yu S, Tan Z, Lai Y, Li Q, Liu J (2023) Nanoparticulate pollutants in the environment: analytical methods, formation, and transformation. Eco-Environ. Health 2(2):61–73
219. Yuvaraj M, Subramanian KS (2018) Development of slow release Zn fertilizer using nano-zeolite as carrier. J Plant Nutr 41:311–320
220. Zeng Y, Zhu Z, Du D, Lin Y (2016) Nanomaterial-based electrochemical biosensors for food safety. J Electroanal Chem 781:147–154
221. Zhang P, Guo Z, Ullah S, Melagraki G, Afantitis A, Lynch I (2021) Nanotechnology and artificial intelligence to enable sustainable and precision agriculture. Nat Plants 7:864–876
222. Zhang Q, Ying Y, Ping J (2022) Recent advances in plant nanoscience. Adv Sci 9:2103414
223. Zhang W, Tang H, Geng P, Wang Q, Jin L, Wu Z (2007) Amperometric method for rapid detection of *Escherichia coli* by flow injection analysis using a bismuth nano-film modified glassy carbon electrode. Electrochem Commun 9(4): 833–838
224. Zhang W, Yang T, Huang D, Jiao K, Li G (2008) Synergistic effects of nano-ZnO/multi-walled carbon nanotubes/chitosan nanocomposite membrane for the sensitive detection of sequence specific of PAT gene and PCR amplification of NOS gene. J Membr Sci 325:245–251
225. Zhang W, Yang T, Huang DM, Jiao K (2008) Electrochemical sensing of DNA Im-mobilization and hybridization based on carbon nano-tubes/nano zinc oxide/chitosan composite film. Chin Chem Lett 19:589–591
226. Zhao L, Hu Q, Huang Y, Fulton AN, Hannah-Bick C, Adeleye AS, Keller AA (2017) Activation of antioxidant and detoxification gene expression in cucumber plant s exposed to a $Cu(OH)_2$ nanopesticide. Environ Sci Nano 4(8):1750–1760
227. Zhao L, Huang Y, Keller AA (2017) Comparative metabolic response between cucumber (*Cucumis sativus*) and corn (*Zea mays*) to a $Cu(OH)_2$ nanopesticide. J Agric Food Chem 66(26):6628–6636
228. Zhao YG, Shen HY, Shi JW, Chen XH, Jin MC (2011) Preparation and characterization of amino functionalized nano-composite material and its application for multi-residue analysis of pesticides in cabbage by gas chromatography–triple quadrupole mass spectrometry. J Chromatogr A 1218(33):5568–5580
229. Zoidis E, Seremelis I, Kontopoulos N, Danezis GP (2018) Selenium-dependent antioxidant enzymes: actions and properties of selenoproteins. Antioxidants 7:1–26

Precision Agriculture and Nanosensor

Vikas Sharma, Animesh Gosh Bag, Nitin M. Changade, and Neeraj Singh Parihar

Abstract This chapter explores the role of precision agriculture in enhancing resource efficiency, crop productivity, and environmental sustainability through the integration of advanced technologies such as GIS, remote sensing, and nanosensors. Nanosensors, with their ability to operate at the nanoscale, offer unprecedented accuracy in measuring soil health, water quality, and environmental variables, providing critical data for optimized farming practices. However, their adoption, particularly in developing countries, is impeded by the need for specialized technical knowledge and concerns over sensor stability compared to traditional laboratory instruments. The chapter highlights the importance of bridging these gaps through education and rigorous validation of nanosensor reliability to ensure their effective integration into sustainable agricultural systems.

Keywords Precision Agriculture · Nanosensor · Soil health · Resource conservation · Crop improvement

1 Introduction

Precision agriculture is an advanced farming approach that emphasizes the sustainable utilization of resources through site-specific management practices to optimize crop productivity and environmental stewardship. Nanosensors have emerged as one of the most exciting advancements in recent years, with their potential spanning across agriculture, water management, environmental protection, and healthcare [3]. Nanosensors operate at the nanoscale, meaning they either use phenomena at this scale or are crafted using nanotechnological methods. Essentially, any sensor whose size, sensitivity, or interaction with the surrounding environment occurs at the nanometer level can be termed a nanosensor. This distinction places nanosensors in a unique position, as they can interact with materials and phenomena at molecular

V. Sharma (✉) · A. G. Bag · N. M. Changade · N. S. Parihar
Department of Soil Science and Agricultural Chemistry, School of Agriculture, Lovely Professional University, Phagwara, Punjab, India
e-mail: vikas.27227@lpu.co.in

A. Husen (ed.), *Nanobiosensors for Agricultural and Other Related Sectors*, Smart Nanomaterials Technology, https://doi.org/10.1007/978-981-96-6499-3_2

and atomic levels, offering insights and capabilities beyond traditional sensors [58]. Probably the most important landmark in global agriculture is the Green Revolution which improved agricultural productivity several times than the pre-green revolution era. The Green Revolution reduced hunger and starvation by a significant amount but side by side increased the dose of chemical fertilizer, herbicide and pesticide application which gradually started to ruin the soil health. After 60 years of the green revolution, we have achieved food security as its blessing, but the curse is also imposed on human beings in the form of soil quality deterioration due to excess chemical use [45]. Productivity started to reduce gradually even after increasing the fertilizer doses 5 times than the recommended dose of fertilizer predominantly due to improper soil health. Thus, it is very important to measure soil health and other quality parameters regularly and if it is declining then actions should be taken to revive it. The ever-green revolution is a modern concept which deals with sustainable production for a prolonged period rather than short-term higher productivity and to achieve this goal determination of soil health frequently should be the topmost priority [20]. Soil health can be assessed by analysing a significant number of physical, chemical and biological properties of soil as shown in Fig. 1. Physical parameters like texture, structure, moisture content, aggregate stability etc. are time-consuming and labour intensive while chemical parameters like pH, electrical conductivity (EC), Organic Carbon (OC), Soil nitrogen (N), phosphorus (P) or potassium (K) require sophisticated instruments and skilled labour [21]. Biological indicators like colony-forming units of bacteria, fungi, actinomycetes or enzymes (Dehydrogenase, phosphatase, urease) require scientific personnel and are sensitive to contamination. So, from the aforesaid statements, it can be concluded that traditional methods of soil health assessment are very much time or labour-consuming and require costly laboratory set-up. Apart from that, continuous assessment of soil health requires daily collection analysis of soil samples which is not a feasible option for larger geographical areas. The use of nanosensors may be the probable solution for all these problems which not only savelabour and capital but also gives amuch quicker response with satisfactory accuracy [42].

This chapter discusses the application of nanosensors in precision agriculture, focusing on their potential to enhance sustainable resource management, optimize crop productivity, and monitor soil health. It highlights the limitations of conventional soil health assessment techniques and presents nanosensors as an efficient, cost-effective solution for real-time, large-scale soil monitoring.

2 Precision Agriculture

Precision agriculture integrates advanced technologies, including Global Positioning System (GPS), remote sensing, and data analytics, to optimize resource allocation and enhance crop productivity while minimizing environmental degradation. Through site-specific management practices, it facilitates the spatially and temporally targeted application of inputs such as water, fertilizers, and pesticides, based

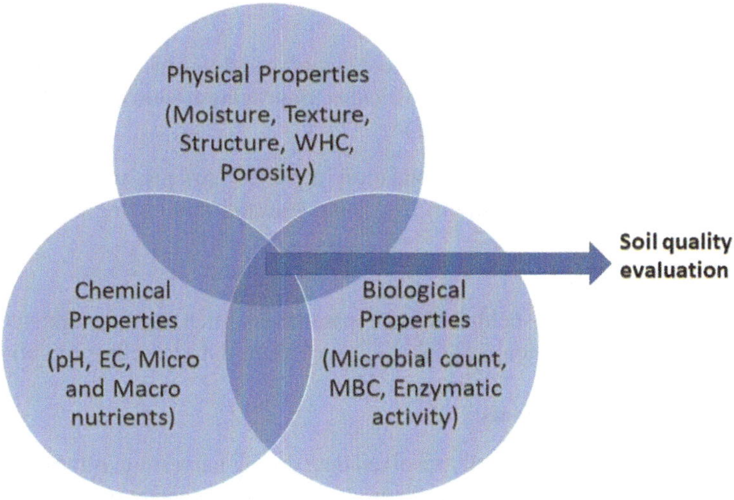

Fig. 1 Determination of soil quality

on field variability. This precision-driven approach improves agronomic outcomes, reduces input wastage, and conserves natural resources. Consequently, precision agriculture fosters sustainability by achieving a balance between maximizing agricultural output, economic viability, and environmental conservation.

Concept of Precision Agriculture: Precision agriculture is grounded in a systematic framework that includes diagnosing issues through scientific methodologies, analyzing the results to assess the severity and spatial variability of the problem, and making informed, data-driven decisions based on the findings to implement targeted interventions.

Precision Agriculture Techniques: Precision agriculture integrates cutting-edge technologies, including geographic information systems (GIS), remote sensing, and data analytics, to address spatial and temporal variability within agricultural fields. This approach focuses on improving resource use efficiency, maximizing crop productivity [60], and reducing environmental impacts by applying inputs such as water, fertilizers, and pesticides in precise quantities and specific locations, guided by real-time field data and site-specific conditions.

The primary techniques include:

- **Geographic Information Systems (GIS)**:

  GIS is utilized to create spatially detailed maps of soil properties, crop yields, and other field parameters, facilitating site-specific management decisions.
- **Remote Sensing**:

  Satellite, aerial, and drone-based remote sensing provide multispectral and hyperspectral data for monitoring crop health, water stress, and pest infestations, enabling timely interventions.

- **Global Positioning System (GPS)**:

  GPS technology supports precision field mapping and ensures accurate application of inputs, such as fertilizers and pesticides, improving resource use efficiency.
- **Variable Rate Technology (VRT)**:

  VRT enables the site-specific application of water, nutrients, and agrochemicals based on field variability, optimizing input use and minimizing environmental impacts.
- **Soil Sensors**:

  Sensors embedded in fields measure parameters such as moisture, temperature, and nutrient content in real-time, providing data to refine irrigation and fertilization strategies.
- **Yield Monitoring Systems**:

  These systems collect spatially resolved data during harvesting, offering insights into crop performance variability to guide future management practices.
- **Data Analytics and Machine Learning**:

  Advanced computational methods analyze large datasets, identify trends, and provide predictive insights to support data-driven decision-making in farm management.
- **Drones and Unmanned Aerial Vehicles (UAVs)**:

  UAVs equipped with imaging sensors capture high-resolution data, facilitating real-time crop monitoring and the early detection of anomalies such as diseases or nutrient deficiencies.
- **Internet of Things (IoT)**:

  IoT-enabled devices, including automated weather stations and smart irrigation systems, allow for continuous environmental and operational monitoring, enhancing system efficiency.
- **Automated Machinery**:

  GPS-guided tractors, robotic planters, and automated weeders enhance precision in planting, cultivation, and harvesting operations, reducing labor dependency and operational costs.

3 Nanosensors

At its core, a nanosensor transforms signals from its environment using nanostructured materials that possess at least one dimension smaller than 100 nm [1]. These structures include familiar names like nanowires, which are ultra-thin wires with diameters under 100 nm, and nanofibers, which are tiny filaments measuring between 50 and 500 nm [29]. Similarly, carbon nanotubes, known for their hollow cylindrical shapes, and quantum dots, nanoscale semiconductor crystals that fluoresce, also fall under this category [19]. Such materials showcase how nanosensors are uniquely suited for operating at a scale where even the smallest interactions matter.

While some nanosensors may not physically be tiny, they still qualify because of their extreme sensitivity at the nanoscale [25]. For example, a displacement sensor might measure movement down to the nanometer level, or a temperature sensor could detect changes as minute as nanokelvin [8]. Additionally, instruments like atomic force microscopy (AFMs), which detect nanoscale forces, fall under this definition because of the small-scale interaction between the sensor and the object it measures [41].

Types of Nanosensors

Nanosensors can be grouped into two broad categories: those that are themselves nanoscale in size, and those that, while not necessarily small, perform measurements with nanoscale precision [25]. The first category includes sensors like quantum dots and thin films, where one or more dimensions are in the nanometer range. The second category covers sensors that might be larger but are designed for ultra-sensitive detection, making them indispensable in applications that require extremely precise measurements.

- ***Active and Passive Nanosensors***: Like their larger counterparts, nanosensors can also be divided into active and passive types [30]. Active nano sensors require an external power source to function [1]. Examples include thermistors, which change resistance with temperature, or strain gauges, which detect pressure or deformation. Passive nanosensors, on the other hand, do not need an external energy source [49]. Devices like thermocouples, piezoelectric sensors, and photodiodes fit into this category. This distinction is essential because it affects how nanosensors are deployed in different environments, with passive sensors often being preferred for autonomous or energy-efficient applications [68].
- ***Absolute and Relative Nanosensors***: Another way to classify nanosensors is by the type of measurement they perform [23]. Absolute nanosensors measure against a fixed physical standard, like a pressure sensor comparing readings to a vacuum (zero pressure) [28]. In contrast, relative nanosensor measures are based on a chosen reference point, such as atmospheric pressure. This distinction allows nanosensors to cater to different measurement needs, depending on whether absolute precision or relative comparison is required.
- ***Physical, Chemical, and Biological Nanosensors***: Nanosensors are highly versatile and can detect different forms of energy or signals [13]. Physical nanosensors are designed to measure parameters like temperature, pressure, stress, strain, or position [48]. Chemical nanosensors focus on detecting specific substances or concentrations, such as ethanol ($C_2H\_{2}OH$) or carbon monoxide (CO) [54]. Biological nanosensors, meanwhile, are sensitive to biological substances, whether they are bacteria, viruses, or even toxins [11]. These sensors can be used to detect harmful biological agents, such as anthrax spores or flu viruses, making them crucial tools in healthcare and safety applications. While biosensors are often considered a subset of chemical sensors, their role in detecting biological entities is distinct and critical for fields like medical diagnostics and environmental monitoring [43].

- ***Other Common Types of Nano Sensors***: Carbon nanotube-based nanosensors represent a breakthrough in this field [59]. These sensors can be functionalized to detect specific molecules, and their properties change when exposed to different gases or vapours, making them highly useful in environmental monitoring [44]. Similarly, nanoscaled thin films—often made from polymers or metal oxides—are capable of concentrating analytes like heavy metals, improving detection limits to levels previously unimaginable (e.g., parts-per-billion). Arrays of nanocantilevers, which bend in response to molecular interactions, have even been used to detect explosives, offering a low-cost alternative to traditional mass spectrometers [33].

4 Existence of Nan Sensors in the Real World

In nature, nanosensors are not new, many animals have evolved their biological nanosensors. For example, dogs have an extraordinary sense of smell that allows them to detect nanosized molecules [64]. Fish sense small vibrations in the water, and insects use nanosensors to detect pheromones [63]. Even plants have nanosensors that help them sense sunlight [5]. In the artificial world, nanosensors have been in use for years, such as in film cameras, where silver ions on photographic film react to light and form nanoscale clusters that create the image [61]. These nanosensors have since evolved to detect a wide range of electromagnetic spectra, from infrared to ultraviolet and even x-rays, highlighting their adaptability [32].

5 Role of Nanosensor in Precision Agriculture

Nanosensors play a pivotal role in advancing precision agriculture by enabling real-time monitoring and management of crop health, soil conditions, and environmental factors. Their applications include:

- Soil Monitoring/Soil Health Monitoring
- Crop Health Monitoring
- Environmental/Weather Monitoring
- Smart Irrigation and Resource Management
- Data Integration and Decision Support

By enabling highly sensitive, accurate, and cost-effective data collection, nanosensors are transforming precision agriculture into a more sustainable and efficient practice. Sensor-based monitoring systems are the cornerstone of contemporary precision agriculture because they provide real-time information on crop status, soil health, and environmental influences [40, 62]. Sensor technology, utilized to monitor and regulate crucial aspects affecting plant health, quality, and stress responses, enables farmers to make informed decisions [34].

6 Soil Monitoring

Soil monitoring systems include soil moisture sensing, soil nutrient sensing, soil temperature sensing, soil health, soil compactions, etc. Soil monitoring sensors are essential in precision agriculture, enabling efficient resource management and sustainable farming practices [69]. Soil moisture sensors such as Tensiometer and capacitive devices measure water availability, critical for optimizing irrigation strategies. Nutrient sensors, like electrochemical devices, assess soil pH, nitrate, and potassium levels, which help in targeted fertilization to enhance yield [7]. Precision agriculture holds promise for the use of flexible temperature and humidity sensors that can detect precise and simultaneous changes in the temperature and humidity of the plant microenvironment [22]. Compaction sensors, such as penetrometers, gauge soil density to improve water infiltration and root penetration. Remote sensing tools e.g., multispectral imaging and LiDAR provide large-scale data on soil properties and erosion, enabling precise land management [2].

The introduction of nanosensors boosts agriculture by detecting critical parameters like pH, EC, organic carbon, N, P, K, enzyme, soil moisture etc. which helps in detecting soil health. Apart from that some nano sensors can detect toxic substances like heavy metal and pesticide residue which have a detrimental impact on soil health. Scientists have developed some nanosensors in recent years by which microorganisms present in the soil can be detected effortlessly. Important nanosensors which contributeto determining and maintaining soil health are represented in Fig. 2.

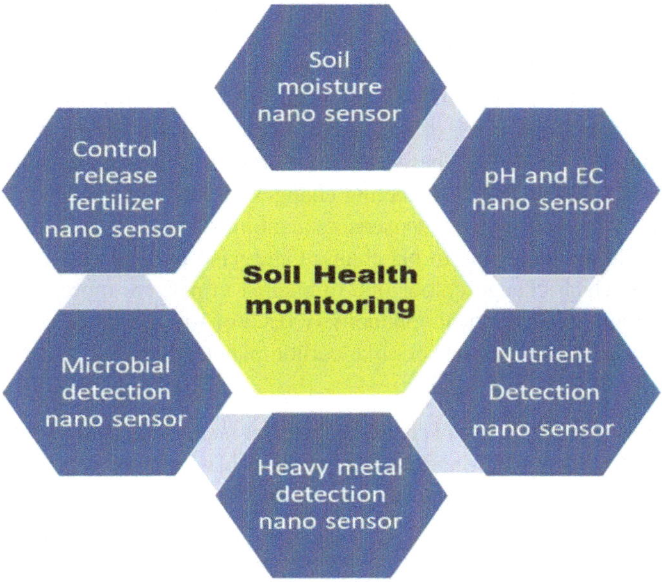

Fig. 2 Nanosensors used to monitor soil healths

Soil Moisture Nano-sensor: Water is a precious natural resource as 4 billion people throughout the globe are facing water scarcity due to inefficient irrigation systems as 70% of global water consumption is used for agricultural purposes [10]. So, it is very critical to detect the moisture content of soil and then schedule irrigation based on their requirement. However, determining soil moisture with the traditional gravimetric or tensiometer method requires a huge amount of skilled labours and money which is difficult to arrange by the small and marginal farmers of India. Moisture nanosensors are introduced as a solution to the problem which is labour or cost-effective and easy to handle. There are several types of nanosensors available in the market which are used to detect soil moisture namely- graphene oxide sensors, carbon nanotube sensors, nanoscale capacitive sensors etc. Graphene oxide has a carbon-based ring structure which naturally does not conduct electricity but when it absorbs moisture the conductivity of the system increases. Two types of graphene oxide are mostly used in nanosensors- ultra-large graphene oxide (lateral size = 47.44 ± 22 μm) and small-size graphene oxide (lateral size = 0.8 ± 0.5 μm) both have high sensitivity towards humidity or moisture [65]. Carbon nanotube sensor is made up of graphene sheets which roll up and form a porous cylindrical structure [72]. Unlike the two-dimensional (2D) structure of graphene sheets, carbon nanotubes have 3D structures which are devoid of functional groups like OH or COOH and detect moisture by analyzing changes in conductivity or resistance due to water absorption. Capacitive nano sensors mainly measure moisture content by accessing dielectric constant with the help of metal oxides, polymers or ceramic plates.

pH and EC nanosensors: Soil pH and EC are generally measured by traditional instruments like glass rod pH and EC meter which are very laborious and time-consuming. Apart from that these are exhaustive methods and continuous reading is not properly collected. So, addressing these problems, the most feasible and realistic approach is the use of nano-sensor-based pH and EC instruments. PH can be determined by different types of nanosensors among which optical or fluorescent, electrochemical and hydrogel-based nanosensors are most important [57]. Optical nanosensors can detect pH by detecting changes in optical properties like absorption or spectra. Among these nanosensorsquantum dot based nanosensors (detect a change in fluorescence due to pH change) and nano-dye sensors (made of silica and polymer which change colour due to change in pH) are more common [16]. Similarly, continuous electrical conductivity (EC) of soil can be measured only by different types of nano EC sensors like carbon nanotube sensors and nanoparticle sensors.

Nutrient Monitoring: The use of chemical fertilizer increasing rapidly globally with a rise in global food demand to increase the overall yield within a short period. This situation leads to a reduction in soil fertility and an increase in nutrient losses ultimately polluting the environment. So, it is very important to check the soil nutrient status after each season from every piece of land. However, due to a lack of facilities and skilled labour, it is very difficult to analyse soil and apply judicious fertilizer based on recommendations. For monitoring essential nutrients (N, P, K, secondary and micronutrients) in soil use of nanosensors are more feasible options over traditional lab-based methods. There are mainly three types of nanosensors used to measure nitrogen content and loss in soil namely optical, electrochemical and carbon tube nanosensors. These sensors detect the concentration of ammonia (NH_4^+)and nitrate (NO_3^-) ions in soil and give precise information about nitrogen status whether it is low, medium or high [53]. Similarly, metal-oxide, nanoporus and optical sensors can measure the phosphate ion concentration in soil [31]. Nano ion selective electrodes and fluorescent and graphene tube-based nanosensors are mainly used to detect potassium ions (K^+).

Detection of Contaminants: Contaminants from different sources like agriculture, industry or household wastewater cause a harmful impact on the soil by deteriorating its quality and killing most of the beneficial microbes. Detection of these toxins, i.e. organic, inorganic contaminants and heavy metals (Pb, Cd, Hg, As etc.) required sophisticated instruments like HPLC, AAS or ICAP-AES which are very costly to establish and maintain [27]. So, for easy detection of heavy metals in soil nanosensors are getting popular because they are cheap and easy to handle. Optical nanosensors are used widely where silver and gold dols are used for detecting heavy metals while carbon nanotubes are used to detect organic and inorganic toxin molecules [51]. Graphane-based nanosensors and metal–organic frameworks are also available in the market which can detect precisely the content of heavy metals [36]. Nanosensors have a wider scope compared to costly instruments to detect contaminants as they have rapid response, high sensitivity, portability and affordable price which make them more feasible.

Improving Soil Health with Nano Sensors

Nanotechnologies discussed above are mostly used to detect different quality parameters of soil and can efficiently improve soil health. Strategies to improve soil health with emerging nanotechnologies are discussed below.

- ***Precision agriculture***—It is a tool to improve agricultural productivity with innovative technologies in a sustainable way where all the inputs are used judicially. Nanotechnologies can be used efficiently to detect the pH, EC or organic carbon content of soil as well as the content of N, P, K and other nutrients so that fertilizers and amendments can be used in a precise manner. If a nutrient deficiency is detected by a nanosensor then nano fertilizers can be used which have higher efficiency and slow nutrient release properties [12]. Nanosensors also help in detecting moisture stress precisely and continuously. Apart from that nano sensors can analyse the overall health of soil or plant and other environmental factors which helps in the decision-making of modern precision agriculture (Table 1).
- ***Site-specific nutrient management***—The Green Revolution deposited huge amounts of chemical fertilizers in soil with the last 60 years which gradually reduced the soil quality and fertility. Scientists are more focused on giving a particular fertilizer when the specific nutrient is deficient which is known as site specific nutrient management (SSNM). SSNM requires continuous detection of primary (N, P, K), secondary (Ca, Mg, S) and micronutrients (Zn, B, Fe) to detect the deficiency and toxicity which can be easily analysed by nanosensors [18].

Table 1 Types of sensors used for monitoring different parameter in precision farming

Sr. no	Sensor types	Technology	Parameter	Application
1	Soil sensor	Tensiometer, Capacitive and resistive type	Soil moisture, PH, EC	Soil health and moisture monitoring
2	Optical sensor	Multispectral, hyperspectral	NDVI, LAI, Spectral data	Crop health monitoring
3	Weather sensor	Thermistors, hygrometers	Temperature, humidity, Rainfall	Climate management
4	Proximal sensor	Gamma-ray, ground penetrating radar	Soil composition, texture	Soil data
5	Remote sensor	LiDAR, RADAR	Satellite, UAV based image	Field data

- ***Detection of soil degradation***—It is very difficult to evaluate soil degradation in the initial phase as analysis of the parameters mostly took place in lab. Nanosensors can detect soil degradation parameters in the standing crop also without disturbing the soil. Nanosensors are efficient in detecting electrical conductivity (EC) at regular intervals and as a result, can detect soil salinity in the initial phase. Information about the heavy metal contamination can be detected effectively so that remediation can be taken if it exceeds its permissible value. Other parameters to detect soil degradation like organic carbon or microbial respiration can be accessed suitably with the help of nanosensors [9].
- ***Improve microbial activity***—Microbial activity is an important parameter to evaluate soil health which includes parameters like microbial count (colony forming unit or cfu), microbial biomass carbon (MBC), respiration quotient, enzymatic activity (Phosphatase, dehydrogenase, urease) etc. Detection of these parameters can be successfully done by modern nanosensors [50]. Early detection of low microbial activity can be improved by applying biofertilizers like Azotobacter or VAM (Vascular arbuscular mycorrhiza). Similarly, low microbial biomass detected by nanosensors can be improved by applying FYM or vermicompost.

7 Crop Health Monitoring

A crop monitoring sensing system includes a group of instruments and technologies intended to collect, examine, and decipher information about the health, development, and environmental circumstances of crops. Precision agriculture relies on these systems to help farmers make well-informed decisions that maximize resource use and yield. NDVI sensors utilize the Normalized Difference Vegetation Index to assess plant health, ecosystem vitality, and productivity and detect stress, aiding early intervention [56, 66]. Multispectral cameras provide data across multiple wavelengths, supporting indices like NDVI to evaluate crop vigor, while hyperspectral sensors offer detailed spectral analysis to identify nutrient deficiencies and stressors [17, 37]. Thermal imaging cameras measure leaf temperature, detect water stress, disease detection and optimise irrigation practices [14]. Drone-based monitoring with UAVs equipped with sensors allows for large-scale surveys, facilitating rapid crop health assessments. Ground-level tools like chlorophyll meters evaluate nitrogen levels via chlorophyll content, and plant health sensors monitor biomass and Leaf Area Index (LAI) for growth tracking [35]. AI-based, IoT and Remote sensing (RS) have emerged as a critical technique for crop disease detection, prediction, and management to reduce the environmental effect of chemical pesticides and facilitate well-informed decision-making in insect pest control [4, 47, 70]. Together, these technologies optimize resource use and improve crop yield potential.

8 Weather Monitoring

Microclimate sensors enabled with IoT are essential for optimizing irrigation and planting by providing real-time, localized weather data [38]. Automated Weather Stations (AWS) are categorised into two types: basic stations, which monitor temperature, humidity, wind speed, and precipitation, and advanced stations which include sensors for solar radiation, soil moisture, and atmospheric pressure. Soil sensors track temperature and moisture at various depths, offering insights into crop-specific conditions, while canopy sensors measure microclimate factors within plant canopies to guide irrigation [15, 24]. Doppler radar helps farmers predict storms and manage risks by providing real-time precipitation data [71]. Remote sensing technologies, such as satellite imagery and drone-based sensors deliver wide-area and localized weather data, assessing cloud cover, vegetation health, and atmospheric conditions [6, 63]. Numerical weather prediction (NWP) models forecast patterns using advanced simulations, while cloud-based platforms integrate data from multiple sources to deliver actionable insights. Farm-specific mobile apps and UAVs equipped with meteorological sensors offer precision data and tailored weather alerts, empowering farmers to make data-driven decisions for better yields.

9 Yield Monitoring

Yield monitoring technology provides precise, often geo-referenced data on field conditions collected with the aid of sensors, drones, tractors, and harvesters, enabling them to analyze crop performance [52]. The information could help to identify possible hazards, maximizing growth prospects, and optimizing yield management. Crop modelling techniques in conjunction with space-born Synthetic Aperture Radar (SAR) technologies could be used to estimate and forecast the growth and yield data [55]. Crop modelling tools like growth simulation models (e.g., APSIM, DSSAT) and phenology models predict crop development stages and yields by incorporating environmental factors, soil properties, and management practices [39, 67]. Field sensors continuously measure soil moisture, nutrient levels, and canopy parameters such as leaf area index (LAI) and chlorophyll content, enabling real-time, site-specific adjustments to improve yield outcomes. Weather forecasting systems and climate models evaluate the impact of climatic variability on crop performance. Geographic Information Systems (GIS) conduct spatial analyses to map yield variability across fields, guiding precision farming. Yield mapping technologies provide real-time harvest data, which informs future planning, while decision support systems (DSS) synthesize diverse data streams to offer actionable insights. Data analytics platforms integrate machine learning and big data, drawing from weather forecasts, sensor networks, and historical records to enhance prediction accuracy. Smart agriculture apps deliver tailored yield forecasts and management recommendations, fostering efficient, sustainable farming practices.

10 Conclusion

Precision agriculture leverages advanced technologies and data analytics to enhance sustainable farming by addressing spatial and temporal field variability. Through tools like GIS and remote sensing, it optimizes resource use efficiency, boosts crop productivity, and reduces environmental impacts. This approach ensures precise input application and supports long-term agricultural sustainability, meeting global food demands while conserving environmental resources. Nanosensors, by harnessing the power of nanotechnology, have opened new possibilities in scientific research and practical applications [46]. Their ability to operate on the nanoscale means they can detect and measure phenomena with unprecedented accuracy. Whether used in biology, chemistry, environmental science, or technology, nanosensors are transforming the way we perceive and interact with the world around us [26]. As research and innovation continue, the scope of nanosensors will undoubtedly expand, driving advancements across diverse fields. Modern developments in agriculture and allied sectors transformed the basic principles of research. Costly and sophisticated instruments used in the laboratory became outdated as cheap and user-friendly nanosensors were already available in the market. But nano sensors are not that popular in developing countries as for installing nano sensors technological knowledge is required. Apart from that, sometimes nanosensors do not give stable reading compared to laboratory instruments which can lead to misinterpretation of the soil health. Though it is a good alternative to lab instruments, due to lack of awareness and a conservative mindsetof farmers are not accepting this in a wider range. Governments and NGOs can break the social bars to enhance the use of nano-sensor-based products for maintaining soil quality. Apart from that there is a huge gap in soil science research as very few sensors have passed through vigorous laboratory tests for their stability and relevancy.

Competing Interests The authors declare that they have no competing interest.

Declaration of Generative AI and AI-Assisted Technologies in the Writing Process During the preparation of this work the book chapter author(s) have not taken any assistance from AI, and the book chapter author(s) reviewed and edited the content as needed and take(s) full responsibility for the content of the publication. All authors read and approved the final version of the manuscript.

References

1. Abdel-Karim R, Reda Y, Abdel-Fattah A (2020) Nanostructured materials-based nanosensors. J Electrochem Soc 167:037554
2. Abdulraheem MI, Zhang W, Li S, Moshayedi AJ, Farooque AA, Hu J (2023) Advancement of remote sensing for soil measurements and applications: a comprehensive review. Sustainability 15(21):15444
3. Adam T, Gopinath SC (2022) Nanosensors: recent perspectives on attainments and future promise of downstream applications. Process Biochem 117:153–173
4. Anam I, Arafat N, Rayed ME, Mridha MF, Shin J (2024, December) PlantNet-Lite: a lightweight CNN approach for plant disease recognition. In: 2024 international conference on decision aid sciences and applications (DASA). IEEE, pp 1–5
5. Ansari MH, Lavhale S, Kalunke RM, Srivastava PL, Pandit V, Gade S, Singh AV (2020) Recent advances in plant nanobionics and nanobiosensors for toxicology applications. Curr Nanosci 16:27–41
6. Arroyo-Mora JP, Kalacska M, Løke T, Schläpfer D, Coops NC, Lucanus O, Leblanc G (2021) Assessing the impact of illumination on UAV pushbroom hyperspectral imagery collected under various cloud cover conditions. Remote Sens Environ 258:112396
7. Bah A, Balasundram SK, Husni MHA (2012) Sensor technologies for precision soil nutrient management and monitoring. Precision Agric 13:501–509
8. Bai W, Tang Z, Gupta S, Diebold GJ (2019) Differential infrared pyrometry for determination of microkelvin temperature variations. Anal Chem 92:2058–2064
9. Bakshi M, Abhilash PC (2020) Nanotechnology for soil remediation: revitalizing the tarnished resource. In: Nano-materials as photocatalysts for degradation of environmental pollutants. Elsevier, pp 345–370. https://doi.org/10.1016/B978-0-12-818598-8.00017-1
10. Belhassan K (2021) Water scarcity management. In: Water safety, security and sustainability: threat detection and mitigation. Springer, pp 443–462. https://doi.org/10.1007/978-3-030-76008-3_19
11. Brahmkhatri V, Pandit P, Rananaware P, D'Souza A, Kurkuri MD (2021) Recent progress in detection of chemical and biological toxins in water using plasmonic nanosensors. Trends Environ Anal Chem 30:e00117
12. Chugh G, Siddique KH, Solaiman ZM (2021) Nanobiotechnology for agriculture: smart technology for combating nutrient deficiencies with nanotoxicity challenges. Sustainability 13:1781. https://doi.org/10.3390/su13041781
13. Darwish MA, Abd-Elaziem W, Elsheikh A, Zayed AA (2024) Advancements in nanomaterials for nanosensors: a comprehensive review. Nanoscale Adv
14. Dong Y, Sloan G, Chappuies J (2024) Open-source time-lapse thermal imaging camera for canopy temperature monitoring. Smart Agric Technol 7:100430
15. Faqir Y, Qayoom A, Erasmus E, Schutte-Smith M, Visser HG (2024) A review on the application of advanced soil and plant sensors in the agriculture sector. Comput Electron Agric 226:109385
16. Gadly T, Chakraborty G, Tyagi M, Patro BS, Dutta B, Potnis A, Ghosh SK (2021) Carbon nano-dot for cancer studies as dual nano-sensor for imaging intracellular temperature or pH variation. Sci Rep 11:24341. https://doi.org/10.1038/s41598-021-03686-x

17. Garcia-Garcia D, la Rosa XR, Bedoya DG, Schrevens E (2021) Linear mixed model analysis of NDVI-based canopy coverage, extracted from sequential UAV multispectral imagery of an open field tomato irrigation experiment. Comput Electron Agric 189:106399
18. Garg S, Rumjit NP, Roy S (2024) Smart agriculture and nanotechnology: technology, challenges, and new perspective. Adv Agrochem 3:115–125. https://doi.org/10.1016/j.aac.2023.11.001
19. Gaur M, Misra C, Yadav AB, Swaroop S, Maolmhuaidh FÓ, Bechelany M, Barhoum A (2021) Biomedical applications of carbon nanomaterials: fullerenes, quantum dots, nanotubes, nanofibers, and graphene. Materials 14:5978
20. Halanaik B (2022) Zero budget-zero wastage natural farming: possible prototype to evergreen revolution in agriculture. Indian Farming 72:5
21. Huang J, Hartemink AE (2020) Soil and environmental issues in sandy soils. Earth Sci Rev 208:103295. https://doi.org/10.1016/j.earscirev.2020.103295
22. Ikram M, Ameer S, Kulsoom F, Sher M, Ahmad A, Zahid A, Chang Y (2024) Flexible temperature and humidity sensors of plants for precision agriculture: current challenges and future roadmap. Comput Electron Agric 226:109449
23. Inobeme A, Adetunji CO, Ajai AI, Inobeme J, Mathew JT, Obar A, Eziukwu C (2023) Chemical nanosensors for monitoring environmental pollution. In: Advanced application of nanotechnology to industrial wastewater. Springer Nature Singapore, pp 93–103
24. Jackson T, Mansfield K, Saafi M, Colman T, Romine P (2008) Measuring soil temperature and moisture using wireless MEMS sensors. Measurement 41:381–390
25. Javaid M, Haleem A, Singh RP, Rab S, Suman R (2021) Exploring the potential of nanosensors: a brief overview. Sens Int 2:100130
26. Johnson MS, Sajeev S, Nair RS (2021, March) Role of Nanosensors in agriculture. In 2021 international conference on computational intelligence and knowledge economy (ICCIKE). IEEE, pp 58–63
27. Kabir MH, Guindo ML, Chen R, Sanaeifar A, Liu F (2022) Application of laser-induced breakdown spectroscopy and chemometrics for the quality evaluation of foods with medicinal properties: a review. Foods 11:2051. https://doi.org/10.3390/foods11142051
28. Klinghammer S, Voitsekhivska T, Licciardello N, Kim K, Baek CK, Cho H, Cuniberti G (2020) Nanosensor-based real-time monitoring of stress biomarkers in human saliva using a portable measurement system. ACS Sens 5:4081–4091
29. Kondawar SB, Pangul CN, Haque MA (2021) Polymer nanofibers via electrospinning for flexible devices. In: Electrospun nanofibers: fabrication, functionalisation and applications, pp 53–86
30. Kumar S, Nidhi Soni S, Sharma KK, Kumar R, Mohan H (2024) Nanosensors: types and customisation. In: Nanotechnology theranostics in livestock diseases and management. Springer Nature Singapore, pp 89–106
31. Lata S, Bhardwaj S, Garg R (2024) Nanomaterials for sensing and biosensing: applications in agri-food diagnostics. Int J Environ Anal Chem 104:4868–4885. https://doi.org/10.1080/03067319.2022.2115895
32. Lay-Ekuakille A, Massaro A, Singh SP, Jabłoński I, Rahman MZU, Spano F (2021) Optoelectronic and nanosensors detection systems: a review. IEEE Sens J 21:12645–12653
33. Leao MB, Grassi MT, Jauris CF, Fernandes AN (2022) Recent nanomaterials development and application in diffusive gradients in thin-film devices. Chem Pap 76:1923–1938
34. Li XH, Li MZ, Li JY, Gao YY, Liu CR, Hao GF (2024) Wearable sensor supports in-situ and continuous monitoring of plant health in precision agriculture era. Plant Biotechnol J 22:1516–1535
35. Lu J, Nie W, Song J, Zhan Q, Wang M, Chen F, Fang W, Chen S, Zhang F, Zhao S, Guan Z (2022) Estimation of nitrogen nutrition index in chrysanthemum using chlorophyll meter readings. Ind Crops Prod 187:115459
36. Mahajan MR, Nangare SN, Patil PO (2023) Nanosize design of carbon dots, graphene quantum dots, and metal–organic frameworks-based sensors for detection of chlorpyrifos in food and water: a review. Microchem J 109056

37. Martins GD, Santos LCS, dos Santos Carmo GJ, da Silva Neto OF, Castoldi R, Machado AIMR, de Oliveira Charlo HC (2023) Multispectral images for estimating morphophysiological and nutritional parameters in cabbage seedlings. Smart Agric Technol 4:100211
38. Mathew TE, Sabu A, Sengan S, Sathiamoorthy J, Prasanth A (2023) Microclimate monitoring system for irrigation water optimization using IoT. Measurement: Sensors 27:100727
39. Mihret YC, Ketsela GM, Mintesinot SM (2024) Implementation and application of APSIM for crop modelling in Ethiopia: a comprehensive review. Heliyon
40. Mishra A, Pant PK, Bhatt P, Singh P, Gangola P (2019) Management of soil system using precision agriculture technology. J Plant Dev Sci 11:73–78
41. Moraille P, Abdali Z, Ramkaran M, Polcari D, Patience GS, Dorval Courchesne NM, Badia A (2022) Experimental methods in chemical engineering: atomic force microscopy—AFM. Can J Chem Eng 100:2778–2806
42. Mustapha UF, Alhassan AW, Jiang DN, Li GL (2021) Sustainable aquaculture development: a review on the roles of cloud computing, internet of things and artificial intelligence (CIA). Rev Aquac 13(4):2076–2091
43. Naresh V, Lee N (2021) A review on biosensors and recent development of nanostructured materials-enabled biosensors. Sens 21:1109
44. Norizan MN, Moklis MH, Demon SZN, Halim NA, Samsuri A, Mohamad IS, Abdullah N (2020) Carbon nanotubes: functionalisation and their application in chemical sensors. RSC Adv 10:43704–43732
45. Nortjé GP, Laker MC (2021) Factors that determine the sorption of mineral elements in soils and their impact on soil and water pollution. Minerals 11(8):821
46. Perdomo SA, Marmolejo-Tejada JM, Jaramillo-Botero A (2021) Bio-nanosensors: fundamentals and recent applications. J Electrochem Soc 168:107506
47. Prasath B, Akila M (2023) IoT-based pest detection and classification using deep features with enhanced deep learning strategies. Eng Appl Artif Intell 121:105985
48. Premnath BJ, Pratima BJ, Ravichandiran R, Srinivasan MK, Nalini N (2024) Recent advancements and applications of nanosensors in various fields. In: Metal oxide–based carbon nanocomposites for environmental remediation and safety. CRC Press, pp 207–229
49. Queiroz DV, Gomes RD, Fonseca IE, Benavente-Peces C, Alencar MS (2020) Using nanosensors in wireless sensor networks. In: Nanosensors for smart cities. Elsevier, pp 515–526
50. Rajput VD, Minkina T, Upadhyay SK, Kumari A, Ranjan A, Mandzhieva S, Verma KK (2022) Nanotechnology in the restoration of polluted soil. Nanomaterials 12:769. https://doi.org/10.3390/nano12050769
51. Rathinavel S, Priyadharshini K, Panda D (2021) A review on carbon nanotube: an overview of synthesis, properties, functionalization, characterization, and the application. Mater Sci Eng B 268:115095. https://doi.org/10.1016/j.mseb.2021.115095
52. Redhu NS, Thakur Z, Yashveer S, Mor P (2022) Artificial intelligence: A way forward for agricultural sciences. In: Bioinformatics in agriculture. Academic Press, pp 641–668
53. Saini R, Devi P (2022) Nanosensors for detecting nutrient losses from soil (as gaseous ammonia and nitrous oxide, and/or soil leachates as nitrate and phosphate). In: Nanosensors for smart agriculture. pp 303–327. https://doi.org/10.1016/B978-0-12-824554-5.00015-X
54. Sargazi S, Fatima I, Kiani MH, Mohammadzadeh V, Arshad R, Bilal M, Behzadmehr R (2022) Fluorescent-based nanosensors for selective detection of a wide range of biological macromolecules: a comprehensive review. Int J Biol Macromol 193:302–319
55. Setiyono T, Nelson A, Holecz F (2014) Remote sensing-based crop yield monitoring and forecasting. In: Crop monitoring for improved food security
56. Shammi SA, Meng Q (2021) Use time series NDVI and EVI to develop dynamic crop growth metrics for yield modeling. Ecol Indic 121:107124
57. Shamsipur M, Molaei K, Molaabasi F, Hosseinkhani S, Taherpour A, Sarparast M, Moosavifard SE, Barati A (2019) Aptamer-based fluorescent biosensing of adenosine triphosphate and cytochrome c via aggregation-induced emission enhancement on novel label-free DNA-capped silver nanoclusters/graphene oxide nanohybrids. ACS Appl Mater Interfaces 11(49):46077–46089

58. Sharma V, Singh PK, Yadav KK, Singh M, Lakhawat SS (2022) Economic feasibility of okra crop grown under sensor based automated drip irrigation system for water conservation. In: Climate smart agriculture (irrigational aspects). Agrotech Publishing Academy, Udaipur, India
59. Sharma V, Bhakar SR, Yadav KK, Singh PK, Lakhawat SS (2020) Effects of advance irrigation technique on crop yield and water use efficiency of Okra Crop. In: National conference on recent advancement in engineering
60. Timmermann C, Gerhards R, Kühbauch W (2003) The economic impact of site-specific weed control. Precision Agric 4(3):249–260
61. Toydemir G, Cekic SD, Ozkan G, Uzunboy S, Avan AN, Capanoglu E, Apak R (2020) Nanosensors for foods. In: Nano-food engineering, vol 1. Springer International Publishing, Cham, pp 327–375
62. Triantafyllou A, Sarigiannidis P, Bibi S (2019) Precision agriculture: a remote sensing monitoring system architecture. Inf 10:348
63. Tyagi R, Pandey PC (2024) Applications of drones in precision agriculture: future of smart and sustainable farming. In: Remote sensing in precision agriculture. Academic Press, pp 429–453
64. Wasilewski T, Brito NF, Szulczyński B, Wojciechowski M, Buda N, Melo ACA, Kamysz W, Gębicki J (2022) Olfactory receptor-based biosensors as potential future tools in medical diagnosis. TrAC Trends Anal Chem 150:116599
65. Wee BH, Khoh WH, Sarker AK, Lee CH, Hong JD (2015) A high-performance moisture sensor based on ultralarge graphene oxide. Nanoscale 7(42):17805–17811
66. Wei Z, Fang W (2024) UV-NDVI for real-time crop health monitoring in vertical farms. Smart Agric Technol 8:100462
67. Wolday K, Hruy G (2015) A review: performance evaluation of crop simulation model (APSIM) in prediction crop growth, development and yield in semi-arid tropics. J Nat Sci Res 5:34–39
68. Yeshe AS, Shinde GU, Gourkhede PH (2023) Application of wireless nano sensors network and nanotechnology in precision agriculture. Int J Adv Agric Sci Technol 10(4):36–65
69. Yin H, Cao Y, Marelli B, Zeng X, Mason AJ, Cao C (2021) Soil sensors and plant wearables for smart and precision agriculture. Adv Mater 33:2007764
70. Yu X, Zheng J (2024) A review on the application of deep learning methods in detection and identification of rice diseases and pests. Comput Mater Contin 78:1
71. Zhong L, Zhang Z, Chen L, Yang J, Zou F (2016) Application of the Doppler weather radar in real-time quality control of hourly gauge precipitation in eastern China. Atmos Res 172:109–118
72. Zhu Z, Cui C, Bai Y, Gao J, Jiang Y, Li B, Wang Y, Zhang Q, Qian W, Wei F (2022) Advances in precise structure control and assembly toward the carbon nanotube industry. Adv Funct Mater 32(11):2109401

Nanosensors, Climate Change and Agricultural Sectors

Maryam Zakavi and Shahnoush Nayeri

Abstract With the world population expected to increase by 10 billion by 2050, food demand will increase, particularly in low-income countries. On the other hand, malnutrition might affect up to 2 billion people, putting the need for urgent agriculture reform at the forefront worldwide. The rapid expansion of populations combined with an unprecedented adoption of unsustainable consumption habits depletes natural resources and causes environmental degradation by eutrophication, disruption of ecology, and global warming. These issues are further driven by the widespread usage of pesticides and fertilizers, destroying ecosystems and health disorders. The solution to these international issues makes nanotechnology attractive, providing novel solutions for achieving sustainable agriculture and food security. Nanomaterials like nanoparticles, nanocomposites, and nano biosensors enable better agricultural performance for better yields and more significant precision farming and environmental monitoring. These technologies have a better likelihood of improving ecological change observation, better exploitation of resources, and reducing environmental impacts, hence contributing to improved and sustainable agriculture. Besides, nanomaterials have been identified to play a role in reducing global warming by capturing greenhouse gases. Despite this potential, agricultural nanotechnology faces expensive costs, technical barriers, and environmental questions. However, further research and development could make nanotechnology a game-changer in agriculture that improves production and resilience to climate while reducing environmental degradation. Indeed, such innovation will help mitigate these challenges and go a long way toward food security and sustainability within the following decades.

Keywords Climate change · Global warming · Nanotechnology · Nanosensors · Sustainable agriculture

M. Zakavi
Department of Cellular and Molecular Biology, Faculty of Life Sciences and Biotechnology, Shahid Beheshti University, Tehran, Iran

S. Nayeri (✉)
Department of Biology, Pennsylvania State University, University Park, PA, United States
e-mail: smn5869@psu.ac.ir

A. Husen (ed.), *Nanobiosensors for Agricultural and Other Related Sectors*, Smart Nanomaterials Technology, https://doi.org/10.1007/978-981-96-6499-3_3

1 Introduction

According to FAO projections, the global population will grow to 10 billion by 2050, increasing food demand by 50%, especially in developing countries. Malnutrition is expected to surge up to 2 billion by 2050; hence, the present structure of global food production needs drastic alteration (FAO 2009). The exponential population growth and consumerist habits quickly deplete natural resources, which has raised great attention as a significant environmental and sustainable development challenge [17]. Widespread use of fertilizers and pesticides increased the chances of eutrophication in both surface water bodies and groundwater, which then lost their essential integrity as ecosystems [1, 19]. Additions of biomass have increased the fertility of the soil. Still, it is ineffective in storing carbon compounds and can't handle nitrogen pollution, causing an overall rise in the ecosystem's temperature, known as global warming [43]. Global warming is the most prevalent environmental impact of high atmospheric CO_2 concentration, and immediate action is required now. "This global temperature rise has been gauged since 1880 at a rate that has more than doubled in the last 40 years" [28]. The presence of greenhouse gases such as CO_2 significantly compromises environmental integrity by depleting the ozone layer, which turns around and affects all various life forms and their health [3]. Global warming causes several issues, such as glacial melting, coastal erosion, flooding, droughts, contamination, and infestations of infectious diseases apart from agricultural growth and productivity [55].

Nanotechnology is a promising tool that provides an opportunity to achieve sustainable agriculture, food productivity, and security approaches. The most potential applications of nanotechnology are in agriculture, environmental science, energy production, food safety, medicine, catalysis, and materials science and industry. Nanotechnology has opened enormous possibilities for preparing sustainable tools that may replace conventional instruments in many fields [24, 37]. These unique attributes of nanomaterials confer advantages in material interactions that may facilitate the transport of clean energy and the adsorption of greenhouse gases [49]. In nanotechnology, nano-sized particles and materials, with at least 1–100 nm length in one dimension, are defined as nanomaterials (NPs) such as nano lubricants, nanoparticles, nanocomposites, nanocoatings, and promising approaches to substitute traditional hazardous material in an eco-friendly manner [13, 25].

Nanomaterials have also been employed to create a variety of devices, such as nanosensors and nano biosensors, which provide groundbreaking answers in areas including environmental monitoring, food safety, and agriculture [9]. Nanosensors can provide immediate information that improves precision farming techniques by detecting specific ecological changes like humidity, temperatures, and nutrient levels. In contrast, nano biosensors utilize biological components to identify diseases, poisons, or pathogens, allowing for early intervention and lowering the need for dangerous chemicals. By reducing resource consumption and environmental effects, these technologies increase agricultural productivity and support safer, more sustainable practices. It assumes a significant increase in agricultural growth rates by about 25% [9, 54].

Moreover, further projections based on its application in agriculture have estimated an increase in economic growth with the use of nano-based strategies in agriculture by up to three trillion USD in recent years. The studies reveal the absorption of greenhouse gases using some promising nanomaterials to reduce the effect of global warming. However, nanomaterials' environmental and health impacts should be cautiously considered [24, 34].

2 Climate Change Effects

Climate change has highly impacted the Earth's climatic system. Some manifestations include increased air, water, and soil temperatures, ocean current changes, ocean acidification, rise in sea levels, hydrological cycle changes, sea ice and glacier melts, and other manifestations. Changes are also related to high incidences of extreme weather events [15]. Here, we discuss the most relevant effects of global warming on environmental and agricultural systems.

2.1 Environmental Impacts

Climate change is one of the most important current topics because it seriously damages the ecological systems. Many scientific studies have recorded the various consequences resulting from the impacts of climate change on the environment. Events associated with global warming have negatively impacted various natural systems. Such changes in the climate system impact coastal flooding, localized flooding, loss of biodiversity, increased thermal stress, impairment of marine ecosystems, and increased frequency of extreme events associated with shifts in seasonality. There is evidence of a substantial increase in the number and area of glacial lakes and the heating of rivers, lakes, and oceans [6].

2.2 Agricultural Impacts

The changes in weather patterns induced by climate change have harmed agriculture productivity. The extreme weather conditions causing severe damage to the quantity and quality of crop yield must be considered to alleviate such weather conditions' effects [5, 34, 37]. Climate change has dramatically influenced the start and duration of growing seasons and the magnitude and duration of heat and water stress in agricultural systems [51]. Drought and increased soil and water salinity are also challenging factors [30].

3 Nanomaterials: Types and Potentials

Nanomaterials, nanoparticles (NPs), and nanostructures represent the materials with dimensions in the nanoscale range (from approximately 1 to 100 nm) and show outstanding physical, chemical, and biological properties. The nanotechnology follows a general overview of the main classes of advanced nanomaterials developed so far, such as super, smart, active, and swarm nanomaterials, and introduces their possible applications [9, 47, 49].

3.1 Super Nanomaterials

Super nanomaterials, or nano pure materials, are defined based on the atomic composition at the optimal geometric arrangement. Such materials possess outstanding strength without defects that advanced nano-engineering techniques could further enhance. Typical examples of super nanomaterials include aircraft wings and diamond bolts; these have shown exceptional properties typical for this material class [40, 57].

3.2 Active Nanomaterials

Sensors, actuators, and computational units are active nanomaterials due to their integration, the abilities possessed by active nanomaterials include environmental perception, the detection of changes in surroundings, and subsequent responses [56].

3.3 Swarms Nanomaterials

Active nanomaterials are swarms of various nanomachines harnessed together to realize some activities; their differences in characteristics and applications are just several reasons why varieties of nanomaterials exist [23].

4 Nano Sensors and Nano Biosensors: General Design and Principle

Nanosensors and nano biosensors devices contain nanostructures to improve output through functional transformation. Various kinds of biosensors are now available (Table 1). Sensors are made from various nanomaterials, such as metallic gold, iron oxide NPs, quantum dots, graphene oxide, and carbon nanotubes [14, 26, 44]. These NPs show unique characteristics, such as nanoconfinement, flexible morphology, and improved optical, mechanical, electrical, and thermal performance of the resulting materials.

4.1 Electrochemical Nanosensors

These are categorized based on their mechanism of transduction, including electrochemical nanosensors with reliance on nanomaterials or carbon nanotube-based electrodes, optical nanosensors based on metallic NPs, up conversion NPs, and quantum

Table 1 Classification of nanobiosensors

Based on detection system	Optical
	Electronic
	Electrical
	Magnetic
	Thermal
	Mechanical
Based on bioreceptors	Enzyme-based
	Aptamer-based
	Whole cells
	Nano-biosensors
	Antibody based
Based on technology	Nano-biosensors
	Biosensors on chip
	Sensor plasmon Resonance
	Electrometers
Based on transducers	Electrochemical
	Acoustic
	Gravimetric
	Thermal
	Electronic
	Optical

dots, and mass nanosensors which make use of the mechanical and piezoelectric properties of microcantilevers and crystals [33].

4.2 *Optical Nanosensors*

Optical biosensors rely on detecting changes in optical signals. They are well suited for numerous spectroscopic techniques, including absorption, fluorescence, phosphorescence, Raman spectroscopy, surface-enhanced Raman scattering, and refraction. Changes that these sensors can detect range over wavelength, phase, time, intensity, and the polarization of light. Many optical methodologies have been applied to biosensors, including fluorescence spectroscopy, surface plasmon resonance, interferometry, and general spectroscopy. Such a phenomenon includes fluorescence and phosphorescence characterized by high sensitivity, fast response times, and high resolution in space. Gold NPs, silver NPs, and quantum dots are essential because their intrinsic fluorescence properties have been widely applied to detect nitrite, reactive oxygen species, pathogenic bacteria, and organophosphates [8, 48].

4.3 *Nanobarcode Technology*

Bio barcode is an outstanding technology that identifies species using a single gene's small segment. Gathering short DNA sequences from various species and making the phylogenetic trees using a distance-based neighbor-joining approach has also been a remarkable application [31]. The most fascinating applications have been the 650-bp segment of mitochondrial cytochrome c oxidase subunit I gene (COI), for taxonomic identification in American birds, Australian fishes, and tropical lepidopterans [12, 36]. This universal DNA-based barcode technology enables sample analysis using PCR-based sequencing analysis. Some other applications include monitoring conditions in food processing, identification of fish species, and adulterant analysis [39].

4.4 *e-NOSE and e-TONGUE*

An e-nose and an e-tongue are functional analogs of human perception of smell and taste, providing essential indications about qualitative and quantitative features of materials in food science, beverage production, agriculture, pharmacology, and even personal care product manufacturing. From the viewpoint of new application fields, it could be considered an alternative to traditional human sensory panels or consumer panels for quality assessment and control at many levels in manufacturing processes to assess consumer satisfaction. The general architecture of the e-nose consists of a headspace sampling system, a sensor array, an electronic data acquisition control system, and pattern recognition software. The electronic tongue is a complementary tool throughout the detection and post-detection analysis phases, with two main differences: selectivity for liquid samples and quantitative evaluations according to taste attributes [50].

4.5 Wireless Nano Sensors and Wireless Sensor Network (WSN)

Using radio-frequency transceivers, sensor nodes, microcontrollers, and power sources, WSNs are intelligent decision-making systems that measure pressure, temperature, and humidity. Being adaptable and low-cost, they have established their practical potential for applications in environmental monitoring, defense, agriculture, and the food industry [41]. WSNs are used in agriculture by deploying in its agronomical inputs control, such as fertilizers and irrigation [29]. WiPAM is an irrigation management system that applies ZigBee IEEE 802.15.4 with the aim of cost-effective and energy-efficient personal area networking. Other uses include greenhouse management, mango and black pepper cultivation, real-time traceability, food chain management systems, and underwater wireless sensor networks for marine fish farming and sustainability monitoring [21].

5 Nanomaterials Used in Nanosensors

Carbon nanotubes represent a family of functional nanomaterials. The unique carbon molecular structure and variable wall construction sizes give these nanomaterials several effective characteristics, such as energy storage, filtration efficiency, and particle packaging properties. Due to this reason, it also finds wide application in influencing growth rates and nutrient transfer in agriculture. Carbon-based nanomaterials usually have high sensitivity in sensing technologies [16]. Indeed, gold, platinum, palladium, and some rare earth metals represent vital components in sensor transducers due to their enhanced sensitivity and potential. Noble metal NPs possess considerable resistance to corrosion and stability. Gold-based electrodes guarantee a fast electron-transfer process between biomolecules and transducers. Silver NPs enhance the sensing upon incorporation into matrices like graphene and poly(amidoamine) dendrimers [4, 11].

6 Nana Sensors: Components and Mechanisms

The nanosensor consists of a probe, transducer, and detector that works together with a bioelement or analyte and provides signals. These are transmitted to farmers, who use the information for further decision-making. Nanotechnology has emerged as an essential discipline for agriculture to enhance the quality of plants and increase the quantity of crop production. One of the most effective methods has been the precision use of nano fertilizers and nano herbicides by delivering nanocarriers or nanocapsules. These developments are part of "Precision Farming," a new strategy that improves yields with minimal effects on soil and water quality [9, 22, 48].

7 NanoParticles

NPs come in the 1–100 nm size range and engender improvements in plant cultivation in growth and development. They may be organic, inorganic, or hybrid and may be presented in such forms as 1-dimensional (D), 2D, and 3D. Green NPs are synthesized using plants and microorganisms that have a higher surface area with active absorption sites that increase the rate of molecular transport [2, 32]. It was observed that adding NPs enhances seedlings' growth, photosynthesis rates, and gas exchange [32]. Precision agriculture has brought information collection methods such as sensors and satellite navigation technologies that could improve yield while reducing costs. Some of the agricultural issues sensor networks face include crop diseases, sustainable irrigation management, and fertilizer application. A sensor-to-server data transmission framework must collect and maintain optimal values to provide farmers with complete reports on various factors [11, 26, 44].

8 Applications in Agriculture

The application of nano biosensors is illustrated in Fig. 1. Accurate detection and quantification of hazardous chemicals in the soil and wastewater are the most important aims of using nanosensors or nanobiosensors, both of which threaten natural resources and agricultural productivity. This class of materials is used for quantitative and qualitative assessment of agricultural soil regarding its fertility and microbial populations [52]. The colloidal gold NPs that were 40 nm in diameter were conjugated with micron-sized polymers through surface modification using 2-amino ethanethiol, yielding micron-sized aptamer-laden gold NPs that functioned as autosensing microspheres. Surface-enhanced Raman spectroscopy has shown a sensitivity threshold of 3.3 $\mu g\ mL^{-1}$. The synthesis of hollow gold NPs with 50 nm diameter was conducted, and their surface-functionalization was done with L-cysteine interacting with the immobilized enzyme AChE and the substrate acetylthiocholine chloride yielded the electroactive product thiocholine. Quantification was obtained based on linear relationships between inhibition percentage and the concentrations of chlorpyrifos and carbofuran. Detection limits were estimated as 0.06 μg dm^{-3} regarding chlorpyrifos and 0.08 μg dm^{-3} in the case of carbofuran. A lateral flow immunochromatographic assay was developed using a gold nanoparticle-labeled antibody for simultaneous detection of pesticides such as carbofuran and triazophos. This allows for rapidly identifying and quantifying these pesticides as environmental contaminants. Metal nanostructures can alter radiative and non-radiative decay rates, changing the fluorescence lifetime and quantum yield [18]. They find applications in agriculture, medicine, and biotechnology. Waterproof sensors from wireless sensors monitor pH in water. Food industry sensors keep products fresh [53]. In medical processes, nanosensors are used to detect diseases at an early stage or to perform drug delivery. They are also used to transfer molecules and modify genes wherein new species are produced [42]. Nanobots employ nanosensors for biotechnology and molecular transportation [20]. Nanosensors are non-destructive and minimally invasive, thus, they offer practical real-time analysis of plant signaling pathways and metabolic processes [46]. The nanosensors that detect the biochemical pathways in plants are presented in Table 2. They can accurately detect food components, soil moisture levels, nutrient requirements, pesticide residues, and crop pests. Their high sensitivity and low detection limits make them ideal sensors for innovative agricultural use [48].

Nanobiosensors have also been applied to determine agricultural inputs like fertilizers, herbicides, insecticides, pathogens, soil texture, and plant hormone regulators. Electrochemical biosensors are gaining favor for their simplicity, sensitivity, specificity, reproducibility, accuracy, ease of use, rapid response time, low cost, and miniaturization [35]. One major disadvantage of optical biosensors is that they sometimes give false positive or negative results and are prone to environmental interference, physical damage, and inconsistencies [38]. Sensing technologies have, until now, enabled the monitoring of multiple plants in every aspect of agricultural settings by employing electrical devices such as mobile phones, hyperspectral cameras, and automated machinery. Chemical signals of plant stress, invisible to human or animal detection, are transduced by sensors into measurable optical, wireless, or electrical signals, thus allowing real-time detection. Wearable sensors have been developed to monitor plant growth, transpiration rate, stress factors, and microclimatic conditions and find a place in pesticide detection. Precision agriculture involves the application of only the quantity of water, nutrients, and agrochemicals for optimal productivity in agriculture. In turn, GIS and GPS interaction allow the remote identification of crop pests and environmental stressors like drought. Nanosensors provide an advantage while monitoring crop growth, analyzing rhizosphere parameters, assessing nutrient content, the effect of toxins in the environment, and microbial infection. In plants, the genome-specific sensors rely on the transgenic materials in intact plant systems while enhancing the fluorescence intensity. Fluorescence resonance energy transfer-based sensors have been applied for sensing sugar levels in leaves and roots of plants under varying abiotic and biotic stresses. Quantum dots have helped to further the optical detection of glucose inside the tissues of plants. MOFs have also been used

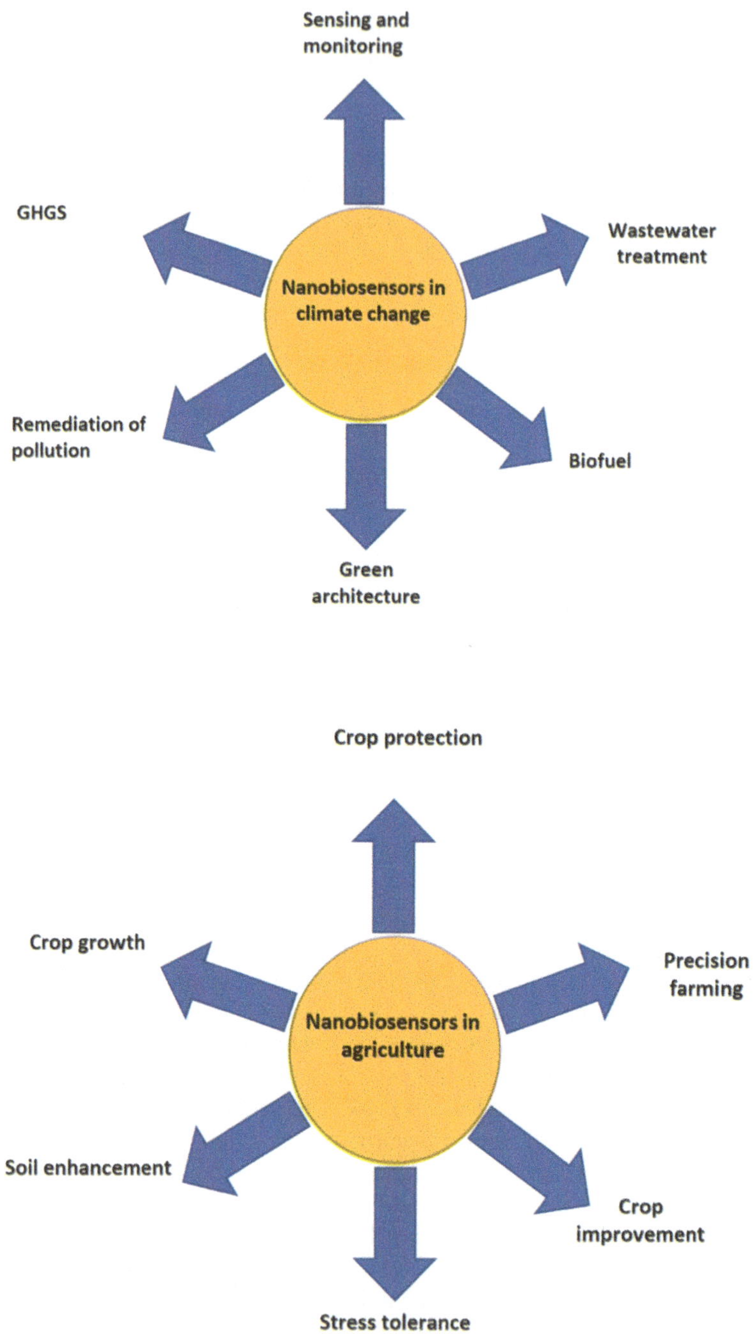

Fig. 1 The applications of Nano biosensors in climate change and sustainable agriculture

Table 2 Nano sensors used for monitoring biochemical pathways in plants

Nano sensors used in monitoring biochemical pathways in plants	Surface-enhanced Raman scattering nanosensors	–
	FRET nanosensors	Genetically encoded FRET nanosensors
		Nanoparticle-based FRET sensors
	Nano lantern-based nanosensors	–
	Nanosensor using corona phase molecular recognition	–
	Electrochemical nanosensors	–
	Piezoelectric nanosensors	–

in plants' in situ fluorescence emission, delivering the acetone concentrations in the atmosphere [20, 22, 27, 46, 48].

9 Benefits and Challenges of Utilizing Nanosensors in Agriculture

Incorporating nanosensors into agriculture offers a transformative potential to boost productivity, optimize resource use, and minimize the negative consequences of climate change. One of the most important advantages of nanosensors is their capacity to facilitate accurate farming. These sensors provide real-time information about crop health, soil conditions, water levels, and environmental elements, helping producers to make informed decisions that maximize efficiency. Nanosensors, for example, can detect nutrient deficits or plant stress at an early stage, allowing for timely interventions to increase crop yields. Furthermore, nanosensors help optimize irrigation and fertilization operations by correctly detecting soil moisture and nutrient content, decreasing resource waste and pollution. This accuracy protects fertilizer and water and reduces the potential of runoff and pollution in surrounding ecosystems [7].

Additionally, nanosensors play a crucial part in controlling diseases and pests. Their remarkable sensitivity enables early identification of diseases and pests, frequently before obvious signs occur. This early detection system minimizes reliance on broad-spectrum insecticides, promotes environmentally acceptable pest control approaches, and saves farmers money. Nanosensors can also be used to measure environmental characteristics such as relative humidity, temperature, and carbon dioxide levels, essential for understanding how climate change affects agricultural systems. Nanosensors assist farmers in adopting climate-resilient techniques, such as selecting crop types suitable for changing circumstances or altering planting schedules, by incorporating this data into prediction models. Nanosensors assist farmers in adopting climate-resilient methods by integrating this data into prediction models. For example, they can help select crop types suited to changing conditions or alter planting schedules to decrease weather risk [32, 35, 45, 48].

Despite these benefits, using nanosensors for agricultural purposes confronts numerous challenges. One significant challenge is the high cost related to nanosensor technology. Nanosensors' intricate design and manufacturing methods frequently make them pricey, limiting their availability, particularly among smallholder farmers in low-income nations. Addressing this issue involves the development of low-cost alternatives as well as government subsidies to encourage widespread adoption. Another difficulty is farmers' lack of information and technical expertise. Nanosensors require a certain amount of technical knowledge and the availability of resources to analyze the data they create and use it effectively. Producers may struggle to fully grasp the promise of these technologies if they do not receive proper training and knowledge [20].

Regulatory and ethical concerns complicate the use of nanosensors in agriculture. The usage of nanoparticles raises concerns about the potential environmental and wellness consequences. For example, issues concerning nanoparticle accumulation in water and soil systems must be addressed by extensive testing and monitoring. Explicit norms and regulations are required to ensure secure nanosensor installation and avoid unforeseen outcomes. Furthermore, integrating nanosensors into agricultural systems frequently requires collecting vast data. This improves privacy and data security issues, particularly when confidential data about farming procedures is shared with other organizations. Farmers require guarantees that their information will be secure and utilized ethically [10, 54].

10 Conclusion and Future Perspectives

Revolutionary nanotechnology can help solve some of the pressing issues that human beings face in the modern era: climate change, sustainable agriculture, and food security. The application of nanomaterials and nanosensors will maximize efficient resource use, reducing ecological deterioration and enhancing agricultural production. Such technologies would lead to improved health of crops, soil condition monitoring, and reduced use of toxic agrochemicals, making agriculture more resilient to climate change. However, there are some serious challenges that must be overcome for the successful adaptation of nanotechnology: high cost, incomplete understanding of technology, and potential health and environmental risks arising from the use of nanoparticles. Cooperation among researchers, legislators, and industry participants is called for in the development of affordable, safe, and widely available nanotechnology. Application ethics and farmers' acceptance will be guaranteed through passing clear legislation and farmer education. Finally, nanotechnology has great potential to impact agricultural and environmental sustainability. Challenges lie ahead, but with further innovation and proper deployment, the security and sustainability of farm production and environmental health in the future can be guaranteed. In summary, while nanosensors provide significant benefits to agriculture, such as increased productivity, resource efficiency, and resilience to climate change, their implementation is not without hurdles. High prices, knowledge gaps, legal barriers, and ethical concerns must all be overcome for this technology to reach its full potential. Collaboration among researchers, politicians, and industry stakeholders is critical to overcoming these obstacles and making nanosensors a viable tool for contemporary agriculture. With the correct support systems, nanosensors can transform farming techniques, improving nutritional safety and environmental sustainability under climate change.

Competing Interests The authors declare that they have no competing interest.

Declaration of Generative AI and AI-Assisted Technologies in the Writing Process During the preparation of this work the book chapter author(s) have not taken any assistance from AI, and the book chapter author(s) reviewed and edited the content as needed and take(s) full responsibility for the content of the publication. All authors read and approved the final version of the manuscript.

References

1. AbuQamar SF, El-Saadony MT, Alkafaas SS et al (2024) Ecological impacts and management strategies of pesticide pollution on aquatic life and human beings. Mar Pollut Bull 206:116613
2. Ahmed S, Chaudhry SA, Ikram S (2017) A review on biogenic synthesis of ZnO nanoparticles using plant extracts and microbes: a prospect towards green chemistry. J Photochem Photobiol B 166:272–284
3. Ajayi F, Adewale AA, Ademola OM Integrating engineering innovations to enhance environmental resilience: evaluating the impact of greenhouse gas emissions, ozone depletion, and aquatic ecosystem degradation on public health. Secur Challenges 1:2
4. Ali Z, Ullah R, Tuzen M et al (2023) Colorimetric sensing of heavy metals on metal doped metal oxide nanocomposites: a review. Trends Environ Anal Chem 37:e00187
5. Arora NK (2019) Impact of climate change on agriculture production and its sustainable solutions. Environ Sustain 2:95–96
6. Babcock RC, Bustamante RH, Fulton EA et al (2019) Severe continental-scale impacts of climate change are happening now: extreme climate events impact marine habitat forming communities along 45% of Australia's coast. Front Mar Sci 6:466674
7. Bharti A, Jain U, Chauhan N (2024) From lab to field: nano-biosensors for real-time plant nutrient tracking. Plant Nano Biol 100079
8. Damborský P, Švitel J, Katrlík J (2016) Optical biosensors. Essays Biochem 60:91–100
9. Darwish MA, Abd-Elaziem W, Elsheikh A, Zayed AA (2024) Advancements in nanomaterials for nanosensors: a comprehensive review. Nanoscale Adv
10. Dubey A, Mailapalli DR (2016) Nanofertilisers, nanopesticides, nanosensors of pest and nanotoxicity in agriculture. Sustain Agric Rev 19:307–330
11. Ekrami E, Pouresmaieli M, Shariati P, Mahmoudifard M (2021) A review on designing biosensors for the detection of trace metals. Appl Geochem 127:104902
12. Elyasigorji Z, Izadpanah M, Hadi F, Zare M (2023) Mitochondrial genes as strong molecular markers for species identification. The Nucleus 66:81–93
13. Findik F (2021) Nanomaterials and their applications. Period Eng Nat Sci 9:62–75
14. Fritea L, Banica F, Costea TO et al (2021) Metal nanoparticles and carbon-based nanomaterials for improved performances of electrochemical (Bio) sensors with biomedical applications. Materials 14:6319

15. Gahlawat IN, Lakra P (2020) Global climate change and its effects. Integr J Soc Sci 7:14–23
16. Geng H, Peng Y, Qu L et al (2020) Structure design and composition engineering of carbon-based nanomaterials for lithium energy storage. Adv Energy Mater 10:1903030
17. Glavič P (2021) Evolution and current challenges of sustainable consumption and production. Sustainability 13:9379
18. Hill LJ, Pinna N, Char K, Pyun J (2015) Colloidal polymers from inorganic nanoparticle monomers. Prog Polym Sci 40:85–120
19. Hossain ME, Shahrukh S, Hossain SA (2022) Chemical fertilizers and pesticides: impacts on soil degradation, groundwater, and human health in Bangladesh. In: Environmental degradation: challenges and strategies for mitigation. Springer, pp 63–92
20. Javaid M, Haleem A, Singh RP et al (2021) Exploring the potential of nanosensors: a brief overview. Sens Int 2:100130
21. John J, Kasbekar GS, Sharma DK et al (2019) Design and implementation of a wireless sensor network for agricultural applications. arXiv preprint arXiv:191009818
22. Johnson MS, Sajeev S, Nair RS (2021) Role of nanosensors in agriculture. In: Proceedings of 2nd IEEE international conference on computational intelligence and knowledge economy, ICCIKE 2021. Institute of Electrical and Electronics Engineers Inc., pp 58–63
23. Jokanović V (2023) The wondrous challenges of the exotic bio-inspired worlds of nanomachines and artificial intelligence. In: Technological prospects and social applications of society 5.0. Chapman and Hall/CRC, pp 47–87
24. Khan F, Pandey P, Upadhyay TK (2022) Applications of nanotechnology-based agrochemicals in food security and sustainable agriculture: an overview. Agriculture 12:1672
25. Khan Y, Sadia H, Ali Shah SZ et al (2022) Classification, synthetic, and characterization approaches to nanoparticles, and their applications in various fields of nanotechnology: a review. Catalysts 12:1386
26. Kumar S, Ahlawat W, Kumar R, Dilbaghi N (2015) Graphene, carbon nanotubes, zinc oxide and gold as elite nanomaterials for fabrication of biosensors for healthcare. Biosens Bioelectron 70:498–503
27. Kwok GAMA, Chong F, Salam A, Makhlouf H Handbook of nanosensors
28. Lindsey R (2021) Climate change: global sea level. Available online: Climate gov (accessed on 14 August 2020)
29. Mowla MN, Mowla N, Shah AFMS et al (2023) Internet of things and wireless sensor networks for smart agriculture applications-a survey. IEEE Access
30. Mukhopadhyay R, Sarkar B, Jat HS et al (2021) Soil salinity under climate change: challenges for sustainable agriculture and food security. J Environ Manage 280:111736
31. Munir S, Ahmed S, Ibrahim M et al (2020) A spellbinding interplay between biological barcoding and nanotechnology. Front Bioeng Biotechnol 8:556291
32. Nayeri S, Dolatyari M, Mouladoost N et al (2023) Ag/ZnO core–shell NPs boost photosynthesis and growth rate in wheat seedlings under simulated full sun spectrum. Sci Rep 13:14385. https://doi.org/10.1038/s41598-023-41575-7
33. Nimal R, Selcuk O, Kurbanoglu S et al (2022) Trends in electrochemical nanosensors for the analysis of antioxidants. TrAC, Trends Anal Chem 153:116626
34. Pandey G (2020) Agri-nanotechnology for sustainable agriculture. Ecol Pract Appl Sustain Agric 229–249
35. Pathak J, Singh DK, Singh PR, et al (2024) Application of nanoparticles in agriculture: nano-based fertilizers, pesticides, herbicides, and nanobiosensors. In: Molecular impacts of nanoparticles on plants and algae. Elsevier, pp 305–331
36. Pentinsaari M (2016) Utility of DNA barcodes in identification and delimitation of beetle species, with insights into COI protein structure across the animal kingdom
37. Prasad R, Bhattacharyya A, Nguyen QD (2017) Nanotechnology in sustainable agriculture: recent developments, challenges, and perspectives. Front Microbiol 8:1014
38. Rasheed S, Kanwal T, Ahmad N et al (2024) Advances and challenges in portable optical biosensors for onsite detection and point-of-care diagnostics. TrAC Trends Anal Chem 117640

39. Rasmussen RS, Morrissey MT (2008) DNA-based methods for the identification of commercial fish and seafood species. Compr Rev Food Sci Food Saf 7:280–295
40. Schiavo L, Cammarano A, Carotenuto G et al (2024) An overview of the advanced nanomaterials science. Inorganica Chim Acta 559:121802
41. Sen S (2023) Wireless sensor networks. IK International Pvt Ltd
42. Shafiee A, Ghadiri E, Kassis J, Atala A (2019) Nanosensors for therapeutic drug monitoring: Implications for transplantation. Nanomedicine 14:2735–2747
43. Sharma HK, Tirpude JB, Meshram YK, Chahande PP (2023) A text book of environmental chemistry and pollution control. Academic Guru Publishing House
44. Sharma P, Kumar S, Patel A et al (2021) Nanomaterials for agricultural and ecological defense applications: active agents and sensors. Wiley Interdiscip Rev Nanomed Nanobiotechnol 13:e1713
45. Sharma P, Pandey V, Sharma MMM et al (2021) A review on biosensors and nanosensors application in agroecosystems. Nanoscale Res Lett 16:1–24
46. Shaw DS, Honeychurch KC (2022) Nanosensor applications in plant science. Biosensors (Basel) 12:675
47. Spence K, Habibi Y, Dufresne A (2011) Nanocellulose-based composites. In: Cellulose fibers: bio-and nano-polymer composites. Springer, pp 179–213
48. Srivastava AK, Dev A, Karmakar S (2018) Nanosensors and nanobiosensors in food and agriculture. Environ Chem Lett 16:161–182
49. Sun Y, Jiang Y, Wei H et al (2024) Nano-enabled strategies for greenhouse gases emission mitigation: a comprehensive review. Nano Today 57:102378
50. Tan J, Xu J (2020) Applications of electronic nose (e-nose) and electronic tongue (e-tongue) in food quality-related properties determination: a review. Artif Intell Agric 4:104–115
51. Teixeira EI, Fischer G, Van Velthuizen H et al (2013) Global hot-spots of heat stress on agricultural crops due to climate change. Agric For Meteorol 170:206–215
52. Thakur A, Kumar A (2022) Recent advances on rapid detection and remediation of environmental pollutants utilizing nanomaterials-based (bio) sensors. Sci Total Environ 834:155219
53. Wang B, Liu K, Wei G et al (2024) A review of advanced sensor technologies for aquatic products freshness assessment in cold chain logistics. Biosensors (Basel) 14:468
54. Yadav A, Yadav K, Ahmad R, Abd-Elsalam KA (2023) Emerging frontiers in nanotechnology for precision agriculture: advancements, hurdles and prospects. Agrochemicals 2:220–256
55. Yadav N, Upadhyay RK (2023) Global effect of climate change on seasonal cycles, vector population and rising challenges of communicable diseases: a review. J Atmos Sci Res 6:21–59
56. Yao S, Ren P, Song R et al (2020) Nanomaterial-enabled flexible and stretchable sensing systems: processing, integration, and applications. Adv Mater 32:1902343
57. Yusaf T, Mahamude ASF, Farhana K et al (2022) A comprehensive review on graphene nanoparticles: preparation, properties, and applications. Sustainability 14:12336

Role of Chemical Nano Sensors for Alleviation of Sustainable Agricultural and Biomass Production

Walaa Yasir, Nadia Sharif, Mona F. A. Dawood, Irfan Aziz, and Zainul Abideen

Abstract Climate change, extreme weather events, and anthropogenic activity pose a serious threat to agriculture and crop productivity. In addition to these, heatwaves, droughts, floods, persistence of leftover pesticide particles, the buildup of heavy metals, and pollution with harmful organic residues negatively influenced the agricultural environment. Additional clues need to be taken in order adapt and protect the agricultural systems against such combination events guarantee food supply security. Thus, it is a current need to ensure global food security. The conventional methods are not adequate enough to overcome such challenges. Hence, alternative approaches are required on an urgent basis. Application of chemical nanosensors could be a long lasting solution towards sustainable agriculture for their accuracy, real-time data on different environmental parameters and crop health indicators, which presents an inventive solution to these problems. These nanosensors possess the capacity to greatly raise the yields of crops, lessen environmental impact, and manage resources more effectively. This chapter aims to discuss various recent applications of chemical nanosensors in sustainable agriculture such as checking the soil health, monitoring the change in soils' pH, soil moisture content, and nutrient levels of the soil, detecting of pollutants and heavy metals, as well as identification of plant diseases. Thus, Smart farming prevents overuse and minimizes runoff that can cause environmental degradation, so in addition to increasing productivity, it also ensures sustainability.

Keywords Agriculture · Growth · Food security · Soil properties · Nanoscience · Sensors · Crop · IOT · Sustainable agriculture

---

W. Yasir · Z. Abideen (✉)
College of Agriculture, University of Al Dhaid, Sharjah, United Arab Emirates
e-mail: zuabideen@uok.edu.pk

N. Sharif
Department of Biotechnology, Women University Mardan, Mardan, KPK, Pakistan

M. F. A. Dawood
Botany and Microbiology Department, Faculty of Science, Assiut University, Assiut, Egypt

I. Aziz · Z. Abideen
Dr. Muhammad Ajmal Khan, Institute of Sustainable Halophyte Utilization, University of Karachi, Karachi, Pakistan

A. Husen (ed.), *Nanobiosensors for Agricultural and Other Related Sectors*, Smart Nanomaterials Technology, https://doi.org/10.1007/978-981-96-6499-3_4

1 Introduction

The constantly rising worldwide demand of food in light of exponential rise in human population besides decreases in crop growth and yield are the major concerns in agriculture. Several biotic and abiotic factors have contributed to limiting food resources hence, conventional approaches have failed to cope up the challenges of food scarcity. Moreover, rapid land degradation and increased production cost have compounded the problem for sustainable food resources in agriculture sectors [74]. The application of synthetic chemicals including pesticides (e.g., organophosphates, carbamates, and organochlorines, pyrethroids etc.) on one hand has enhanced agriculture but, on the other hand it has caused serious public health concerns. Since most of the pesticides are not biodegradable, it has culminated in increased environmental pollution instead of boosting agriculture [21]. Thus, there is a need of using alternative approaches that are both conventional and innovative to minimize the usage of harmful pesticides.

The agroecosystem has largely been compromised due to heavy metal accumulation besides increasing crop diseases [101, 33]. Contaminated foodstuff has serious implications for the consumers [130]. In view of increasing use of harmful chemicals, pathogens, metals, etc., food safety has become the need of the hour. A number of advanced lab procedures that includes spectroscopy, chromatography, stable isotope analysis, proteomics, metabolomics, etc., are in practice to augment farming practices for sustainable agriculture [130] yet, they require more skilled operators and are expensive and time-consuming. As a result, trustworthy, easy-to-use, and on-site screening techniques are essential. Global initiatives to promote sustainability could greatly benefit from the commercialization of nanotechnology. Recent advances in nanotechnology have shown great potential in overcoming such problems [106]. The development of advanced biosensors has significantly boosted nanotechnology in various scientific disciplines [21].

The use of nanosensors has provided greater adaptability and accuracy compared to the conventional methods for their resilient operation in agriculture. Nanosensors may resolve some issues along with conventional methods as they offer consistent evaluation with minimal time and cost commitment, as well as accurate and reliable identification [54]. The use of nanotechnology-based farming methods has resulted in higher yields [9, 54]. Additional research on the use and deployment of sensors has been demonstrated.

Nanosensors are non-invasive devices that combine sensors with nanomaterials and are made especially for monitoring a wide range of environmental samples [149]. Nanosensors for target detection use metal nanoparticles (Au, Ag, etc.), metal oxides (ZnO, TiO_2, MgO, etc.), magnetic nanoparticles (CNPs), dendrimers, QDs nanoparticles, etc. [139]. These portable sensors have great promise for on-site monitoring of soil and plant physiological variables in the agriculture industry and are simple to use [33]. Because of their low detection limit and sensing capabilities, nanosensors are more widely used in sustainable farming methods [33].

Nanosensor for their accuracy to monitor, process, and identify changes, have typically been employed to assist in the advancement of "precise farming" and/or "smart agriculture" [96]. By maximising crop production without the use of artificial fertilisers and pesticides, this technique offers a comprehensive explanation of the data collected from soil or landscape conditions. Furthermore, intelligent farming systems provide the timely assessment and efficient management of costly agrochemicals, abiotic conditions, and phytopathogens—all of which significantly reduce crop output.

Crop monitoring is done in a more accurate way by using intelligent nanosensors [99]. Advanced nanosensors integrated with computers have been proposed by [8] for the enhancement of rural areas' agricultural output. The detection method may be enhanced by connecting the nanosensor to the global positioning system, which will also help farmers in remote areas regulate the usage of pesticides, herbicides, the amount of water used as well as physical and biochemical stessors besides pathogens to hamper plant disease [134].

Based on the aforementioned knowledge, this chapter addresses a brief outline of climate change impacts, and the types of nanosensors that could be used for agriculture sustainability and to keep an eye on the quality of food grains, plant development conditions, and soil health.

2 Climate Change and Agriculture

The agriculture sector is mostly affected by the raw material, feed and harvests resources. In view of rise in global population to boost food production by 35–56% to reach 9.7 billion by 2050, will be needed in the coming decades [36]. Climate change would likely cause serious threats to agriculture and food production systems throughout the world. Heatwaves, droughts, and frequent floods are likely to happen in the coming years which would severely affect the crop production and livestock output. Of them, drought stress has been attributed to more than 65% of crop loss since 2007 and estimates suggest an increase in the overall cost of up to $3. 8 trillion dollars on crop production as well as livestock over the last three decades. Such economic losses have not only resulted in food scarcity but have has also afflicted millions of farmers globally. Moreover, agroecosystem is also facing other dilemmas, like sporadic crop infections, pollution and heavy metal contagion in the agricultural soils [33]. The frequency of such events is continuously rising that has added enormous pressure on agricultural systems [43], therefore, these problems must be solved by adapting proper policy making.

Approximately 874 million individuals, or 27% of the world's workforce, are actively involved in agriculture worldwide. Heavy industrialization and limiting natural resources have resulted in an all-time quality and quantity of food resources. Estimates suggest that between 2000 and 2019, 127 million hectares of agricultural land were degraded. Malnourishment has also increased drastically with each passing year. About 650 million people have been reported to suffer from undernourishment between 2014 to 2019 and by the end of 2020 the projected figures have shown a dramatic increase of 770 million malnourished human population. By 2050, when the world's population is expected to reach about 10 billion people, researchers anticipate that food demand would increase by up to 98%.

In the modern era of industrialization, the widespread application of inorganic fertilisers and injurious substances have significantly contributed to a great imbalance in agroecosystem and serious public health problems. The usage of chemical pesticides increased by 36% between 2000 and 2019, and about 190 million tonnes of inorganic fertiliser were spread on agricultural areas in 2019 alone. Furthermore, greenhouse gas emissions have caused global warming, which has made plants more vulnerable to illnesses and epidemics [169]. The already precarious global food security system may be seriously threatened by climate change, which is predicted to not only reduce the effectiveness of traditional disease control but also directly cause crop growth to decline. However, conventional agricultural practices usually use excessive amounts of chemicals to maximise crop yields [46]. At the moment, 400 million tonnes of fertiliser and 3 million tonnes of pesticide are consumed year worldwide [153]. On the other hand, these agrochemicals have extremely poor utilisation and transportation efficiency [26]. The biogeological cycles of crops have been significantly impacted, and the worldwide mean N:P ratio has changed, since more than half of the nitrogen (N) and 85% of the phosphorus (P) are not absorbed by crops but instead discharged into the environment [116]. Furthermore, according to [46], less than 10% of pesticides are successful in reaching their intended targets. Significant energy and economic losses result from the loss of agrochemicals [124], and soil, surface water, and ground water are all severely contaminated. Therefore, improving the use of agrochemicals by plants or investigating environmentally friendly substitutes that do not jeopardise ecosystems or human health become essential tactics for attaining sustainable agriculture. Careful monitoring of the usage of injurious substances, pathogens, trace metals, etc., in agricultural by-products would ensure food security. Sophisticated lab techniques that are steadfast in detecting unwanted stuff may be used in farming [130] but all such techniques are monotonous, expensive and require highly skilled operators. Thus, it is essential to provide quick, dependable, easy-to-use, and on-site screening techniques.

3 Sustainable Agriculture

Achieving sustainable agriculture and global food security while reducing the negative environmental effects of agriculture is one of the biggest issues of the new millennium [36, 153]. The goal of sustainable agriculture is to satisfy society's present needs for food and textiles without endangering the capacity of future generations to satisfy their own [94].

The literature now in publication has a variety of definitions and metrics for sustainable agriculture. For instance, sustainable agriculture is described as an agricultural method that consistently satisfies fundamental requirements. These circumstances ought to align with socially just farming systems, economic sustainability, and ecological stability. While environmental topics may garner more emphasis, the evaluation of agricultural sustainability stresses the convergence of environmental, economic, and social factors. The need for a more comprehensive approach to agricultural sustainability monitoring and the multifaceted and multidisciplinary nature of sustainable agriculture are reflected in the differences in definitions and metrics of sustainable agriculture. The efforts to define and quantify agricultural sustainability are exemplified by the creation of SDG Indicator 2.4.1. Sustainable agriculture is defined by the FAO using a multifaceted perspective as agricultural practices that can sustainably meet the demands of both current and future generations of people. In addition to not degrading the environment, these agricultural practices should be socially acceptable, commercially feasible, and technically sound.

Sustainable agriculture is paramount for ensuring long-term food security, preserving the environment, and maintaining the health and well-being of communities worldwide. This approach is made to satisfy current demands without endangering the capacity of future generations to satisfy their own. It emphasizes the efficient use of natural resources, including soil and water, and biodiversity, to produce food in a manner that is both ecologically safe and generally liable. According to the Food and Agriculture Organization (FAO), sustainable agricultural practices can significantly enhance productivity, improve resilience to climate change, and reduce greenhouse gas emissions [43]. The FAO emphasizes the need for climate-smart agriculture, or the application of sustainable practices that combine soil health, biodiversity, and overall agricultural system resilience to climate-related stresses in order to adapt to and mitigate the effects of climate change on agriculture.

Meeting the nutritional needs of people worldwide and enhancing economic growth are two major benefits of agriculture. Sustainable agriculture is impacted by a wide range of issues, including population, economic situation, weeds, and insects and pests. To satisfy the demands and obstacles, it is crucial to improve the agricultural aspect's grade. To reach the desired agricultural output, a significant amount of chemicals are needed in pesticides and fertilisers (30–40%). With only 20–40% efficiency, potassium, nitrogen, and phosphorus may not be long-term beneficial to the soil. Therefore, this unfertilised soil in and around the plateau regions affects the overall yield of food crops. Farmers deal with a number of issues, such as hazardous materials or compounds, the economic downturn, the consequences of climate change, and low yield. Managing many risks is more important than ever due to the challenges facing the agriculture sector as a result of the world's population growth, dietary changes that raise the demand for foods derived from animals, and climate change. Several technical innovations must be selected to increase agricultural yields and improve farmer standards in order to overcome these challenging conditions. Such technology involves the use of nanosensors based on the nanotechnology implemented in every concept of agriculture which will be discussed in the following sections.

4 Nanosensors: Definitions, Biosynthesis, and Types

A sensor is a device that tracks a signal and reacts to it by sending out a signal. Physical, chemical, or biological signals are all possible. Non-electrical signals are converted to electrical signals using sensors. According to their need for external energy, sensors can be classified as either passive or active [114]. In order to meet the demands of increased sensitivity, compactness, and intelligence, sensor technology has advanced [21].

Numerous implications have been made regarding the use of sensors for soil monitoring analysis [93], including assessing the soil's state with regard to its organic matter, pH, texture, salt conditions, and nutrient availability and content. In this context, acoustic sensors have been used to measure the bulk density (compaction), soil depth variations (depth of topsoil, depth to hardpan), and soil texture (sand, silt, and clay) [144].

A chemical sensor is a tool that converts chemical data into a signal that can be used in analysis. This data can range from the concentration of a particular sample component to an analysis of the sample's overall composition. A receptor portion and a transducer part are the two fundamental functional units found in most chemical sensors. A membrane known as a separator may be a part of certain sensors. It gives details about the environment's chemical makeup. Usually, it is composed of a chemically selective layer and a physical transducer [66].

Due to their affordability and ease of use, chemical sensors and microbial biosensors have attracted a lot of attention. Chemical sensors and biosensors have a lot of potential for the detection of metallic cations, anions, organic dyes, medications, pesticides, and other harmful pollutants and microorganisms due to their ease of fabrication, rapid detection, high sensitivity, and selectivity, as well as their ease of naked-eye sensing [69, 80, 142]. A performing analytical tool that can analyse soils and has broad applications in environmental monitoring is created by combining chemical sensors with a transducer that can generate a signal proportional to the target analyte concentration [168].

The most advanced and adaptable soil sensors are electrochemical ones [3]. They make use of ion-selective membranes, which react to the activity of particular ions by producing a voltage output. They are separated into a number of categories, including conductometric (which measures conductivity), amperometric (which measures current), and potentiometric (which measures voltage). When an analyte is present, metal-based electrodes are usually employed as a transducer element in all of these sensors. Selectivity and sensitivity towards target chemicals have increased thanks to the development of novel electrode materials. The literature claims that because sensors can identify a set of criteria that are helpful for soil certification, their application in soil analysis has become increasingly significant.

The application of nano sensors in agriculture can augment productivity and overall farm management in addition to decreasing environmental effects and increasing resource efficiency. Larger farm management systems can incorporate the data gathered by nanosensors to give farmers comprehensive insights into the condition and productivity of their crops. With this data, planting schedules can be optimized, crop growth can be tracked, and yield predictions can be made with greater accuracy. Furthermore, agriculture becomes more tolerant of severe weather conditions and the effects of long-term climate change due to the capability to monitor environmental conditions in real-time and better adapt to changing climate conditions. The sustainability and productivity of farming operations can be further increased by combining nanosensors with other technologies, such as drones and satellite imagery, to create a comprehensive precision agriculture system.

Types of Chemical Sensors

Depending on the converter that transforms the chemical change into a processable signal, chemical sensors can be classified as thermal, mass, electrochemical (potentiometric, aerometry, conductivity), or optical [2].

Heat sensors: A good physical factor for sensing, detecting, and measuring temperature changes produced during a reaction is proportional to changes in analytic concentration since heat is a prevalent aspect of chemical reactions.

Mass sensors: Measuring mass change is another appropriate metric for chemical sensors, much as measuring the heat produced by a reaction. This characteristic can be applied to reactions where the release of a selective catalytic reactor causes a change in net mass. These sensors have two key characteristics: first, they can be employed in the liquid phase; second, they are used in immunoassay applications because to their selectivity in the gas phase. Mass sensors can be divided into two main categories: piezoelectric oscillators and surface acoustic waves [143].

Electrochemical sensors: Potentiometric (cell voltage measurement), amphometer (cell current measurement), and conductivity (conductivity measurement) sensors are the three types of electrochemical sensors [44, 84]. Electrochemical sensors are the oldest and largest group of chemical sensors, the response generated in these sensors is due to the interaction between chemistry and electricity. Many of these sensors are currently commercially available and available on the market, and many are in development.

Chemical optical sensors: One of the newest types of chemical sensors is the chemical optical sensor. These sensors should be closely monitored for a number of reasons: The optical components needed for these sensors have previously been created and are readily adaptable to chemical sensors. Their use is more secure because they have numerous uses for remote process control. They can even be positioned on the tip of an optical fibre and are made in tiny sizes [44, 84]. Similar to electrochemical sensors, optical-chemical sensors rely on a wealth of spectroscopic information that is readily transferable to distant sensors. There are numerous applications for optical sensors, including the measurement of luminescence and absorption fluorescence across a broad spectrum of wavelengths, Chemical-optical sensors are divided into several categories based on their application, such as: Immunosensor, pH sensors, gas optical sensors, humidity sensors, ionic optical sensors, sensors used in petrochemistry, etc.

Materials used to make sensors: There are two types of materials used to create the sensors: polymers and classical materials [44, 84].

5 Chemical Versus Biological Nanosensors

As they provide accurate and real-time biological and environmental condition monitoring, nanosensors are a crucial technological advancement in agriculture. Chemical and biological nanosensors are the two main categories of nanosensors. Understanding the differences between the two types is essential to maximizing their use in sustainable agricultural practices, as they have different mechanisms and applications. Chemical nanosensors are intended to identify chemical substances and variations in their ambient concentrations. These nanosensors frequently depend on chemical processes to generate signals that can be detected, like variations in electrical characteristics, color, or fluorescence. For example, by identifying ions or molecules unique to each of these nutrients, chemical nanosensors can be used to track the amounts of nutrients in soil, such as potassium, phosphorus, and nitrogen. This minimizes waste and its negative effects on the environment by enabling precise fertilizer application. Chemical nanosensors can also identify toxins and pollutants in soil and water, allowing farmers to quickly take corrective action to reduce contamination and guarantee the health of their crops (FAOHome) (Federal Reserve Bank of Chicago).

Biological nanosensors, or biosensors, on the other hand, use biological molecules, like enzymes, antibodies, or nucleic acids, to identify biological entities or modifications in biological processes. These highly specific sensors can identify pests, diseases, or plant health-related biomarkers. Biosensors, for instance, can recognize pathogen proteins or DNA sequences, enabling prompt intervention with targeted treatments and early disease outbreak detection. Because of its specificity, fewer broad-spectrum pesticides are used, which lowers chemical inputs and encourages more environmentally friendly pest management techniques. Moreover, biological nanosensors can track physiological alterations in plants, such as stress reactions to nutrient shortages or drought, providing important information for adaptive management plans [128].

Both chemical and biological nanosensors have special benefits, but they also have drawbacks. The accuracy of readings from chemical nanosensors, for example, can occasionally be affected by environmental factors like pH and temperature, necessitating calibration. However, they are extremely useful agricultural uses due to their sturdiness and capacity to measure a broad range of chemical parameters. Although very specific, biological nanosensors can be more difficult to develop and may not be stable in some environments. Notwithstanding these obstacles, developments in nanotechnology are consistently enhancing the functionality and dependability of both kinds of sensors (Federal Reserve Bank of Chicago) (Fifth National Climate Assessment).

Chemical and biological nanosensors can work better together when combined into a single system, offering a more complete monitoring solution. For example, integrating biological nanosensors to track plant health with chemical nanosensors to measure soil nutrient levels can provide a comprehensive picture of the agricultural environment. This coordinated strategy can boost overall sustainability, enhance crop management, and maximize resource utilization. The data produced by these nanosensors will also enable farmers to make informed decisions based on predictive analytics and real-time data, which can also be fed into sophisticated analytical tools and decision-support systems.

A nanosensor is a sensitive, non-invasive device that combines nanomaterials and biosensors. It is specifically made to monitor a wide range of environmental samples [64, 149]. For target detection, nanobiosensors employed magnetic nanoparticles, CNPs, dendrimers, QDs nanoparticles, metal nanoparticles (Au, Ag, etc.), and metal oxides (ZnO, TiO_2, MgO, etc.) [140]. Three parts make up a nanobiosensor: a transducer, a detector, and a biological probe or recognition elements (such as aptamers, antibodies, peptides, and enzymes) [104]. These portable, user-friendly devices have enormous potential in the agricultural industry and have been used to monitor soil and plant physiological conditions on-site, including soil fertility, pH, moisture content, mineral concentrations, levels of toxic chemicals, and early pest and pathogen detection [103, 33]. Because of their low detection limit and sensing capabilities, nanobiosensors have a wide range of applications in sustainable farming methods [33]. The comparison between chemical and bionanosensor has been applied in Table 1.

Table 1 Comparison Between Chemical and Bionanosensor

Type of nanosensor	Characteristics	Applications	Example	Key references
Chemical nanosensors	– Detects chemical substances	– Soil nutrient monitoring	– Sensors for measuring nitrogen, phosphorus, and potassium	[14]
	– Measures variations in ambient concentrations	– Detection of toxins and pollutants	Levels in soil	[17]
		– Precision agriculture	– Sensors for detecting heavy metals in water	Jasim et al. (2016)
Biological nanosensors	– Uses biological molecules (enzymes, antibodies)	– Early disease detection	– Biosensors for detecting specific pathogen proteins or DNA	[71]
	– Highly specific to biological targets	– Monitoring plant health	Sequences	[100]
		– Pest managemet	– Sensors for tracking physiological changes in plants	[82]

6 Applications of Chemical Nanosensors in Agriculture

The development of sensor devices is a new approach, which has many merits over traditional methods, such as simplicity, sensitivity, selectivity and applicability in the field, which remains very desirable for pesticide contamination monitoring (Dutta et al. 2014). Furthermore, using nanomaterials to build sensors and biosensors turns out to be a very effective analytical method for tracking pesticides and other analytes. It contrasts with traditional methods, which have many drawbacks, such as matrix interference, sample decomposition, a limited number of samples in a given time period, and a lengthier test duration. Sensors open the door to more accurate, cost-effective, continuous monitoring, and improved detection capabilities in the detection of pesticides [151] (Fig. 1).

6.1 Soil Health

6.1.1 Detection of Soil Nutrient

Nutrients such as nitrogen (N), phosphorous (P), and potassium (K), or NPK, are essential for plant growth and can be found in large quantities in oil. In addition to essential elements like sulfur, calcium, and magnesium. Determining the soil's nitrogen availability is essential for detecting element-deprived soil. Since these nutrients are the most crucial for plant growth, N, P, and K fertilisers are used to increase agricultural yields. High concentrations, however, have detrimental consequences on the environment and economy. Chemical nanosensors are changing agriculture

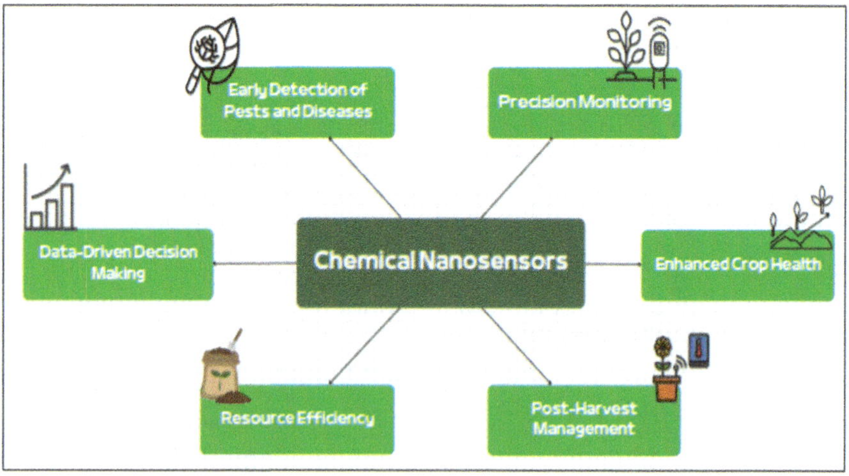

Fig. 1 Role of chemical nanosensors in sustainable agricultural practice

by making it possible to precisely monitor a range of environmental elements that have an impact on crop growth. These sensors can identify the amount and type of nutrients—such as soil nitrogen, phosphorous, and potassium. They can also provide farmers with real-time data that helps them plan their fertilization techniques. This focused approach lowers fertilizer overuse, which can pollute the environment, while simultaneously increasing crop yields. By making certain plants receive the right amount of nutrients at the right time, for example, precision nutrient management enabled by nanosensors promotes ideal growth conditions and reduces runoff of nutrients into bodies of water.

The availability of nitrogen in the soil for plants has been sensed using laser-induced graphene and NIR spectrophotometry technology, among other sensors, to assess vital nutrients in soils [45]. The levels of ammonia nitrogen, nitrate nitrogen, available phosphorus, accessible iron, exchangeable manganese, and exchangeable calcium are determined by photometric detection of soil nutrients using optical sensors [162]. Due to their great sensitivity, single-walled carbon nanotube-based chemi-resistive sensors and silicon nitride microcantilever-based optical sensors have been employed extensively. Chlorophyll meters and other indirect techniques that monitor plant growth and responses are also used to identify nutrients. Phosphorus may also be measured efficiently and affordably using UV and NIR. To determine phosphorus availability in situ, an NIR and cobalt electrochemical approach was proposed, creating a data fusion method to assess phosphorus content [61].

P has also been detected in soil using an on-chip microsensor S or with planar cobalt microelectrodes and an Ag/AgCl reference electrode. According to Jochen et al. (2018), planar cobalt exhibited a selective potential response to phosphorus in an acidic media for both organic and inorganic phosphate molecules (ATP and KH_2PO_4, respectively). This disposal is less costly and simpler to utilise. The Raman sensor, a portable field sensor used to measure P concentration, is another inexpensive sensor. Nutritional measures are typically taken in lab settings using ion-selective electrodes. Ion-selective electrodes are frequently used for quick nutritional analysis by integrating them into microfluidic architectures at the micro- and pico-levels. Phosphate may be accurately and quickly detected in situ using flow loop microfluidic devices combined with sensor-based ion-selective electrodes [30].

Additionally, a variety of membrane materials have been employed to chemically and/or electrochemically measure the amounts of nutrients in soil; the most often used membranes are based on molecularly imprinted polymers and polyvinyl chloride (PVC) ligand membranes. Doped polymers were used to detect microelements like nitrate using tetradecyl ammonium-based o-nitrate, nitrogen using nitrogen-doped polypyrrole (N-doped PPy), potassium using electrodes based on valinomycin-bis(2-ethylhexyl) sebacic acid (DOS), and phosphate using electrodes based on chip microsensors with planar cobalt rods [4]. Similarly, [170] created a miniature phosphate sensor with a phosphate-sensitive response. Furthermore, [13] showed that adding polypyrrole nanowires to ion-selective electrodes for soil nitrate detection increased their sensitivity.

6.1.2 Detection of pH

A measure of the alkalinity or acidity of soil, pH takes into account the entire range of soil-forming elements, including organisms, climate, and organic matter. According to [83], the ideal pH range for soil that supports the growth of the majority of plants and nutrient availability is between 5.5 and 8. Because increased amounts of manganese and aluminium reduce microbial activity and decompose soil organic matter, a shift in soil pH outside of its normal range to an acidic level may be harmful. Conversely, the phosphorous content and other fundamental microelements are reduced when the level is shifted to the basic level [131]. The standard techniques for determining the pH of soil need several steps to collect the specimen and can be costly [107]. Ion-selective electrodes and ion-selective field-effect transistors (ISFETs) are two examples of optical, electrochemical, and chemical sensors that have been used to measure the pH of soil. Colorimetric and photometric indicator names, dyes, and pH test strips are examples of optical techniques that rely on colour changes. The pH level is ascertained using pH-color charts and pH-sensitive organic pigments [79]. In comparison to colorimetric techniques, optical fibre pH sensors, fluorometric pH sensors, holographic pH sensors, and CCD camera full range pH sensors offer significantly greater resolution and accuracy, photometric methods. However, these sensors do not align to in-site work because of the complexity of optical system [79]. Lemos et al. [83] used a composite glass electrode to create an ion-selective electrode sensor for pH analysis. Furthermore, for dysprosium (III), Zamani and Mohaddeszadeh [165] substituted a PVC membrane electrode for the proton-selective glass membrane in order to improve pH sensitivity. At 55 Mv, this ion-selective electrode membrane demonstrated a great sensitivity to pH values between 2 and 10. Furthermore, the performance of ISFET materials can also serve as a substitute for pH sensing, as demonstrated by the potentiometric response to hydrogen ion concentration of silicon nitride (Si_3N_4), silicon oxide (SiO_2), and aluminium oxide [138, 163].

Thus, chemical nanosensors can track the pH change of the soil, guaranteeing that crops are grown under ideal circumstances. Microbial activity and nutrient availability are both impacted by soil pH and are essential for the growth of healthy plants. Farmers can maintain a balanced growing environment by monitoring pH levels and taking appropriate action to amend the soil. This feature allows farmers to better manage soil health overall and customize soil treatments for specific crop needs, which is especially helpful for crops that are sensitive to pH fluctuations. Certain crops, like potatoes and blueberries, need more acidic soils, while other crops, like asparagus, do better in more alkaline environments. To meet these unique needs, soil management can be adjusted with the help of nanosensors.

6.1.3 Detection of Soil Moisture

One of the most crucial factors for determining the amount of water in the soil that influences weather patterns, plant development, groundwater storage, etc. is soil moisture [123]. Soil moisture sensors indirectly measure the volumetric water and express it in volume or weight because direct gravimetric measurement of free-soil moisture requires the removal, drying, and weighing of a sample. In order to maintain appropriate soil conditions, such as selecting fertiliser rates for crops and plants to maintain their physiological activities, it is also crucial to check the moisture content of the soil [113]. The water content of soil and its associated properties, such as dielectric permittivity, matric potential, and volumetric water content, are typically determined using soil moisture sensors (Selma et al. 2022). Examples of sensors has been used for estimating soil moisture, capacitive soil moisture sensors which detect the water content of soil (Jesus et al. 2020), electromagnetic induction-based moisture sensors which allow rapid, high accuracy, and easy detection of water in deep soil layers (Rushikesh et al. 20), ultrasonic-based moisture sensors, and optical fiber sensors (Selma et al. 2022). In contrast to alternative methods, [34] created a tiny sensor that measures and detects soil water using a microelectromechanical system (MEMS). Sulaiman et al. (2009) created a Zigbee-related fringing electric field based on a moisture sensor that can measure soil water content in wide range from 1 to 80%.

Determining the moisture content of soil is another essential use. Effective irrigation management is made possible by the accurate measurements of soil water content that nanosensors can provide. This is especially crucial in areas where there is a shortage of water because it guarantees that water is used wisely, cutting down on waste and encouraging sustainable farming methods. In turn, adequate irrigation promotes robust root growth and increases the general resilience of plants. Nanosensors assist in maintaining ideal soil moisture conditions, which are necessary for nutrient uptake and the avoidance of water-related stress, by preventing over- or under-watering. For example, crops that are extremely sensitive to water stress include tomatoes and cucumbers, and accurate irrigation management can greatly increase both their yield and quality.

6.1.4 Detection of Soil Contaminants and Heavy Metals

The remains of plants, animals, and microbes that break down at different phases are included in the soil's organic composition. Additionally, heavy metals and other harmful compounds can enter soil naturally or as a result of human activity. Pesticides, including herbicides like atrazine, acetochlor, and bifenox; fungicides, including lindane, metalaxyl, procymidone, and penconazole; insecticides, including benomyl, endosulfan, heptachlor, and endrin; PAHs; chlorinated compounds, phenolic compounds, and monomeric aromatic hydrocarbons (such as benzene, toluene, ethylbenzene, and xylene, collectively known as BTEX); and PAHs.

Anthropogenic petroleum hydrocarbon leaks have caused soil contamination, which is now a serious environmental problem, so the detection of hydrocarbon pollution in soil is also crucial. Kaur et al. [70] suggested a sensor for sensing hydrocarbon pollution in the soil. By detecting the amount of CO_2 as contamination indicators, which emitted in soil microbial activity and biota respiration in hydrocarbon-contaminated soils. Following the addition of glucose, a reasonably priced nondispersive infrared CO_2 sensor was applied to measure CO_2 emission in contaminated soils.

A large number of chemicals are released into the water and food supply as a result of the extensive use of pesticides and herbicides, first contaminating the soil (Ragaey et al. 2023). A modified sensor based on a double-template molecularly imprinted polymer gold nanofilm-modified pencil graphite electrode was created by researchers as an effective way to identify herbicides such as glyphosate and glufosinate that include phosphorus-containing amino acids [120]. Dong et al. [39] developed a multiwalled carbon natotubeCeO_2Au nanocomposite electrochemical sensor as tool to detect the methyl parathion. In order to detect picloram-herbicide, Tang et al. [148] created an electrochemical immunosensor based on a conductive chitosan/gold nanoparticle composite membrane. In order to detect herbicide isoproturon, Baskeyfield et al. [18] developed a membrane-based immunosensor that uses a heterogenous competitive enzyme-linked immunosorbent assay with an immobilised isoproturon ovalbumin conjugate. Additionally, a voltametric competitive immunosensor that detects atrazine by immobilising the conjugate atrazine-bovine serum albumin on a gold nanostructured substrate functionalised with poliamidoaminic dendrimers.

However, excessive nitrogen fertilisation with N or new complexes can impact negatively soil and plant growth. Grell et al. [47] created readily available electrical Point-of-Use measurements for NH_4 that enable real-time nitrate level prediction. Additionally, the membrane sensors based on poly (vinyl chloride) and neutral ionophores allow the detection of dysprosium, and better performance has been achieved using sodium tetraphenylborate and various plasticizers [48]. Bojdi et al. [22] created a palladium (II)-selective electrode with remarkable selectivity by using palladium (II) imprinted polymer nanoparticles and sensitivity to measure palladium (II). Alizadeh et al. [7] developed an electrode-based carbon paste to detect 2,4,6-trinitrotoluene (TNT). Additionally, polyphenols were detected using voltammetric and amperometric sensors, which are very sensitive to detect 4-phenoxyphenol and 4-methoxyphenol [108] and their antioxidant activity and electron-giving potential (Selma et al. 2022). Similarly, [152] used a very inexpensive pencil graphite of 6B grade to detect phenols.

To assess the amount of metals like aqueous mercuric, silver, and lead ions, the acrylic mould and selective colorimetric method were created [27]. Individual or more heavy metals were monitored using electrochemical electrodes, this technique can be enhanced by altering the electrode surface to increase sensitivity. Label-free gold nanoparticles (AuNPs)-alkanethiols based on selective sensors were developed by [59] to detect Pb, silver, and mercuric in aqueous solutions. In order to increase the selectivity of Pb (II) ions, [158] also created an improved sensor consisting of two PVC membrane ion-selective electrodes based on two bis-thioureas: 1,3-bis(N′-benzoylthioureido) benzene and 1,3-bis(N′-furoylthioureido) benzene as ionophores. Additionally, a Cu-selective PVC membrane electrode based on a 1-phenyl-2-(2-hydroxyphenylhydrazo) butane-1,3-dione (H2 L) ionophore has been used to perform accurate potentiometric measurement of Cu ions in soils [75]. Using benzothiazole-modified chelating ionophores (1,3-bis[2-(1,3-benzothiazol2-yl)-phenoxy] propane (L1) and 1,20-bis [2-(1,3-benzothiazol2-yl)-phenoxy]2-ethoxyethane (L2)), Gurjaspreet et al. (2021) enhanced an ion-selective electrode sensor for improved Cu detection. Numerous researchers enhanced and combined techniques on the electrode surface, including ion-selective membrane coatings, DNA addition, and nanoparticle coatings. Pb and Hg were found in soil samples using a DNA-based thrombin-binding aptamer probe equipped with the donor carboxyfluorescein (FAM) and 4-([4-(dimethylamino)phenyl] azo) benzoic acid (DABCYL) sensor [88]. DNA aptamer was also employed by [23] as a molecular recognition for Hg ions at the nanomolar level. Furthermore, for quicker combined sample collection and analysis of materials with high sensitivity to Cr and Pb, real-time microfluidic paper-based analytical devices employing passive capillary force have been created [121].

6.2 *Nanosensors for the Detection of Pesticide Residue*

Nanosensors have emerged as powerful tools for detecting pesticides that pose risks to crops and human health. For instance, imazapyr, an herbicide, has been detected using fluorescent nanosensors made by coating ytterbium oxide (Yb_2O_3) with 3-aminopropyl-triethoxysilane via a hydrothermal technique [77]. Atrazine can be detected with atrazine-imprinted nanoparticles produced via emulsion polymerization and integrated with a surface plasmon resonance (SPR)-based affinity sensor [164]. Similarly, fenitrothion detection has been achieved using fiber-optic SPR nanosensors made from tantalum pentoxide (Ta_2O_5) nanoparticles embedded in a reduced graphene oxide matrix and adhered to silver-coated probes [68].

Glyphosate detection employs fluorescence sensors, such as copper oxide/multi-walled carbon nanotubes (CuO/MWCNT), synthesized by precipitating copper nitrate with aqueous NaOH [25]. Electrochemical luminescence sensors, created using graphene-gold nanoparticles and luminol-gold nanoparticle composites, are also effective for glyphosate detection [87]. For methyl parathion, an electrochemical sensor utilizing copper oxide-titanium dioxide (CuO-TiO_2) nanocomposites coated onto a glass carbon electrode has been developed through a liquid-phase precipitation process [150].

Electrochemical nanosensors have also been designed for malathion detection. One approach involves immobilizing aptamers on iron oxide nanoparticles integrated into a chitosan-coated fluorine tin oxide (FTO) electrode using streptavidin. Another uses a copper oxide nanocatalyst on 3D graphene, synthesized via a hydrothermal process, to enhance detection sensitivity [159]. Optical nanosensors for dimethoate detection are based on silver nanodendrites fabricated using a laser-assisted photochemical method and mounted onto optical fibers. Meanwhile, nanosensors for metribuzin utilize lanthanide metal-EDTA complexes integrated with PVC polymer, dioctyl phthalate, and near-infrared dyes to create sensor films [127].

Electrochemical Sensors:

Electrochemical pesticide detection is contingent on the reaction between the analyte and a working electrode that serves as a transducer. Modified electrodes, such as those using nanomaterials, have shown great potential. For instance, an electrode made with poly(3,4-ethylenedioxythiophene) (PEDOT) and carbon nanotubes (CNT) successfully detected carbamates in water using cyclic voltammetry (CV), demonstrating high electrocatalytic activity and economic advantages [166]. Other innovations include carbon nanosphere-based screen-printed electrodes for detecting carbaryl, carbofuran, and isoprocarb in agricultural samples [35], as well as gold nanoparticle electrodes for selective detection of carbofuran in fruits and vegetables [147]. Electrochemical nanosensors using fluorine tin oxide (FTO) decorated with gold nanoparticles have proven effective for chlorpyrifos detection due to their simplicity, stability, and miniaturization capabilities [146]. Similarly, glyphosate has been detected with high sensitivity using layered double hydroxide (LDH) electrodes composed of nickel-aluminum hydroxides [73].

Optical Nanosensors:

Optical nanosensors, including those founded on surface plasmon resonance (SPR) and fluorescence techniques, are widely utilized due to their high sensitivity. SPR-based immunosensors allow real-time detection of pesticide residues by leveraging molecular interactions. Fluorescence-based sensors, such as carbon quantum dots, have been established for carbofuran finding in food samples with impressive sensitivity [24]. Additionally, surface-enhanced Raman spectroscopy (SERS) coupled with nanomaterials has shown exceptional signal enhancement, making it ideal for pesticide detection in fruits and vegetables [89].

Chemiluminescence Sensors:

Chemiluminescence sensors rely on light emission from electron transfer reactions at the electrode surface. For example, silver nanoparticles functionalized with luminol have been used for carbamate pesticide detection, achieving a low detection limit [167]. Electrochemiluminescence (ECL) sensors, incorporating zinc sulfide quantum dots in mesoporous carbon, have demonstrated high sensitivity for glyphosate detection in vegetable juice samples [111].

Piezoelectric Sensors:

Piezoelectric sensors, like quartz crystal microbalance (QCM)-based immunosensors, are effective for detecting pesticides like carbaryl, parathion, and chlorpyrifos in water and agricultural products. These sensors offer high thermal stability and sensitivity, making them suitable for on-site inquiry [1, 97].

Advancements in nanosensor technology have significantly enhanced pesticide detection, offering rapid, sensitive, and cost-effective solutions. From electrochemical and optical approaches to chemiluminescence and piezoelectric sensors, these technologies provide real-time monitoring capabilities critical for environmental protection and food safety. Further research into improving sensitivity, selectivity, and portability will ensure their broader application in agriculture and environmental monitoring.

6.3 *Detection of Pest/Pathogen in the Plants*

The development of numerous sensor platforms has been fuelled by the need for quick, affordable, and user-friendly technologies. These platforms allow for the label-free detection of target pathogens with high sensitivity and specificity, surpassing the limitations of conventional diagnosis methods and the need for qualified scientists. Using a quartz crystal microbalance (QCM) to create an immunosensor for the detection of maize chlorotic mottle virus (MCMV), an example of label-free detection is provided. This is because a change in mass causes a variation in the crystal's resonance frequency [58]. Surface plasmon resonance (SPR) was used by [85] as a label-free detection method to track variations in the sensor surface's refractive index brought on by the interaction of the analyte in solution with an immobilized ligand. They employed gold nanorods (AuNRs) functionalised with antibodies against either the Odontoglossum ringspot virus (ORSV) or the Cymbidium mosaic virus (CymMV), two viruses that infect orchids. Another pertinent method for detecting Alternaria mycotoxins in pear fruit is the use of silver nanoparticles (AgNPs) and surface-enhanced Raman spectroscopy (SERS) [112].

Additionally, crop sciences rely on electrochemical impedance spectroscopy sensors to identify plant viruses or pathogens [62]. The ability to respond quickly without destroying the sample and the availability of portable readers that can track changes in the device impedance upon the specific recognition of the target analytes make EIS a useful technique for on-field analysis in addition to its simplicity and sensitivity. Citrus tristeza virus (CTV) can be detected selectively using a DNA hybridisation sensor based on screen-printed carbon electrodes modified with gold nanoparticles (AuNPs) even when other non-specific DNAs are present [72]. This feature is especially helpful in cases of mixed infections, which are fairly common for cultivated plants. Rather, utilising colloidal gold nanoparticles for antibody immobilisation in extracts from tobacco (*Nicotiana benthamiana*) and plum (*Prunus domestica*) leaves, a selective electrochemical immunosensor was created for the detection of Plum Pox virus (PPV). Samples of healthy plants and those that included 0.01% extract from infected plant material could be distinguished by this sensor. Using glassy carbon electrodes as platforms and transducers, the same group then developed an evolution of this platform for the detection of Prunus necrotic ringspot virus (PNRSV) [60]. In addition to viruses, *X. fastidiosa* subsp. pauca bacterial strain De Donno can be detected by an electrochemical impedance biochip in both naturally infected and asymptomatic olive trees [31].

Because they are produced by plants and released as a defence mechanism against pathogen attack, volatile organic compounds (VOCs) serve as an indirect method of plant pathogen detection [129]. Gas chromatography–mass spectrometry (GC–MS)-based techniques are typically used to detect volatile organic compounds (VOCs).

These techniques are complicated, time-consuming, costly, large, and require a great deal of training to use correctly [136]. Electronic noise (EN), one of the rapid, simple, and alternative techniques, appears to be a good way to identify volatile organic compounds. A multimodal array, an artificial neural network-based information processing unit, software with digital pattern recognition algorithms, and reference library databases make up this system [157]. T serves a variety of purposes in plants, such as forestry and agriculture [156], food quality [11, 117], plant pest monitoring [80], and the automotive industry [63]. However, due to atmospheric interference, detection in open fields is challenging, necessitating additional advancements [32]. A completely automated portable GC apparatus for in situ analysis was developed, overcoming the majority of the restrictions [136]. Aphids contaminated half of the ten milkweed (*Asclepias syriaca*) plants used in the studyWhen compared to VOC changes found in other research, the ability to distinguish between healthy and infected plants with an accuracy of 90–100% within 48–72 h of attack and 3–4 days earlier was demonstrated by the separation and detection of 35 VOC peaks in eight minutes. In the asymptomatic stage of Huanglongbing-infected citrus trees, four volatile organic compounds (VOCs)—ethylhexanol, linalool, tetradecene, and phenylacetaldehyde—can be detected using a novel electrical biosensor array based on single-walled carbon nanotubes (SWNTs) functionalised with single-stranded DNA (ssDNA) (Wang et al., 2019). Several mathematical models were used to distinguish between different species and concentrations of volatile organic compounds. With quick reaction and recovery durations, the functionalisation with ssDNA makes up for the lack of selectivity displayed by SWNT-based sensors (Staii et al. 2005) by distinguishing between a range of odoursn comparison to bare SWNTs, the ssDNA-SWNT devices demonstrated superior sensitivity and outstanding repeatability while taking into account the concentration's validated range of analytes. Because of its low specificity, this method is still challenging to use in the field. To improve its suitability for real-time detection, the scientists suggested concentrating on combinations of volatile organic compounds (VOCs) released by diseased citrus trees.

In order to enable non-invasive diagnosis of late blight, which is caused by Phytophthora infestans, a smartphone-based fingerprinting of leaf volatiles was also recently reported (Li et al. 2019). Plasmonic nanomaterials that target green leafy aldehyde and (E)-2-hexenal are employed as chemical sensors and transducersprimary late blight marker), demonstrating high sensitivity down to the sub-ppm threshold of LOD. With a disease detection accuracy of over 95%, the system was able to distinguish between 10 distinct plant volatiles, enabling an earlier diagnosis prior to the onset of symptoms in both laboratory-inoculated and field-collected tomato leaves. VOCs can serve as an indirect warning system for the existence of a disease, but they are unable to pinpoint the pathogen that is causing it. The following sections provide examples of wearable sensors and Internet of Things (IoT) technologies to address the demand for online plant disease monitoring. Different applications of the chemical nanosensors have been provided in Table 2.

Table 2 Different applications of the chemical nanosensors

Study	Nanosensor	Target	Result	Key references
Detection of nitrogen levels in soil for precision farming	Zinc oxide (ZnO) nanosensor	Soil nitrogen content	Achieved an exposure limit of 0.5 mg/L and enhanced nitrogen use efficiency by 25%	[14]
Nanosensors for early detection of drought stress in crops	Silver nanoparticle sensor	Water content in plant leaves	Detected water stress 5 days earlier than visual symptoms with 92% accuracy	[17]
Real-time monitoring of soil pH using carbon-based Nanosensors	Graphene oxide nanosensor	Soil pH levels	pH levels monitored with 98% accuracy in variable field conditions	[6]
Detection of pesticide residue in tomato plants using gold nanoparticles	Gold nanoparticle sensor	Organophosphate pesticide residue	Detected pesticide residue at concentrations as low as 0.005 mg/L	[19]
Nanosensors for monitoring soil moisture in smart irrigation	Silicon-based nanosensor	Soil moisture levels	Improved irrigation efficiency by 30%, reducing water usage by 15%	[105]
Detection of heavy metals in soil using nanocomposites	Iron oxide nanoparticle sensor	Lead and cadmium levels in soil	Detected lead and cadmium concentrations at 0.1 mg/L with 95% accuracy	Nhan et al. (2015)
Real-time detection of nutrient deficiency in crops	Quantum dot nanosensor	Essential nutrient levels	Detected nutrient deficiencies with 90% precision and enabled timely fertilizer application	[141]
Monitoring crop disease outbreaks	Carbon nanotube (CNT) sensor	Pathogen presence in crops	Identified fungal pathogens with 93% accuracy before visible symptoms appeared	[20]

(continued)

Table 2 (continued)

Study	Nanosensor	Target	Result	Key references
Detection of airborne pollutants in greenhouses	Titanium dioxide (TiO2) nanosensor	Volatile organic compounds (VOCs)	Monitored VOC levels and provided early warning for potential plant stress with 97% accuracy	
Detection of soil microbial activity	Zinc sulfide (ZnS) quantum dots	Enzyme activity in soil	Measured soil enzyme levels with 94% accuracy, indicating microbial health and soil fertility	[145]
Continuous monitoring of water quality in aquaponics	Carbon dot nanosensor	Nutrient and contaminant levels in water	Detected contaminant levels as low as 0.01 mg/L with 96% accuracy, ensuring optimal fish and plant growth	[42]
Real-time detection of crop maturity	Copper oxide (CuO) nanoparticle sensor	Ripeness indicators in fruit	Provided accurate ripeness data with 95% reliability, enabling optimized harvest timing	[161]
Monitoring fertilizer efficiency	Magnesium oxide (MgO) nanosensor	Nutrient uptake efficiency in soil	Enhanced fertilizer use by 20% with real-time data on nutrient absorption	[100]
Detection of soil salinity for precision farming	Platinum nanoparticle sensor	Salt concentration in soil	Measured soil salinity with 97% accuracy to prevent crop damage due to high salt levels	[55]
Real-time plant health monitoring	Nano-diamond sensor	Chlorophyll content in leaves	Achieved a 98% accuracy in detecting plant health status based on chlorophyll fluorescence	[119]

(continued)

Table 2 (continued)

Study	Nanosensor	Target	Result	Key references
Monitoring heavy metal uptake in plants	Cadmium sulfide (CdS) quantum dots	Heavy metal absorption in plant tissue	Detected cadmium uptake in plants with 92% precision, aiding in crop safety assessments	[67]
Early detection of root rot	Silicon nanowire sensor	Root infection markers	Detected root rot 7 days before visible symptoms, achieving a 91% accuracy rate	[81]
Detection of microplastic contamination in soil	Polystyrene nanoparticle sensor	Microplastic particles	Identified microplastic contamination at concentrations as low as 0.01% with 95% accuracy	[91]
Rapid detection of viral pathogens in plants	Gold-silver bimetallic nanoparticle sensor	Viral DNA/RNA in crops	Detected plant viruses within 30 min with 94% sensitivity, enabling faster intervention	[109]
Monitoring temperature stress in crops	Thermosensitive polymer-based nanosensor	Heat stress markers in leaves	Detected temperature fluctuations with 96% accuracy, allowing for timely interventions	[137]

7 IoT and Remote Sensing Technologies

The agricultural sector and food industry face significant challenges, including population growth, climate change, and emerging plant diseases. Nanotechnology and the Internet of Things (IoT) offer sustainable solutions to address these issues [95]. Key advancements like "real-time communication" and "wireless sensing" are central to modern agricultural innovations. "Smart farming" incorporates advanced information and communication technologies (ICT), such as remote sensing, cloud computing, and IoT, enabling farmers to monitor field conditions remotely or with in-field support [56]. Additionally, robotics plays a crucial role in tasks such as seedling management, fruit harvesting, plant protection, and weed control, further enhancing agricultural efficiency and precision [10].

Remote sensing collects data on objects or environments via satellites, UAVs, or probes, enabling non-destructive monitoring of agricultural systems to reduce resource depletion and environmental impacts. It analyzes crop stress, biophysical traits, and early stress-induced changes like chlorophyll content and surface temperature (Mola et al. 2013). Since the 1980s, remote sensing in precision farming has advanced from visible and NIR bands to hyperspectral imaging, which monitors plant events across spectral regions, such as pathogen detection in VIS, chlorophyll degradation in VIS/red-edge, and transpiration changes in TIR [155]. Hyperspectral radiometers can also estimate photosystem II efficiency (Fv/Fm), correlating stress with NIR reflectance changes [115].

Agricultural drones (UAVs) assist in planting, pest control, and crop monitoring. The eBee SQ drone integrates multispectral imaging and cloud-based analysis, covering vast areas [133]. Aircraft and satellites provide remote sensing data for land use, nitrogen flow, and terrain mapping via LiDAR and hyperspectral imaging [110, 125]. IoT applications in agriculture leverage sensors, RFID, and cloud computing for crop monitoring, traceability, and pest detection [15, 92]. Wireless networks collect data on temperature, humidity, and greenhouse conditions. RFID aids in plant tracking and phytosanitary management. Robotic systems process crop images in real time for pest control [126], while platforms like VineSens monitor vineyard conditions and prevent diseases. Tools like "Arable" and "Semios" deliver real-time stress and pest data via IoT [12, 132]. Remote sensing identifies pathogens like *Xylella fastidiosa* using hyperspectral and thermal imagery [57, 118]. Advances include nano-sensors for precise data collection in inaccessible areas [16]. Big data applications in agriculture continue to evolve, enhancing ecosystem monitoring and crop management [65].

8 Limitations of Nanosensors in Agriculture

The practice of nanosensors in agriculture has certain disadvantages despite their many advantages. The price of these cutting-edge technologies is one of the main issues. For small-scale and resource-constrained farmers, the high cost of producing and deploying nanosensors can be a barrier. This financial obstacle prevents nanosensors from being widely used, which could worsen productivity disparities in agriculture. This could lead to unequal access to cutting-edge farming equipment since smallholder farmers, especially those in developing nations, might find it difficult to afford these technologies. The overall advancement of agricultural sustainability and productivity may be hampered by this economic disparity [5].

The possible effects of nanoscale materials on the environment are a significant additional worry. There are concerns about the safety of nanoparticles because it is unclear how they will affect agricultural ecosystems in the long run. These substances run the risk of building up in the soil or water and having unanticipated ecological effects. Although studies are being conducted to gain a deeper understanding of these interactions, the environmental risks continue to be of great concern until complete safety data is obtained. Cautious consideration should be given to the persistence of nanoparticles in the environment and their potential to disturb aquatic ecosystems and soil microbial communities. For example, some nanoparticles may disrupt the processes involved in naturally remediating soil or have an adverse effect on the health of beneficial soil organisms like earthworms and mycorrhizal fungi [135].

Threats to non-target organisms are also posed by the problem of bioaccumulation. Plants could absorb nanosensors and their byproducts, which can then enter the food chain and have an impact on human health and wildlife. To make sure that nanoparticles do not endanger living things, it is important to thoroughly investigate their toxicity at every stage of their lifecycle, from production to degradation. In order to develop safe and sustainable nanosensor technologies, it is essential to comprehend the pathways through which nanoparticles can accumulate and their potential effects on the health of humans and animals. For instance, nanoparticles used in nanosensors may be absorbed by plants and subsequently passed on to herbivores, building up in higher trophic levels and presenting hazards to humans and predators who eat these plants and animals.

Finally, considerable modifications to infrastructure and practices are needed for the incorporation of nanosensors into current agricultural systems. Effective use of these technologies by farmers requires training, and implementing new practices may encounter resistance. To guarantee that the advantages of nanosensors are fully realized without interfering with conventional farming operations, a significant amount of support and education will be required during the potentially difficult transition period. To overcome adoption barriers and achieve widespread utilization, it is imperative to guarantee that farmers receive sufficient information about the advantages and safe application of nanosensors. This could involve extensive training courses, pilot projects, and ongoing technical assistance to ensure farmers are able to seamlessly incorporate nanosensors into their operations.

Strong regulatory frameworks are also required to control the use of nanosensors in agriculture. The safe development, use, and disposal of nanosensors should be covered by these regulations to protect the environment and public health. In the absence of appropriate regulations, the improper or excessive use of nanosensors may have unforeseen consequences that pretense a risk to human health and the environment. To reduce these risks and encourage the responsible utilization of nanosensors in agriculture, standard operating procedures for their use must be developed.

Although chemical nanosensors have many advantages, it is crucial to use them carefully to reduce any risks. Nanosensors must be handled and disposed of properly to avoid contaminating the environment. It is important that farmers and other agricultural workers receive training on how to use these devices safely, including knowing when and under what circumstances to use them and the possible consequences of misusing them. The development of knowledge and proficiency in the safe handling of nanosensors can be facilitated by training programs and educational initiatives. To teach farmers how to properly install, calibrate, and maintain nanosensors, for example, workshops and training sessions can be arranged.

To prevent overuse and possible toxicity, it's also critical to set guidelines for the dosage and frequency of nanosensor applications. Overuse of nanosensors may cause nanoparticles to build up in the soil, which could be harmful to microbial communities and the health of the soil. The efficient and sustainable utilization of nanosensors can be ensured by creating standardized protocols for their use. To optimize the advantages while lowering risks, farmers should be urged to adhere to suggested usage guidelines and best practices. For instance, recommendations for the ideal density and positioning of nanosensors in various fields can aid in avoiding over-concentration and guaranteeing precise gathering of data.

To guarantee the safe application of nanosensors in agriculture, regulatory oversight is also essential. It is recommended that governments and regulatory agencies create frameworks incorporating safety assessments, environmental impact assessments, and safe usage guidelines. These rules can ensure that nanosensors are not misused and that there are no major menaces to human health or the environment when they are deployed. To create comprehensive regulatory standards, researchers, legislators, and stakeholders in agriculture must work together. For example, regulatory bodies and academic institutions can collaborate to carry out environmental risk assessments and set safe thresholds for the concentrations of nanoparticles in agricultural soils [53, 154].

Consequently, continued investigation is necessary to assess the long-term impacts of nanosensors on human health and the environment. Programs for monitoring the effects of nanosensors on the environment should be put in place so that any new issues can be quickly addressed. The agricultural industry can take advantage of chemical nanosensors whereas conserving the environment and public health by remaining vigilant and revising safety protocols considering new scientific discoveries. To gather important information for safe usage practices, long-term field studies can be carried out to track the effects of nanosensors on soil health, crop yield, and biodiversity.

9 The Future Prospects of Chemical Nanosensors

Chemical nanosensors in agriculture appear to have a bright future thanks to ongoing breakthroughs in nanotechnology and growing public awareness of sustainable farming methods. More advanced nanosensors that have greater sensitivity and specificity and can detect a wider range of parameters are being developed by researchers. The increased precision with which soil and plant health can be monitored thanks to these next-generation sensors will further optimize agricultural inputs. It is anticipated that advances in sensor designs and materials will improve the robustness and durability of nanosensors, increasing their suitability for extended field applications. For instance, research is being done to enhance the sensitivity and durability of nanosensors using developments in nanomaterials like graphene and carbon nanotubes [160].

The combination of Internet of Things (IoT) technology and nanosensors is one of the exciting prospects. Farmers can get real-time data on their computers or smartphones through the network connection of nanosensors, which allows for automated decision-making and remote monitoring. Precision agriculture enabled by the Internet of Things (IoT) has the probable to greatly increase productivity and sustainability by improving crop management techniques, lowering labor costs, and increasing efficiency. Agricultural practices can be further improved by applying machine learning algorithms to analyze data from nanosensors, providing predictive insights and optimizing resource allocation. For example, IoT-integrated nanosensors can offer predictive analytics for pest outbreaks, enabling farmers to minimize crop damage and take preventative measures.

To make nanosensors more affordable for small-scale farmers, efforts are also being made to lower their cost. Lower production costs due to advancements in materials and manufacturing techniques should enable a broader adoption of nanosensor technology among various farming communities. By bridging the gap between large and small-scale farmers, this democratization of technology can advance agricultural productivity equity. Government support initiatives and subsidies may also be very important in lowering the cost of nanosensors for all farmers and guaranteeing that everyone has access to cutting-edge farming equipment. Public–private partnerships have the potential to enhance the production and distribution of nanosensors, thereby rendering them more cost-effective for farmers operating in developing regions.

What's more, one important area of continuing research is the effect of nanosensors on the environment. To reduce any possible ecological risks, scientists are investigating biodegradable nanoparticles and eco-friendly materials. It is possible to promote the widespread use of nanosensors without jeopardizing the ecological health of the ecosystem by guaranteeing their environmental safety. Regulatory framework development will be indispensable to guaranteeing the liable and innocuous application of nanosensors in agriculture. Regulations can reduce risks and guarantee that the use of nanosensors complies with environmental preservation objectives. Using biodegradable nanomaterials, for instance, can lower the chance of long-term environmental contamination. To fully comprehend the probable of nanosensors in agriculture, interdisciplinary collaborations between environmental scientists, agronomists, and nanotechnologists are imperative. These partnerships may result in the creation of novel responses to the unique requirements and difficulties faced by various agricultural systems. Collaborative research, for instance, can result in the development of nanosensors specifically designed for organic farming systems, which limit the use of synthetic inputs. These developments can aid in the integration of nanosensors into diverse agricultural techniques, advancing effective and sustainable agriculture across the globe.

10 Conclusion

The growing challenges of damaged and variable soils have highlighted the need for advanced soil assessment technologies. This chapter discussed efficient, low-cost sensors for monitoring soil nutrients, pollutants, moisture, pH, and other parameters, widely applied in agriculture and soil managementSmart nanosensor devices have garnered a lot of attention due to their high sensitivity and capacity to deliver fast, accurate, and predictive information. They are beginning to establish themselves as a significant tool for promoting soil evaluation in the near future. It has been discovered that using nanosensors boosts agricultural productivity. Temperature, soil health, moisture content, soil microbiological/microenvironment, and nutritional status can all be measured by these real-time sensors. Additionally, nanotechnology-based biosensors hold potential for advancing precision agriculture. However, further improvements in sensitivity and specificity are essential to enhance their capability to detect diverse soil elements and organisms effectively.

The incorporation of chemical nanosensors into farming methods presents a viable approach to tackle the complex issues brought about by resource scarcity, climate change, and the demand for sustainable farming techniques. Chemical nanosensors help farmers maximize resource use, improve crop yields, and lessen their impact on the environment by giving accurate, real-time data on soil health, nutrient levels, and ecological conditions. These sensors make it easier to implement targeted interventions, which reduces the overuse of pesticides and fertilizers—two main causes of pollution in the soil and water. Hence, chemical nanosensors' accuracy and effectiveness support more robust and sustainable agricultural systems. It is impossible to overestimate the significance of such precision instruments, which provide a route toward more productive and ecologically friendly farming methods, given the growing worldwide demand for agricultural products.

By offering information about soil conditions, chemical nanosensors greatly enhance agricultural growth by empowering farmers to make decisions that promote the health and growth of their plants. They improve productivity and make effective use of water and fertilizer by keeping an eye on moisture, nutrient concentrations, and pH levels. Identifying possible problems early minimizes crop damage. It is anticipated that chemical nanosensors will transform agriculture by enhancing precision, accessibility, and user-friendliness. They will be combined with cutting-edge technologies like drones and Internet of Things (IoT) gadgets as technology develops, offering complete precision farming solutions. As a result, farmers will be able to produce more with fewer resources, enhancing both environmental sustainability and global food security. Innovation and deployment require cooperation between policymakers, industry stakeholders, and researchers. With chemical nanosensors playing a key role in addressing food production, environmental stewardship, and economic viability, the future of agriculture appears brighter and more sustainable. Chemical nanosensors have many benefits, but there are drawbacks to their use in agriculture. It is necessary to address issues like the cost of deployment, the requirement for technical know-how, and the possible environmental effects of the sensors themselves. Small-scale farmers may find it difficult to adopt nanosensor technology due to its high initial cost, which may prevent it from being widely available. Furthermore, environmental factors may have an impact on the efficacy of nanosensors, requiring ongoing calibration and maintenance. These obstacles should be removed, though, by continued progress in nanotechnology and greater funding for research, which will enable chemical nanosensors to become more widely available and dependable for agricultural applications. To safeguard that the advantages of this technology reach a wider range of farming operations, efforts to lower production costs and increase sensor durability will be essential.

Competing Interests The authors declare that they have no competing interest.

Declaration of Generative AI and AI-Assisted Technologies in the Writing Process During the preparation of this work the book chapter author(s) have not taken any assistance from AI, and the book chapter author(s) reviewed and edited the content as needed and take(s) full responsibility for the content of the publication. All authors read and approved the final version of the manuscript.

References

1. Abad JM, Pariente F, Hernández L, Abruna HD, Lorenzo E, Abad EL (1998) Determination of organophosphorus and carbamate pesticides using a piezoelectric biosensor. Anal Chem 70:2848–2855
2. Abdelbasset WK, Jasim SA, Bokov DO, Oleneva MS, Islamov A, Hammid AT, Mustafa YF, Yasin G, Alguno AC, Kianfar E (2022) Comparison and evaluation of the performance of graphene-based biosensors. Carbon Lett 32(4):927–951. https://doi.org/10.1007/s42823-022-00338-6
3. Adamchuk VI, Hummel JW, Morgan MT, Upadhyaya SK (2004) On-the-go soil sensors for precision agriculture. Comput Electron Agric 44:71–91. https://doi.org/10.1016/j.compag.2004.03.002
4. Adamchuk VI, Lund ED, Sethuramasamyraja B, Morgan MT, Dobermann A, Marx DB (2005) Direct measurement of soil chemical properties on-the-go using ion-selective electrodes. Comput Electron Agric 48:272
5. Ahmad Z, Niyazi S, Firdoos A, Wang C, Manzoor MA, Ramakrishnan M, Upadhyay A, Ding Y (2024) Enhancing plant resilience: nanotech solutions for sustainable agriculture. Heliyon. 10:e40735
6. Akhtar MN, Shaikh AJ, Khan A, Awais H, Bakar EA, Othman AR (2021) Smart sensing with edge computing in precision agriculture for soil assessment and heavy metal monitoring: a review. Agriculture 11(6):475
7. Alizadeh T, Zare M, Ganjali MR, Norouzi P, Tavana B (2010) A new molecularly imprinted polymer (MIP)-based electrochemical sensor for monitoring 2,4,6-trinitrotoluene (TNT) in natural waters and soil samples. Biosens Bioelectron 25:1166–1172
8. Alvarado K, Bolaños M, Camacho C, Quesada E, VegaBaudrit J (2019) Nanobiotechnology in agricultural sector: overview and novel applications. J Biomater Nanobiotechnol 10(02):120
9. Amin Bora K, Hashmi S, Abideen Z (2025) Applying chemical and bio-mediated synthesis of manganese oxide nanoparticles for enhancing growth and nutrient fluxes in Arka anamika plants under saline conditions. J Plant Nutr 1–24
10. Ampatzidis Y, De Bellis L, Luvisi A (2017) IPathology: robotic applications and management of plants and plant diseases. Sustainability 9:1010
11. Ampuero S, Bosset JO (2003) The electronic nose applied to dairy products: a review. Sens Actuators, B Chem 94(1):1–12. https://doi.org/10.1016/S0925-4005(03)00321-6
12. Arable—Decision Agriculture. Available online: https://www.arable.com/. Accessed on 7 Oct 2020
13. Aravamudhan S, Bhansali S (2008) Development of micro-fluidic nitrate-selective sensor based on doped-polypyrrole nanowires. Sens Actuators B 132:623
14. Arora AK, Devi S, Jaswal VS, Singh J, Kinger M, Gupta VD (2014) Synthesis and characterization of ZnO nanoparticles. Orient J Chem 30(4):1671–1679
15. Ashton K (2009) That, "Internet of Things" Thing. RFID J 22:97–114
16. Balasubramaniam S, Kangasharju J (2013) Realizing the internet of nano things: challenges, solutions, and applications. Computer 46:62–68
17. Barrena R, Casals E, Colón J, Font X, Sánchez A, Puntes V (2009) Evaluation of the ecotoxicity of model nanoparticles
18. Baskeyfield DEH, Davis F, Magan N, Tothill IE (2011) A membrane-based immunosensor for the analysis of the herbicide isoproturon. Anal Chim Acta 699:223–231

19. Bedi A (2016) Bioleaching-an approach for synthesizing functionalized nanoformulations for agricultural use Deakin University
20. Bilkiss M, Shiddiky MJ, Ford R (2019) Advanced diagnostic approaches for necrotrophic fungal pathogens of temperate legumes with a focus on Botrytis spp. Front Microbiol 10:1889
21. Bodkhe M, Chalke T, Kulkarni S, Goswami A (2024) A review on sustainable applications of nanobiosensors in various fields and future potential. BioNanoScience 1–21
22. Bojdi MK, Behbahani M, Sahragard A, Amin BG, Fakhari A, Bagheri A (2014) A palladium imprinted polymer for highly selective and sensitive electrochemical determination of ultra-trace palladium ions. Electrochim Acta 149:108–116
23. Brenneman KL, Sen B, Stroscio MA, Dutta M (2010) Aptamer-based optical bionano sensor for mercury(II) ions. In: Proceedings of the IEEE nanotechnology materials and devices conference. https://doi.org/10.1109/nmdc.2010.5652331
24. Campos BB, Contreras-Cáceres R, Bandosz TJ, Jiménez-Jiménez J, Rodríguez-Castellón E, Da Silva JCE, Algarra M (2017) Carbon dots coated with vitamin B 12 as selective ratiometric nanosensor for phenolic carbofuran. Sens Actuators B: Chem 239:553–561
25. Chang Y-C, Lin Y-S, Xiao G-T, Chiu T-C, Hu C-C (2016) A highly selective and sensitive nanosensor for the detection of glyphosate. Talanta 161:94–98. https://doi.org/10.1016/j.talanta.2016.08.029
26. Chen S, Pan Z, Zhao W, Zhou Y, Rui Y, Jiang C et al (2023) Engineering climate resilient rice using a nano biostimulant-based "stress training" strategy. ACS Nano 17:10760–10773
27. Chen C, Shao C, Meng L, Yang J (2019) High-strength, self-adhesive, and strain-sensitive hydrogels for flexible sensors. ACS Appl Mater Interfaces 11(42):39228–39237. https://doi.org/10.1021/acsami.9b15817
28. Chen H, Zhang W, Yang S, Wang J, Zhang Y (2024) The application of nanotechnology in the development of agrochemicals. Chemosphere 250:130567. https://doi.org/10.1016/j.chemosphere.2023.13056
29. Chen Y, Zou H, Yan B, Wu X, Cao W, Qian Y et al (2022) Atomically dispersed Cu nanozyme with intensive ascorbate peroxidase mimic activity capable of alleviating ROS-mediated oxidation damage. Adv Sci 9:2103977
30. Chen Y, Guo X-L, Yan J-C, Zhao Y-F, Pang Y, Jian J-M et al (2018) Toward an in situ phosphate sensor in natural waters using a microfluidic flow loop analyzer. J Electrochem Soc 165(B737)
31. Chiriacò MS, Luvisi A, Primiceri E, Sabella E, De Bellis L, Maruccio G (2018) Development of a lab-on-a-chip method for rapid assay of *Xylella fastidiosa* subsp. *pauca* strain CoDiRO. Sci Rep 8:7376. https://doi.org/10.1038/s41598-018-25798-9
32. Cui S, Ling P, Zhu H, Keener HM (2018) Plant pest detection using an artificial nose system: a review. Sensors 18(2):378. https://doi.org/10.3390/s18020378
33. Dadach M, Abideen Z, Bhatt A (2025) In-vitro evaluation of resistance to trace metals in the early development stage of the Mediterranean endemic Ballota hirsuta Benth. A critical approach for large-scale cultivation and environmental safety. Euro-Mediterr J Environ Integr 1–14
34. Dean RN, Rane AK, Baginski ME, Richard J, Hartzog Z, Elton DJ (2012) IEEE Trans Instrum Meas 61:1105
35. Della Pelle F, Angelini C, Sergi M, Del Carlo M, Pepe A, Compagnone D (2018) Nano carbon black-based screen printed sensor for carbofuran, isoprocarb, carbaryl and fenobucarb detection: application to grain samples. Talanta 186:389–396
36. van Dijk M, Morley T, Rau ML, Saghai Y (2021) A meta-analysis of projected global food demand and population at risk of hunger for the period 2010–2050. Nat Food 2:494–501
37. Domínguez-Niño JM et al (2020) Differential irrigation scheduling by an automated algorithm of water balance tuned by capacitance-type soil moisture sensors. Agric Water Manag. https://doi.org/10.1016/j.agwat.2019.105880
38. Dong NM, Risgaard-Petersen N, Sørensen J, Brandt KK (2011) Rapid and sensitive Nitrosomonas europaea biosensor assay for quantification of bioavailable ammonium *sensu stricto* in soil. Environ Sci Technol 45(3):1048–1054. https://doi.org/10.1021/es1030036

39. Dong J, Wang X, Qiao F, Liu P, Ai S (2013) Sens Actuators, B Chem 186:774
40. Dutta RR, Puzari P Amperometric biosensing of organophosphate and organocarbamate pesticides utilizing polypyrrole entrapped acetylcholinesterase electrode. Biosens Bioelectron 52:166–172
41. Federal Reserve Bank of Chicago (2023) Climate change and risks to Midwest agriculture. Retrieved from https://www.chicagofed.org/research/publications/economic-perspectives/2021/3-2021/climate-change-and-risks-to-midwest-agriculture
42. Foo WH, Chia WY, Ende S, Chia SR, Chew KW (2024) Nanomaterials in aquaculture disinfection, water quality monitoring, and wastewater remediation. J Environ Chem Eng 113947
43. Food and Agriculture Organization of the United Nations (2023) FAO calls for scaling up action to address loss and damage in agrifood systems. Accessed from https://www.fao.org/newsroom/detail/fao-calls-for-scaling-up-action-to-address-loss-and-damage-in-agrifood-systems/en
44. Freeman MH, Hall JR, Leopold MC (2013) Monolayer-protected nanoparticle doped xerogels as functional components of amperometric glucose biosensors. Anal Chem 85(8):4057–4065. https://doi.org/10.1021/ac3037188
45. Garland NT, McLamore ES, Cavallaro ND, Mendivelso-Perez D, Smith EA, Jing D et al (2018) Flexible laser-induced graphene for nitrogen sensing in soil. ACS Appl Mater Interfaces 10(45). https://doi.org/10.1021/acsami.8b10991
46. Gilbertson LM, Pourzahedi L, Laughton S, Gao XY, Zimmerman JB, Theis TL et al (2020) Guiding the design space for nanotechnology to advance sustainable crop production. Nat Nanotechnol 15:801–810
47. Grell M et al (2020) Determining and predicting soil chemistry with a point-of-use sensor toolkit and machine learning model. bioRxiv. https://doi.org/10.1101/2020.10.08.331371
48. Gupta VK, Pal MK, Singh AK (2009) Comparative evaluation of Dy(III) selective poly(vinyl) chloride based membrane electrodes. Talanta 79:528–533
49. Hamimed S, Chatti A (2022) Olive pomace as a source of nutraceuticals. In: Food and agricultural byproducts as important source of valuable nutraceuticals. Springer, pp 88–96. https://doi.org/10.1007/978-3-030-98760-2_6
50. Hamimed S et al (2022) Bacterial cellulose nanofibers. In: Handbook of nanocelluloses. Springer. https://doi.org/10.1007/978-3-030-62976-2_15-1
51. Hamimed S et al (2022) Nanocelluloses for removal of heavy metals from wastewater. Handbook of nanocelluloses. Springer. https://doi.org/10.1007/978-3-030-62976-2_51-1
52. Hamimed S et al (2022) Photocatalytic metal bionanocomposites for biomedical applications. Bionanotechnol: Emerg Appl Bionanomater 227–258. https://doi.org/10.1016/B978-0-12-823915-5.00011-3
53. Hanif M, Munir N, Abideen Z, Yong JWH, El-Keblawy A, El-Sheikh MA (2024) Synthesis and optimization of nanoparticles from *Phragmites karka* improves tomato growth and salinity resilience. Biocatal Agric Biotechnol 55:102972
54. Hanif M, Munir N, Abideen Z, Dias DA, Hessini K, El-Keblawy A (2023) Enhancing tomato plant growth in a saline environment through the eco-friendly synthesis and optimization of nanoparticles derived from halophytic sources. Environ Sci Pollut Res 30(56):118830–118854
55. Haris M, Hussain T, Mohamed HI, Khan A, Ansari MS, Tauseef A, Khan AA, Akhtar N (2023) Nanotechnology–A new frontier of nano-farming in agricultural and food production and its development. Sci Total Environ 857:159639
56. Hashem IAT, Yaqoob, I, Anuar NB, Mokhtar S, Gani A, Ullah Khan S (2015) The rise of "big data" on cloud computing: review and open research issues. Inf Syst 47:98–115
57. Hornero A, Hernandez Clemente R, North PRJ, Beck PSA, Boscia D, Navas Cortés J, Zarco-Tejada P (2020) Monitoring the incidence of *Xylella fastidiosa* infection in olive orchards using ground-based evaluations, airborne imaging spectroscopy and Sentinel-2 time series through 3-D radiative transfer modelling. Remote Sens Environ 236:111480
58. Huang X, Xu J, Ji H-F, Li G, Chen H (2014) Quartz crystal microbalance-based biosensor for rapid and sensitive detection of maize chlorotic mottle virus. Anal Methods 6:4530–4536. https://doi.org/10.1039/c4ay00729c

59. Hung YL, Hsiung TM, Chen YY, Huang YF, Huang CC (2010) J Phys Chem C 114:16329
60. Jarocka U, Radecka H, Malinowski T, Michalczuk L, Radecki J (2013) Detection of Prunus Necrotic Ringspot Virus in plant extracts with impedimetric immunosensor based on glassy carbon electrode. Electroanalysis. https://doi.org/10.1002/elan.201200685
61. Jung D-H, Kim H-J, Kim HS, Choi J, Kim JD, Park SH (2019) Fusion of spectroscopy and cobalt electrochemistry data for estimating phosphate concentration in hydroponic solution. Sensors 19(11):2596. https://doi.org/10.3390/s19112596
62. Jócsák I, Végvári G, Vozáry E (2019) Electrical impedance measurement on plants: a review with some insights to other fields. Theor Exp Plant Physiol 31:359–375. https://doi.org/10.1007/s40626-019-00150-1
63. Kalman E-L, Löfvendahl A, Winquist F, Lundström I (2000) Classification of complex gas mixtures from automotive leather using an electronic nose. Anal Chim Acta 403(1–2):31–38. https://doi.org/10.1016/S0003-2670(99)00610-6
64. Kalyani N, Goel S, Jaiswal S (2021) On-site sensing of pesticides using point-of-care biosensors: a review. Environ Chem Lett 19(1):345–354
65. Kamilaris A, Andreas K, Francesc XP-B (2017) A review on the practice of big data analysis in agriculture. Comput Electron Agric 143:23–37
66. Kangas MJ, Burks RM, Atwater J, Lukowicz RM, Williams P, Holmes AE (2017) Colorimetric sensor arrays for the detection and identification of chemical weapons and explosives. Crit Rev Anal Chem 47(2):138–153. https://doi.org/10.1080/10408347.2016.1233805
67. Kanhed P, Birla S, Gaikwad S, Gade A, Seabra AB, Rubilar O, Duran N, Rai M (2014) In vitro antifungal efficacy of copper nanoparticles against selected crop pathogenic fungi. Mater Lett 115:13–17
68. Kant R (2020) Surface plasmon resonance-based fiber–optic nanosensor for the pesticide fenitrothion utilizing Ta_2O_5 nanostructures sequestered onto a reduced graphene oxide matrix. Microchim Acta 187:8. https://doi.org/10.1007/s00604-019-3935-4
69. Karim F, Fakhruddin ANM (2012) Recent advances in the development of biosensor for phenol: a review. Rev Environ Sci Bio/Technol 11:261–274. https://doi.org/10.1007/s11157-012-9268-9
70. Kaur J, Adamchuk VI, Whalen JK, Ismail AA (2015) Development of an NDIR CO_2 sensor-based system for assessing soil toxicity using substrate-induced respiration. Sensors 15(3):4734–4748. https://doi.org/10.3390/s150304734
71. Kerry RG, Ukhurebor KE, Kumari S, Maurya GK, Patra S, Panigrahi B, Majhi S, Rout JR, del Pilar Rodriguez-Torres M, Das G (2021) A comprehensive review on the applications of nano-biosensor-based approaches for non-communicable and communicable disease detection. Biomat Sci 9(10):3576–3602
72. Khater M, de la Escosura-Muñiz A, Quesada-González D, Merkoçi A (2019) Electrochemical detection of plant virus using gold nanoparticle-modified electrodes. Anal Chim Acta 1046:123–131. https://doi.org/10.1016/j.aca.2018.10.051
73. Khenifi A, Derriche Z, Forano C, Prévot V, Mousty C, Scavetta E, Ballarin B, Guadagnini L, Tonelli D (2009) Glyphosate and glufosinate detection at electrogenerated NiAl-LDH thin films. Anal Chim Acta 654:97–102
74. Khoshru B, Fallah Nosratabad A, Mahjenabadi VAJ, Knežević M, Hinojosa AC, Fadiji AE, Mitra D (2024) Multidimensional role of Pseudomonas: from biofertilizers to bioremediation and soil ecology to sustainable agriculture. J Plant Nutr 1–27
75. Kopylovich MN, Mahmudov KT, Pombeiro AJL (2011) Copper-selective electrode based on PVC membrane. J Hazard Mater 186:1154–1162
76. Kumar S, Nehra M, Dilbaghi N, Marrazza G, Hassan AA, Kim KH (2023) Nano-based smart pesticide formulations: emerging opportunities for agriculture. J Control Release 304:144–161. https://doi.org/10.1016/j.jconrel.2023.04.015
77. Kumar S, Sachdeva S, Chaudhary S, Chaudhary GR (2020) Assessing the potential application of bio-compatibly tuned nanosensor of Yb_2O_3 for selective detection of imazapyr in real samples. Colloids Surf, A 593:124612. https://doi.org/10.1016/j.colsurfa.2020.124612

78. Kumar THV, Sundramoorthy AK (2018) Non-enzymatic electro-chemical detection of urea on silver nanoparticles anchored nitrogen-doped single-walled carbon nanotube modified electrode. J Electrochem Soc 165(B3006)
79. Lamar B et al (2020) Review—the "Real-Time" revolution for in situ soil nutrient sensing. J Electrochem Soc 167:037569
80. Laothawornkitkul J, Moore JP, Taylor JE, Possell M, Gibson TD, Hewitt CN, Paul ND (2008) Discrimination of plant volatile signatures by an electronic nose: a potential technology for plant pest and disease monitoring. Environ Sci Technol 42(22):8433–8439. https://doi.org/10.1021/es801738s
81. Lastochkina O, Aliniaeifard S, SeifiKalhor M, Bosacchi M, Maslennikova D, Lubyanova A (2022) Novel approaches for sustainable horticultural crop production: advances and prospects. Horticulturae 8(10):910
82. Lee G, Wei Q, Zhu Y (2021) Emerging wearable sensors for plant health monitoring. Adv Funct Mater 31(52):2106475
83. Lemos SG, Nogueira ARA, Torre-Neto A, Parra A, Alonso J (2007) Soil calcium and pH monitoring sensor system. J Agric Food Chem 55:4658–4663
84. Lin J, Cai X, Liu Z, Liu N, Xie M, Zhou BP, Wang H, Guo Z (2020) Anti-liquid-interfering and bacterially antiadhesive strategy for highly stretchable and ultrasensitive strain sensors based on Cassie-Baxter wetting state. Adv Func Mater 30(23):2000398. https://doi.org/10.1002/adfm.202000398
85. Lin HY, Huang CH, Lu SH, Kuo IT, Chau LK (2014) Direct detection of orchid viruses using nanorod-based fiber optic particle plasmon resonance immunosensor. Biosens Bioelectron 51:371–378. https://doi.org/10.1016/j.bios.2013.08.032
86. Liu H, Chen P, Liu Z, Liu J, Yi J, Xia F, Zhou C (2020) Electrochemical luminescence sensor based on double suppression for highly sensitive detection of glyphosate. Sens Actuators, B Chem 304:127364. https://doi.org/10.1016/j.snb.2019.127364
87. Liu Y, Guo M, Du R, Chi J, He X, Xie Z et al (2020) A gas reporting whole-cell microbial biosensor system for rapid on-site detection of mercury contamination in soils. Biosens Bioelectron 170:112660. https://doi.org/10.1016/j.bios.2020.112660
88. Liu C-W, Huang C-C, Chang H-T (2009) Highly selective DNA-based sensor for lead(II) and mercury(II) ions. Anal Chem 81:2383–2387
89. Liu B, Zhou P, Liu X, Sun X, Li H, Lin M (2013) Detection of pesticides in fruits by surface-enhanced Raman spectroscopy coupled with gold nanostructures. Food Bioprocess Technol 6:710–718
90. Liu B, Zhuang J, Wei G (2020) Recent advances in the design of colorimetric sensors for environmental monitoring. Environ Sci Nano 7:2195–2213. https://doi.org/10.1039/d0en00449a
91. Luo Y, Su W, Xu D, Wang Z, Wu H, Chen B, Wu J (2023) Component identification for the SERS spectra of microplastics mixture with convolutional neural network. Sci Total Environ 895:165138
92. Luvisi A, Ampatzidis Y, De Bellis L (2016) Plant pathology and information technology: opportunity for management of disease outbreak and applications in regulation frameworks. Sustainability 8:831
93. Lvova L, Nadporozhskaya M (2017) Chemical sensors for soil analysis: principles and applications. Elsevier Inc. https://doi.org/10.1016/b978-0-12-804299-1.00018-7
94. Lykogianni M, Bempelou E, Karamaouna F, Aliferis KA (2021) Do pesticides promote or hinder sustainability in agriculture? The challenge of sustainable use of pesticides in modern agriculture. Sci Total Environ 795:148625
95. Maksimovic M, Omanovic-Miklicanin E (2017) Green Internet of Things and Green nanotechnology role in realizing smart and sustainable agriculture. In: Kovacevic D (ed)Proceedings of the VIII international scientific agriculture symposium "AGROSYM 2017", Jahorina, Bosnia. University of East Sarajevo: Sarajevo, Bosnia, pp 2290–2295
96. Manjunatha SB, Biradar DP, Aladakatti YR (2016) Nanotechnology and its applications in agriculture: a review. J Farm Sci 29(1):1–13

97. March C, Manclús J, Jiménez Y, Arnau A, Montoya A (2009) A piezoelectric immunosensor for the determination of pesticide residues and metabolites in fruit juices. Talanta 78:827–833
98. Mishra RK (2020) Magnetic induction-based wireless sensor network for control and monitoring of underground pipeline. Resincap J Sci Eng 4
99. Mittal D, Kaur G, Singh P, Yadav K, Ali SA (2020) Nanoparticle-based sustainable agriculture and food science: recent advances and future outlook. Front Nanotechnol 2:579954
100. Morsy M, Helal M, El-Okr M, Ibrahim M (2014) Preparation, purification and characterization of high purity multi-wall carbon nanotube. Spectrochim Acta Part A: Mol Biomol Spectrosc 132:594–598
101. Mujeeb A, Abideen Z, Aziz I, Sharif N, Hussain MI, Qureshi AS, Yang HH (2023) Phytoremediation of potentially toxic elements from contaminated saline soils using *Salvadora persica* L.: seasonal evaluation. Plants 12(3):598
102. Mulla DJ (2013) Twenty five years of remote sensing in precision agriculture: key advances and remaining knowledge gaps. Biosyst Eng 114:358–371
103. Nadporozhskaya M, Kovsh N, Paolesse R, Lvova L (2022) Recent advances in chemical sensors for soil analysis: a review. Chemosensors 10(1):35
104. Naresh V, Lee N (2021) A review on biosensors and recent development of nanostructured materials-enabled biosensors. Sensors 21(4):1109
105. Neethirajan S, Ragavan V, Weng X, Chand R (2018) Biosensors for sustainable food engineering: challenges and perspectives. Biosensors 8(1):23
106. Neme K, Nafady A, Uddin S, Tola YB (2021) Application of nanotechnology in agriculture, postharvest loss reduction and food processing: food security implication and challenges. Heliyon 7(12)
107. Nielsen SS (2010) Food analysis, 4th edn. Springer Science and Business Media
108. Nissim R, Compton RG (2014) Introducing absorptive stripping voltammetry: wide concentration range voltammetric phenol detection. Analyst 139(22):5911–5918
109. Nizamani MM, Hughes AC, Zhang H-L, Wang Y (2024) Revolutionizing agriculture with nanotechnology: innovative approaches in fungal disease management and plant health monitoring. Sci Total Environ 172473
110. Omasa K, Oki K, Suhama T (2006) 5.2 remote sensing from satellites and aircraft. ASABE: St. Joseph, MI, USA
111. Ouyang H, Tu X, Fu Z, Wang W, Fu S, Zhu C, Du D, Lin Y (2018) Colorimetric and chemiluminescent dual-readout immunochromatographic assay for detection of pesticide residues utilizing g-C_3N_4/$BiFeO_3$ nanocomposites. Biosens Bioelectron 106:43–49
112. Pan T-T, Sun D-W, Pu H, Wei Q (2018) Simple approach for the rapid detection of alternariol in pear fruit by surface-enhanced Raman scattering with pyridine-modified silver nanoparticles. J Agric Food Chem 66:2180–2187. https://doi.org/10.1021/acs.jafc.7b05577
113. Panhwar QA, Ali A, Naher UA, Memon MY (2019) Fertilizer management strategies for enhancing nutrient use efficiency and sustainable wheat production. Organic Farming. https://doi.org/10.1016/b978-0-12-813272-2.00002-1
114. Patel BC, Sinha GR, Goel N (2020) Introduction to sensors. Advances in modern sensors: physics, design, simulation and applications. IOP Publishing, p 1
115. Peng Y, Zeng A, Zhu T, Fang S, Gong Y, Tao Y, Zhou Y, Liu K (2017) Using remotely sensed spectral reflectance to indicate leaf photosynthetic efficiency derived from active fluorescence measurements. J Appl Remote Sens 11:026034
116. Penuelas J, Sardans J (2023) Human-driven global nutrient imbalances increase risks to health. Eco-Environ Health 2:246–251
117. Peris M, Escuder-Gilabert L (2009) A 21st century technique for food control: electronic noses. Anal Chim Acta 638(1):1–15. https://doi.org/10.1016/j.aca.2009.02.009
118. Poblete T, Camino C, Beck PSA, Hornero A, Kattenborn T, Saponari M, Boscia D, Navas-Cortes JA, Zarco-Tejada PJ (2020) Detection of *Xylella fastidiosa* infection symptoms with airborne multispectral and thermal imagery: assessing bandset reduction performance from hyperspectral analysis. ISPRS J Photogramm Remote Sens 162:27–40

119. Power AC, Gorey B, Chandra S, Chapman J (2018) Carbon nanomaterials and their application to electrochemical sensors: a review. Nanotechnol Rev 7(1):19–41
120. Prasad BB, Jauhari D, Tiwari MP (2014) Biosens Bioelectron 59:81
121. Rattanarat P, Dungchai W, Cate D, Volckens J, Chailapakul O, Henry CS (2014) Analytical developments in chemical sensing. Anal Chem 86:3555
122. Regan F (2019) Sensors | overview, 3rd edn. Elsevier Inc. https://doi.org/10.1016/B978-0-12-409547-2.14540-8
123. Robinson DA, Campbell CS, Hopmans JW, Hornbuckle BK, Jones SB, Knight R et al (2008) Soil moisture measurement for ecological and hydrological watershed-scale observatories: a review. 7:358
124. Rodrigues SM, Demokritou P, Dokoozlian N, Hendren CO, Karn B, Mauter MS et al (2017) Nanotechnology for sustainable food production: promising opportunities and scientific challenges. Environ Sci Nano 4:767–781
125. Rudd JD, Roberson GT, Classen JJ (2017) Application of satellite, unmanned aircraft system, and ground-based sensor data for precision agriculture: a review. In: Proceedings of the 2017 ASABE annual international meeting, St. Joseph, MI, USA, p 1
126. Rupanagudi SR, Ranjani BS, Nagaraj P, Bhat VG, Thippeswamy G (2015) A novel cloud computing based smart farming system for early detection of borer insects in tomatoes. In: Proceedings of the 2015 international conference on communication, information & computing technology (ICCICT), Mumbai, India, pp 1–6
127. Saleh SM, Alminderej FM, Ali R, Abdallah OI (2020) Optical sensor film for metribuzin pesticide detection. Spectrochim Acta Part A Mol Biomol Spectrosc 229:117971. https://doi.org/10.1016/j.saa.2019.117971
128. Samal S, Mohanty RP, Mohanty PS, Giri MK, Pati S, Das B (2023) Implications of biosensors and nanobiosensors for the eco-friendly detection of public health and agro-based insecticides: a comprehensive review. Heliyon. 9:e15848
129. Scala A, Allmann S, Mirabella R, Haring M, Schuurink R (2013) Green leaf volatiles: a plant's multifunctional weapon against herbivores and pathogens. Int J Mol Sci 14:17781–17811
130. Schieber A (2018) Introduction to food authentication. In: Modern techniques for food authentication. Academic Press, pp 1–21
131. Schirrmann M, Gebbers R, Kramer E, Seidel J (2011) Soil pH mapping with an on-the-go sensor. Sensors 11:573–598
132. Semios. Available online: https://semios.com/. Accessed on 7 Oct 2020
133. SenseFly-The professional's mapping drone. Available online: https://www.sensefly.com/. Accessed on 7 Oct 2020
134. Shang Y, Hasan MK, Ahammed GJ, Li M, Yin H, Zhou J (2019) Applications of nanotechnology in plant growth and crop protection: a review. Molecules 24(14):2558
135. Sharma R, Sharma N, Prashar A, Hansa A, Asgari Lajayer B, Price GW (2024) Unraveling the plethora of toxicological implications of nanoparticles on living organisms and recent insights into different remediation strategies: a comprehensive review. Sci Total Environ 906:167697
136. Sharma R, Zhou M, Hunter MD, Fan X (2019) Rapid in situ analysis of plant emission for disease diagnosis using a portable gas chromatography device. J Agric Food Chem 67:7530–7537
137. Sheikh ZN, Raina S, Bhat UH, Sharma N, Saffeullah P, Mohammad F, Al-Andal A, Islam S (2024) Deciphering the potential role of nano-based biosensors in growth, development and stress management of crop plants. S Afr J Bot 167:542–553
138. Simic M, Manjakkal L, Zaraska K, Stojanovic GM, Dahiya R (2017) TiO_2-based thick-film pH sensor. IEEE Sens J 17(2):248–255
139. Singh G (2021) Benzothiazole encapsulated silane and its nano composites for sequential detection of copper ions and cysteine. Chem Select 6(9). https://doi.org/10.1002/slct.202002948
140. Singh RP (2021) Recent trends, prospects, and challenges of nanobiosensors in agriculture. Biosens Agric Recent Trends Future Perspect 3–13

141. Singh R, Choudhary P, Kumar S, Daima HK (2024) Mechanistic approaches for crosstalk between nanomaterials and plants: plant immunomodulation, defense mechanisms, stress resilience, toxicity, and perspectives. Environ Sci: Nano
142. Stradiotto NR, Yamanaka H, Zanoni MVB (2003) Electrochemical sensors: a powerful tool in analytical chemistry. J Braz Chem Soc 14(2):159–173. https://doi.org/10.1590/S0103-50532003000200003
143. Sun C, Pan L, Zhang L, Huang J, Yao D, Wang C, Zhang Y, Jiang N, Chen L, Yuan C (2019) A biomimetic fluorescent nanosensor based on imprinted polymers modified with carbon dots for sensitive detection of alpha-fetoprotein in clinical samples. Analyst 144(22):6760–6772. https://doi.org/10.1039/C9AN01065C
144. Susha Lekshmi SU, Singh DN, Shojaei Baghini M (2014) A critical review of soil moisture measurement. Measurement 54:92–105. https://doi.org/10.1016/j.measurement.2014.04.007
145. Syu Y-y, Hung J-H, Chen J-C, Chuang H-w (2014) Impacts of size and shape of silver nanoparticles on Arabidopsis plant growth and gene expression. Plant Physiol Biochem 83:57–64
146. Talan A, Mishra A, Eremin SA, Narang J, Kumar A, Gandhi S (2018) Ultrasensitive electrochemical immuno-sensing platform based on gold nanoparticles triggering chlorpyrifos detection in fruits and vegetables. Biosens Bioelectron 105:14–21
147. Tan X, Hu Q, Wu J, Li X, Li P, Yu H, Li X, Lei F (2015) Electrochemical sensor based on molecularly imprinted polymer reduced graphene oxide and gold nanoparticles modified electrode for detection of carbofuran. Sens Actuators B: Chem 220:216–221
148. Tang L, Zeng G-M, Shen G-L, Li Y-P, Zhang Y-I, Huang D-L (2008) Environmental applications of environmental science and technology. Environ Sci Technol 42:1207
149. Thakur M, Wang B, Verma ML (2022) Development and applications of nanobiosensors for sustainable agricultural and food industries: Recent developments, challenges and perspectives. Environ Technol Innov 26:102371
150. Tian X, Liu L, Li Y, Yang C, Zhou Z, Nie Y, Wang Y (2018) Nonenzymatic electrochemical sensor based on CuO-TiO_2 for sensitive and selective detection of methyl parathion pesticide in ground water. Sens Actuators B Chem 256:135–142
151. Verma N, Bhardwaj A (2015) Biosensor technology for pesticides—a review. Appl Biochem Biotechnol 175:3093–3119
152. Vishnu N, Kumar AS (2015) A preanodized 6B-pencil graphite as an efficient electrochemical sensor for mono-phenolic preservatives. Anal Methods 7(5):1943–1950
153. Wang D, Saleh NB, Byro A, Zepp R, Sahle-Demessie E, Luxton TP et al (2022) Nano-enabled pesticides for sustainable agriculture and global food security. Nat Nanotechnol 17:347–360
154. Waqif H, Munir N, Farrukh MA, Hasnain M, Sohail M, Abideen Z (2024) Algal macromolecular mediated synthesis of nanoparticles for their application against citrus canker for food security. Int J Biol Macromol 263:130259
155. West JS, Bravo C, Oberti R, Lemaire D, Moshou D, McCartney HA (2003) The potential of optical canopy measurement for targeted control of field crop diseases. Annu Rev Phytopathol 41:593–614
156. Wilson AD (2013) Diverse applications of electronic-nose technologies in agriculture and forestry. Sensors 13(2):2295–2348. https://doi.org/10.3390/s130202295
157. Wilson AD, Baietto M (2009) Applications and advances in electronic-nose technologies. Sensors 9(7):5099–5148. https://doi.org/10.3390/s90705099
158. Wilson D, Arada M, de los A, Alegret S, del Valle M (2010) Lead(II) ion-selective electrodes with PVC membranes. J Hazard Mater 181:140–146
159. Xie Y, Yu Y, Lu L, Ma X, Gong L, Huang X, Liu G, Yu Y (2018) CuO nanoparticles decorated 3D graphene nanocomposite as a non-enzymatic electrochemical sensing platform for malathion detection. J Electroanal Chem 812:82–89. https://doi.org/10.1016/j.jelechem.2018.01.032
160. Yadav N, Garg VK, Chhillar AK, Rana JS (2023) Recent advances in nanotechnology for the improvement of conventional agricultural systems: a review. Plant Nano Biol 4:100032

161. Yan T, Hu C, Que Y, Song Y, Lu D, Gu J, Ren Y, He J (2023) Chitosan coating enriched with biosynthetic CuO NPs: effects on postharvest decay and quality of mango fruit. Int J Biol Macromol 253:126668
162. Yokota YM, Okada T, Yamaguchi I (2007) An optical sensor for analysis of soil nutrients using LED light sources. Meas Sci Technol 18(7):2197–2201. https://doi.org/10.1088/0957-0233/18/7/052
163. Yuqing M, Jianrong C, Keming F (2005) New technology for the detection of pH. J Biochem Biophys Methods 63(1):1–9
164. Yılmaz E, Özgür E, Bereli N, Türkmen D, Denizli A (2017) Plastic antibody-based surface plasmon resonance nanosensors for selective atrazine detection. Mater Sci Eng, C 73:603–610. https://doi.org/10.1016/j.msec.2016.12.117
165. Zamani HA, Mohaddeszadeh M (2008) Monitoring of iron(III) ions with a Fe^{3+}-PVC membrane sensor based on 4,4′-dimethoxybenzil bisthiosemicarbazone. Anal Lett 41:2710
166. Zamora R (2018) School of chemistry, Costa Rica institute of technology, Cartago, Costa Rica development of poly(3,4-ethylenedioxythiophene(PEDOT)/carbon nanotube electrodes for electrochemical detection of Mancozeb in water. Int J Electrochem Sci 13:1931–1944
167. Zhang Q, Xu G, Gong L, Dai H, Zhang S, Li Y, Lin Y (2015) An enzyme-assisted electrochemiluminescent biosensor developed on order mesoporous carbons substrate for ultrasensitive glyphosate sensing. Electrochim Acta 186:624–630
168. Zhang K, Barhoum A, Xiaoqing C, Li H (2019) Handbook of nanofibers. https://doi.org/10.1007/978-3-319-53655-2
169. Zhao Q, Yu P, Mahendran R, Huang W, Gao Y, Yang Z et al (2022) Global climate change and human health: pathways and possible solutions. Eco-Environ Health 1:53–62
170. Zou Z, Han J, Jang A, Bishop PL, Ahn CH (2007) A disposable on-chip phosphate sensor with planar cobalt microelectrodes on polymer substrate. Biosens Bioelectron 22(9–10):1902–1907. https://doi.org/10.1016/j.bios.2006.08.004

Nanosensors and Their Role in Soil Quality, Moisture, pH, Contaminants, and Other Molecules Detection

Patriciani Estela Cipriano, Matias Siueia Júnior, Deivisson Ferreira da Silva, Geslin Mars, Rodrigo Fonseca da Silva, and Flávio Henrique Silveira Rabêlo

Abstract With the current climate change and population growth, agricultural practices from the developed green revolution are becoming unsustainable. As a result, creative methods are urgently needed to meet the growing global need for food. The new scientific discipline of nanoscience and technology has transformed numerous fields. Nanoparticles are more suitable for utilization than their bulk counterparts due to their unique characteristics and larger surface-to-volume fractions. This technology has enormous potential in areas such as nanosensors and nanofertilizers. In order to boost agricultural productivity and guarantee that plants receive the proper amount of nutrients, nanosensors are used to detect pests, diseases, soil pH, humidity, and nutrients in plants and soil. Nanosensors assist in safeguarding crops from outside influences and identify negative impacts on soil and water quality as well as environmental pollutants. These provide information about changes in the environment by using dynamic shifts in the behavior of molecules at the nanoscale as a signal. Here, we looked at how nanotechnology might help plants fight off environmental

P. E. Cipriano (✉) · D. F. da Silva
Department of Soil Science, Lavras School of Agriculture, Federal University of Lavras, Lavras, Minas Gerais, Brazil
e-mail: patricianiestela@gmail.com

P. E. Cipriano · F. H. S. Rabêlo
Luiz de Queiroz College of Agriculture, University of São Paulo, Piracicaba 13418-900, Brazil

M. Siueia Júnior
Laboratory of Plant Nutritional Physiology, Graduate School of Integranted Sciences for Life, Hiroshima University, Higashi-Hiroshima, Japan

College of Agricultural Sciences, Aquila University, Maputo, Mozambique

D. F. da Silva
Instituto Federal Catarinense—Campus Araquari, Santa Catarina, Brazil

G. Mars
Department of Soil Science and Agricultural Engineering, Ponta Grossa State University, Ponta Grossa, Brazil

R. F. da Silva
Department of Agricultural Sciences, Federal University of Piauí, Piauí, Brazil

A. Husen (ed.), *Nanobiosensors for Agricultural and Other Related Sectors*, Smart Nanomaterials Technology, https://doi.org/10.1007/978-981-96-6499-3_5

and biological stress. We describe the use of nanosensors and how they can be used to detect contaminants, moisture, pH, soil quality, and other chemicals.

Keywords Nanotechnology · Sustainable agriculture · Nanofertilizer · Plant nutrition

1 Introduction

Produce food to feed approximately 9.9 billion people on this planet by the year 2050 is one of the biggest challenge that we are facing, especially in a scenario of climate change. With climate change, tapering resources, and soil degradation, the agricultural system is incessantly being pushed toward the wall. Thus, innovations and implementation of sustainable technologies must be made urgently to augment food production while being sensitive to environmental integrity. It would indicate a severe need for agricultural production and integrated technology to ramp up the sustainable use of land resources and minimize environmental deterioration [157].

Some advanced technologies include precision farming, which measures real-time the crop health, soil conditions, and atmospheric variables, with certain highlighted technologies allowing productivity from the land to improve in several parts of the world [37]. However, advancing agricultural practices pose a significant threat to human health and biodiversity due to the chemical pesticide applications, among others factors resulting in adverse effects on the environment throughout the globe [153].

Excessive fertilizer and pesticide applications have resulted in significant soil and water contamination. Recent studies have shown a shocking fact: over 64% of global agricultural soils are at risk of being contaminated by various pesticide-active ingredients, and more than 31% are at high risk [172]. Besides, this not only threatens human and environmental health but also mostly disrupts the fragile biological balance of farmlands. This ultimately results in adverse effects on the population of important microorganisms responsible for maintaining soil health and fertility [199]. Soil degradation caused by improper irrigation and cultivation practices undermines soil's ecological functions. These include losses in the soil organic carbon pool, soil acidification, nutrient loss due to depletion or low availability of macronutrients, and many others [119].

In this respect, nanotechnology appears to be an innovative solution to enhance agricultural productivity in terms of sustainability and precision technologies. Researchers are utilizing the superior capability of nanoparticles to increase the delivery of nutrients and improve crop resistance against environmental stressors. In this manner, the crop yield is improved while the negative impact on the environment is depressed.

It is possible to define nanotechnology as manipulating matter at the nanoscale, thus, nanomaterials have particle sizes between 1 and 100 nm [185] (Fig. 1). This technology is an advance in agricultural sciences that allows real-time monitoring

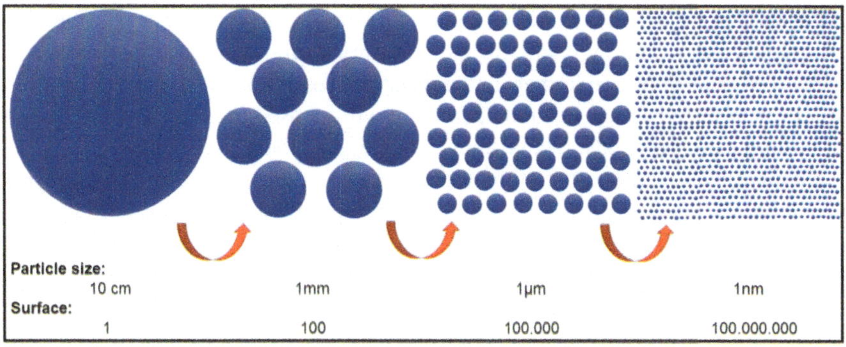

Fig. 1 Example of particle size and surface

of soil quality. While analytical and stimulated sensors already exist and have been used in certain aspects of food supply for their quick performance and disposition for unperturbed analysis, critics think this could never attract massive attention because it assumes the amount of agricultural fertilizers and pesticides expended shall not cross an arduous threshold [88]. Moreover, this is the period when agriculture shall be considered rather environment-friendly.

The nanosensors would counter traditional agriculture predictions by accurately monitoring plant health and soil conditions in real-time. This would enable the assessment of site-specific management practices relevant to individual crops [38]. The functioning of nanosensors infused and blended into an agricultural phenomenon was a complete systematic transition from traditional to sustainable. An ingenious piezoelectric nanosensor that could respond to mechanical stimuli within the plant's environment was present to allow differences in the physical and chemical properties of the soil to be distinguished from structural alterations in the plant. Employing nanosensor technologies might help identify specific nutrient imbalances in soil or during plant development, eliminating imminent disease outbreaks [97].

The aggressive application of nanosensors speeds the progression towards intelligent agriculture, with management choices made based on more dependable information about the agricultural environment. This approach increases agricultural productivity while reducing environmental damage by providing the inputs when and only where needed [56]. In situ, detecting nutrients and soil parameters in real-time creates the possibility of precision agriculture, wherein losses are reduced, and soil quality is improved by maneuver of maximizing fertilizer and pesticide application for increased production efficiency on the environment.

Nanotechnology also plays an important role in diagnosing biotic and abiotic stress factors affecting plant health. The growing resistance of weeds and pathogens to traditional chemicals gives an immediate reason to develop a technology that will monitor these biochemical changes and lead to less dependence on toxic chemicals [107]. Traditional methods typically respond only after challenges emerge, whereas nanosensors have the potential to take a proactive stance. These advanced tools can detect early signs of stress in plants and soil, promoting a healthier ecosystem. This new monitoring technology will change how agricultural challenges are addressed, offering crucial support to farmers and industry professionals. The induced functionality probably envisioned in nanotechnology will be used to create novel solutions for optimizing crop yields, increasing soil fertility, and sustaining practices.

Changes in agriculture that minimize crop losses and assure food security should be quickly invented to allow the minimization of production losses caused by these factors [123]. Agricultural sustainability is immensely improved when soil and water quality is monitored through nanosensors. Advanced sensors using various nanomaterials responsible for identifying pesticide residues in soils and those capable of giving estimates of nutrient levels with more accuracy when compared with traditional methods, as per recent studies, are financially beneficial. This has triggered the responsible use of inputs into degraded soils. Developmental breakthroughs made by nanotechnology have produced devices with impressive sensitivity and stability specifically engineered to monitor soil contaminants and nutrient levels.

Nanomaterial-based biosensors are more advantageous than conventional detection methods because they have rapid response times and require minimal or no reagents, which are vital in food safety and environmental health [107]. Incorporating biosensors into agriculture is one of the most thrilling breakthroughs in pushing for sustainability. Differentiated technologies capture much potential for addressing contemporary global challenges in food production and distribution while nurturing the goodness of Mother Nature's ecosystem.

This chapter describes innovative applications of nanosensors in agriculture, emphasizing soil quality monitoring. It describes how those sensors help determine critical nutrients that plants require, moisture, pH, and harmful contaminants in the soil. Monitoring could be achieved more flexibly and accurately with the help of nanosensors, thus enabling the birth of sustainable ways of production. Technologies based on nanoscale modulations are rapidly becoming the underlying platforms for developing ultimately resilient systems that provide much-needed efficiency in agriculture. This plays a salient role in ensuring amicable environmental protection through productivity enhancement and addressing other challenges that lie ahead.

2 Nanosensors and Soil Quality

Nanosensors can accurately measure various critical soil parameters, including pH levels, essential nutrients availability for plant growth, residual particles, temperature fluctuations, and moisture content. It can detect many substances, like insecticides, glucose, and urea, making it a valuable tool for monitoring and analysis. Farmers and horticulturists can implement targeted irrigation strategies that enhance water efficiency to its fullest potential by accurately measuring the soil's moisture content. To effectively diagnose soil diseases caused by a range of pathogens, including bacteria, fungi, and viruses, it is essential to conduct a comprehensive soil analysis to pinpoint these harmful contaminants. One-dimensional potassium niobate nanofibers have been effectively employed in the innovative measurement of soil humidity [53]. The interactions between plants and their associated microbiomes can be effectively observed and analyzed by applying advanced nanosensors.

By strategically identifying and modifying the specific bacteria within the plant microbiome, researchers have the potential to improve soil health greatly. This groundbreaking method significantly decreases our dependence on chemical fertilizers, which can harm the environment and disrupt plants' essential functions. Interactions among plants, fungi, and soil microorganisms are vital in promoting soil health and enhancing beneficial activities for plants, making them essential for sustainable agriculture.

Nanosensors can identify and analyze the diversity of microorganisms present in the soil, providing insightful data on the various species and their interactions. Carbon nanotubes (CNTs), metal nanoparticles (mNPs), quantum dots (QDs), silica nanoparticles, and upconversion nanoparticles (UCNPs) represent a diverse range of nanomaterials that have gained significant attention for their applications in the detection of microorganisms. These advanced materials offer unique properties that enhance the sensitivity and specificity of microbial detection methods, making them valuable tools in fields such as medical diagnostics and environmental monitoring. Recent research highlights their effectiveness in identifying pathogens, showcasing their potential to improve public health and safety measures [151]. By analyzing and comparing the amount of oxygen that various types of bacteria utilize during their respiratory processes, certain nanosensors can effectively discern which bacteria are beneficial for soil health and growth and which may have detrimental effects [88].

Sensors designed explicitly for microelectromechanical systems, which utilize advanced nanotechnology, are increasingly being employed to accurately monitor critical environmental parameters such as soil temperature and moisture content. Certain innovative moisture-level-detecting nanosensors ingeniously integrate the Wheatstone bridge piezo resistor circuit with the sophisticated design of microcantilever beams, which are meticulously coated in a highly water-sensitive nanopolymer. The nanoresistor plays a crucial role in detecting variations on the surface of the microcantilever, which are intricately linked to the moisture content of the surrounding soil [38]. In the agricultural field, the strategic placement of nanosensors has become increasingly vital for detecting and analyzing various soil conditions. These advanced sensors can efficiently categorize soil quality and health, providing real-time data that farmers can utilize [170] created a method for rapidly detecting and measuring soil moisture content in the field using an integrated capacitive sensor. This method enhances the efficiency of soil moisture measurements while delivering real-time data that supports effective agricultural management. Nanosensors significantly enhance management decisions by offering insights into factors like moisture levels, pH balance, and nutrient content.

The presence of potentially toxic elements (PTEs) like lead (Pb^{2+}), mercury (Hg^{2+}), cadmium (Cd^{2+}), and copper (Cu^{2+}) in soil and water at levels exceeding the permissible thresholds poses serious risks not only to human health but also to the environment [64]. The reuse of industrial wastewater can lead to the PTEs accumulation in agricultural soils. This accumulation can result in diminished soil fertility and potential food chain contamination. Consequently, it is imperative to implement stringent regulations and effective treatment methods for wastewater before it can be safely reused in agriculture. The cadmium detection, poses significant health risks, is effectively accomplished using advanced technologies such as ICTS nanosensors, which utilize innovative signal processing alongside optical nanosensors that harness light for measurement and electrochemical nanosensors that analyze electrical changes in response to Cd presence. To effectively detect the Hg presence in various environments, we can employ a combination of cutting-edge techniques, including advanced colorimetric methods, multimodal detection systems, and highly sensitive magnetic fluorescent nanosensors. These technologies work together to enhance the accuracy and reliability of Hg detection, allowing for precise measurements. Lead detection has been significantly advanced by integrating various innovative technologies, including colorimetric sensors, which offer visual indications of Pb presence, and ICTS technology, which enhances detection capabilities. Additionally, surface plasmon resonance nanosensors are crucial in identifying Pb at extremely low concentrations, demonstrating their sensitivity and precision. Together, these approaches form a comprehensive suite of methods that provide effective and reliable solutions for Pb detection [159].

Delgadillo-Vargas et al. [35] highlight the application of nano-biosensors in quantifying urease and nitrite levels in soil and water through innovative colorimetric and microfluidic impedimetric assays. These advancements in sensor technology pave the way for more precise monitoring of nutrient dynamics. According to Usman et al. [180] nano-biosensors can effectively capture nutrient spatial and temporal fluctuations in agricultural fields. This capability allows for precise tracking of nutrient concentrations, facilitating thorough fertilizer analysis and optimizing applications in precision agriculture. This technology improves fertilizer application efficiency, boosts crop yield, and minimizes environmental harm. Antonacci et al. [13] assert that the only biosensors utilizing nanomaterials capable of measuring organic matter, total carbon, sodium chloride, phosphate, and residual nitrate in soil are deployed in aqueous solutions and in vitro standards. These innovations demonstrate the potential of nanobiosensors to revolutionize soil health assessment and nutrient management practices.

3 Current Technologies and Potential of Nanosensors in Measuring Soil Moisture

Maintaining optimal soil moisture is essential for ecosystem health and plant growth, as it directly influences critical physiological processes such as nutrient uptake, photosynthesis, and transpiration [47, 140]. Insufficient soil moisture can limit these processes, reducing plant yields and compromising agricultural productivity. Conversely, excessive moisture creates anaerobic conditions that increase the risk of root diseases and promote the production of greenhouse gases like nitrous oxide, a potent contributor to climate change [34, 129]. Achieving a balanced soil moisture level is thus critical for sustainable agriculture and mitigating environmental impacts.

In this context, accurate soil moisture measurement is indispensable for optimizing agricultural practices, conserving water resources, and monitoring environmental conditions. Recent advancements in sensor technologies have led to the development of highly precise and reliable tools [2, 50, 129, 155]. These innovations employ diverse approaches and are extensively applied in irrigation management, precision agriculture, and environmental research, enabling more efficient resource utilization and improved sustainability across various sectors.

Various methods and sensors have been developed and implemented to address the growing need for accurate soil moisture monitoring [129]. The thermogravimetric method is considered the gold standard due to its precision. However, its labor-intensive and time-consuming nature limits its practicality for real-time applications [45]. In contrast, modern soil moisture sensors offer real-time measurements and utilize advanced techniques such as neutron thermalization, time-domain reflectometry (TDR), time-domain transmissometry (TDT), electrical capacitance, and impedance analysis [2, 50, 129, 155]. These innovations provide more efficient and accessible solutions, meeting the demands of contemporary agriculture and environmental management.

The simplest and most widely used soil moisture measurement technologies include capacitance sensors [96] and electrical impedance sensors [24]. These devices assess soil moisture by detecting changes in electrical properties, such as current or capacitance, influenced by the presence of water in the soil. Their popularity stems from their robustness, affordability, and ease of installation, making them accessible for various applications. However, their accuracy can be compromised by soil composition, texture, and salinity, which may alter the electrical properties independently of moisture content.

Sensors utilizing TDR technology are regarded as some of the most advanced tools for soil moisture measurement [129]. These sensors emit electromagnetic pulses and analyze their propagation time through the soil, offering highly precise moisture readings. While their accuracy and reliability make them ideal for research and advanced agricultural applications, their high cost and technical complexity pose significant barriers to widespread adoption, particularly for smallholder farmers and resource-constrained operations.

Frequency-Domain Reflectometry (FDR) sensors offer a practical alternative to TDR sensors. They use radio frequencies to measure changes in the soil's dielectric constant, providing reasonably accurate data on soil moisture levels. FDR sensors are generally more cost-effective than TDR sensors, making them a popular choice for many applications. However, their implementation in complex network systems can be challenging. A broader range of interactions within the system can form numerous standing waves caused by the overlap of incident and reflected signals, which may compromise measurement accuracy and reliability [18].

Several other innovative technologies have significantly advanced soil moisture measurement. For instance, optical sensors leverage infrared technology to analyze light reflection and absorption patterns, providing precise estimates of soil water content [103]. These sensors are particularly useful for non-invasive measurements and surface-level analysis. Similarly, thermal sensors measure variations in soil temperature induced by moisture levels, relying on the relationship between specific heat capacity and humidity. This approach offers a unique perspective on soil conditions, particularly in scenarios where temperature gradients are relevant [36]. These emerging technologies complement traditional methods, expanding the range of tools available for efficient and accurate soil moisture monitoring.

Remote sensing systems, including drones equipped with multispectral cameras, have become increasingly popular for soil moisture monitoring [126]. Although these systems do not involve direct ground-based sensors, they offer a valuable aerial perspective, enabling the collection of comprehensive data on soil moisture distribution across large areas. This capability efficiently identifies and analyzes spatial patterns and temporal trends, providing critical insights for precision agriculture, resource management, and environmental monitoring.

In recent years, nanotechnology has revolutionized soil moisture measurement by developing nanosensors [179]. These advanced devices utilize nanoscale materials to detect changes in soil water content with exceptional sensitivity, capturing minute variations that traditional technologies often miss. Furthermore, nanosensors can be seamlessly integrated into Internet of Things (IoT) systems, enabling real-time data transmission to digital platforms. This connectivity enhances remote monitoring and management of irrigation systems, reducing water waste, optimizing resource use, and promoting more sustainable and efficient agricultural practices.

In addition to their exceptional sensitivity, nanosensors are notable for their durability, as they are engineered to endure harsh conditions, including temperature fluctuations and chemical variations in the soil. Their multifunctionality further enhances their value, allowing them to measure not only soil moisture but also other critical parameters, such as nutrient levels and the presence of contaminants. This capability provides a more comprehensive understanding of soil health and composition, enabling more informed decision-making for agricultural and environmental management [179].

In this context, it is crucial to emphasize the need for innovative approaches to sustainable soil management. These strategies focus on improving soil quality, reducing chemical inputs, and promoting the conservation of natural resources. While nanotechnology still faces challenges, such as high costs and environmental concerns related to disposal, it holds significant potential to transform water and agricultural management. By offering sustainable and innovative solutions, nanotechnology can help address the pressing challenges of modern agriculture, paving the way for more efficient and, environmentally friendly practices.

4 Nanotechnology and Soil pH

Soil quality and agricultural productivity are greatly affected by the pH level of soil. Acidic soils (pH < 7) result from high levels of protons (H^+), often caused by organic matter decomposition, root exudates, or the decomposition of clay minerals [182]. The pH level affects chemical, physical, and biological properties of the soil, playing a pivotal role in plant nutrient availability, microbial activity, and overall soil health [4, 131]. To sustain agriculture and ecosystem health, it is essential to understand and manage soil pH.

Soil pH directly affects nutrient solubility and availability. For example, the macronutrients such as nitrogen (N), phosphorus (P), and potassium (K) are most available in soils with a pH range of 6–7 and the micronutrients, like iron (Fe), manganese (Mn), and zinc (Zn), become less available in alkaline soils but may reach toxic levels in acidic soils [90, 138]. Regarding the microbial activity, extreme pH levels can suppress microbial diversity and activity, reducing organic matter decomposition and N fixation [4, 144]. Optimal pH levels promote the aggregation of soil particles through enhanced microbial activity and organic matter stabilization [10]. Soil pH can regulate the mobility and availability of PTEs, because acidic conditions enhance the solubility of metals like aluminum (Al) and Pb, which can be toxic to plants and microorganisms [90, 131]. Understanding soil pH is important for enhancing soil quality and agricultural productivity by improving nutrient use efficiency (NUE), reducing environmental risks, and promoting healthy soil ecosystems.

Improving agricultural productivity and sustainability can be achieved through effective monitoring and management of soil pH. To address the challenges of global food security, new research should emphasize the integration of precision agriculture techniques for real-time pH monitoring and management. For instance, the use of nanosensors to measure and manage soil pH. Nanosensors help farmers monitor and adjust soil pH efficiently, leading to better crop yields and resource management and play a significant role in environmental monitoring by detecting changes in soil acidity or alkalinity due to pollution or other factors [112].

The development of nanotechnology has introduced innovative tools like nanosensors for monitoring and managing soil pH, contributing to enhanced soil quality. Nanosensors are devices capable of detecting and transmitting information at the nanometer scale, offering unparalleled precision, sensitivity, and real-time data for soil management. Nanosensors designed with carbon-based nanocomposites, such as polyaniline-coated carbon nanotubes, have demonstrated high sensitivity in detecting pH changes, making them suitable for agricultural [109]. Nanosensors can accurately measure pH levels in soil, providing a highly sensitive and efficient alternative to traditional methods [112].

Nanosensors have shown excellent photostability and reversible responses across a range of pH levels, making them suitable for long-term monitoring. For instance, nanosensors designed with carbon-based nanocomposites exhibit significant signal variations in response to pH changes, which enhances their ability to detect small fluctuations in pH accurately [109]. The Veris pH Manager™, an on-the-go sensor, utilizes ion-selective electrodes to measure soil pH directly while moving through fields. Nanosensors can be based on various principles, including potentiometric and optical methods. Potentiometric sensors measure the electrical potential difference between electrodes, while optical sensors detect color changes in indicators responsive to pH variations [168]. Both methods have been effectively employed in soil applications, ensuring accurate readings [26].

Initial evaluations on field indicated that this system provided reliable measurements, with deviations from standard reference values being minimal average deviations of 0.37 pH for $CaCl_2$ and 0.68 pH for H_2O [152]. Nanosensors have transformative applications in soil pH management, particularly in precision agriculture. By enabling the precise mapping of soil pH variability across fields, nanosensors assist farmers in applying lime, fertilizers, and soil amendments more efficiently. This precision minimizes resource waste, reduces costs, and mitigates environmental impacts, such as nutrient runoff. In addition, nanosensors play a pivotal role in soil remediation by offering continuous feedback during efforts to correct soil acidification or alkalinity. This ensures interventions remain effective, with real-time adjustments preserving soil functionality and productivity.

Nanosensors offer several advantages over traditional methods for measuring soil pH, leading to increased accuracy and efficiency. Here are the key factors that contribute to the superior performance of pH-sensitive nanosensors: Higher Sensitivity, Real-Time Monitoring, Robustness and Stability, Miniaturization and Portability, Diverse Measurement Techniques and Field-Specific Calibration [26, 109, 152, 168].

Nanosensors can detect minute changes in pH levels due to their small size and high surface area-to-volume ratio. For instance, studies have shown that certain nanosensors exhibit a nearly nine-fold signal variation in response to pH changes, allowing them to distinguish small fluctuations that traditional methods might miss [109]. Unlike conventional laboratory methods, which often involve time-consuming sampling and analysis, nanosensors can provide real-time data on soil pH. This capability enables farmers and researchers to make immediate decisions based on current soil conditions, enhancing precision in agricultural practices [152].

Many nanosensors demonstrate excellent photostability and reversibility, maintaining their performance over extended periods and under varying environmental conditions. For example, nanosensors designed with carbon nanotubes or conductive polymers have shown reliable responses in complex samples, such as polluted water or soils with varying ionic strengths [26, 109]. The compact nature of nanosensors allows for their integration into portable devices, facilitating on-the-go measurements directly in the field. This is exemplified by systems like the Veris pH Manager™, which can map soil pH while moving across fields, providing spatially detailed data that traditional grid sampling cannot achieve [152].

Nanosensors utilize various measurement techniques, including potentiometric and optical methods, which can be tailored to specific applications. Optical sensors, for instance, can measure extreme pH values and operate effectively in colored or turbid solutions where traditional glass electrodes may fail [168]. Nanosensors can be calibrated specifically for the field conditions they are used in, reducing systematic errors associated with standard protocols. This adaptability enhances their accuracy compared to traditional methods that may not account for local variations in soil chemistry [152].

The integration of nanosensors into soil pH measurement represents a significant advancement over traditional methods due to their higher sensitivity, real-time monitoring capabilities, robustness, portability, and diverse measurement techniques, making them invaluable tools for precision agriculture and environmental monitoring, while their continuous technological evolution enhances the accuracy and reliability of these sensors, solidifying their role in modern agricultural practices and promoting ecological sustainability through precise, site-specific interventions.

5 Nanotechnology and Controlled Release Fertilizers

Agrochemicals are frequently used to improve crop production technology and play a key role in agriculture. Chemical pesticides are widely used to prevent and control insects, pests, pathogenic fungi, and weeds, playing a crucial role in improving the productivity of agroecosystems [121]. Pesticides can pose significant health risks when their concentrations in soil exceed established limits. Human pesticide exposure causes various adverse effects, including neurotoxicity, mutagenicity, genotoxicity, and carcinogenicity. Several studies on pesticide use have shown that intensive application of these chemicals can impact the contamination of vegetables, fruits, and water sources [121].

Therefore, monitoring and assessing the amount of pesticides present in the agricultural sector, in food, and in water samples is crucial. An efficient and accurate system for pesticide detection and monitoring must be developed. Traditional instruments such as spectroscopic and chromatographic techniques—including high-pressure liquid chromatography, mass spectrometry, gas chromatography, and antibody-based immunoassays—are widely used to detect the presence of pesticides in various agricultural samples [57, 121, 160].

These traditional analytical methods face several limitations, including the need for expensive equipment and skilled professionals to perform the analyses. In addition, the pre-sampling required is time-consuming [85]. Most traditional pesticide detection methods are time-consuming and impractical for real-time on-site testing. However, these limitations can be overcome by utilizing nanoscale sensing techniques that provide real-time on-site monitoring. Rapid, sensitive, and affordable sensors have been developed to detect pesticide residues [183].

Nanostructure biosensors represent an innovative sensing technology that is both affordable and efficient and delivers results.

Several types of biosensors and nanobiosensors are currently in use, including enzyme-based optical biosensors [87], electrochemical biosensors [63], voltammetric biosensors [111], and biosensors using carbon nanomaterials [201]. These innovative devices are essential for detecting pesticides in food and other agricultural products. Some modern sensors integrate a biological recognition system with a physical transduction method. Nanosensors made of TiO_2 nanotubes have been used to detect the presence of atrazine in soil [194].

Amino acid-based herbicides containing P in soil can be detected using nanosensors, which use a pencil lead electrode modified with a nanofilm. The nanofabrication of biosensors has significantly increased the effectiveness of these devices, improving their sensitivity, selectivity, and multiplexing capacity. Electrochemical nanosensors incorporating the cholinesterase enzyme have shown promising results in detecting organophosphate pesticides and carbamate residues, standing out for their extraordinary sensitivity [204].

Tyrosine-based biosensors demonstrate remarkable sensitivity in detecting pesticides, especially those based on carbamate and atrazine. Biosensors using alkaline phosphatase have proven highly effective in detecting organochlorine pesticides, such as carbamate and fenitrothion [54]. Biosensors using organophosphate hydrolase and glutathione S-transferase (GST) are highly effective in detecting organophosphate pesticides, specifically for compounds such as parathion and methyl parathion, as demonstrated by [41]. In addition, these biosensors have also been successfully used to identify paraoxon, an active metabolite of pesticides that can be highly toxic [101].

The synergistic combination of acetylcholinesterase (AchE) and choline oxidase (ChOx) enzymes showed remarkable performance, achieving a detection limit of only 4.49 nM, significantly effectively identifying organophosphate pesticides. Biosensors using acetylcholinesterase stabilized by Cd QDs with poly(N-vinyl-2-pyrrolidone) (PVP) have shown promising performance in identifying trichlorfon, a potentially toxic substance [106].

The enzyme acetyl coenzyme A carboxylase was instrumental in a study in which an innovative nano biosensor, developed using atomic force microscopy (AFM) technology, was used to identify the presence of herbicides in the environment [9]. The interaction between an enzyme and the analyte is analyzed through a complex process that uses directed molecular dynamics. This approach allows simulating and observing how these molecules behave and interact under specific conditions. This recently developed nanobiosensor demonstrated the ability to identify the herbicide at deficient concentrations, a significant advance in detecting agricultural contaminants. The interaction between biomolecules and analytes was studied at the molecular level using an enzyme carefully positioned on the surface of an atomic force microscope, allowing detailed analysis of molecular dynamics.

The disposable enzymatic nanobiosensor developed for detecting the herbicide Chlortoluron was identified using a screen-printed carbon electrode that underwent modifications in addition to incorporating zinc oxide nanoparticles (ZnO NPs). Chlortoluron is widely recognized as one of the most widely used herbicides in modern agriculture, being especially effective in inhibiting the growth of weeds that compete with crops. This developed nanobiosensor can inhibit the action of tyrosinase, which is a crucial enzyme in several biochemical reactions. This biosensor has a detection capacity with a limit of 0.47 nM. The level of inhibition of the enzyme tyrosinase, which plays a crucial role in biological processes such as melanin synthesis, is influenced by the presence of the herbicide chlortoluron, whose concentrations range from 1 to 100 nM. This range of inhibition suggests that Chlortoluron may significantly impact tyrosinase activity [60].

6 Nanotechnology and Plant Nutrition

Nutrient management is essential to maximize agricultural production, reduce environmental pollution, and preserve soil health [46]. Plant nutrients are divided into two groups: the macronutrients nitrogen (N), phosphorus (P), potassium (K), calcium (Ca), magnesium (Mg), and sulfur (S), which are required in large quantities, and the micronutrients boron (B), chlorine (Cl), copper (Cu), iron (Fe), manganese (Mn), molybdenum (Mo), nickel (Ni), and zinc (Zn). Accurate measurement of these nutrients is essential in agriculture, and nanotechnology-based sensors offer advantages such as real-time monitoring, high spatial accuracy, low energy consumption, and compatibility with wireless technologies [147]. Among the nanoparticles used to produce these nanosensors, carbon nanotubes, nanocrystals, some metal oxides, and graphene have been widely studied in recent research (Table 1). Carbon nanotubes (CNTs) are nanoscale structures composed of graphene sheets rolled into a cylindrical shape, with diameters ranging from 0.4 to 100 nm and reaching lengths of several micrometers [184]. The high surface-to-volume ratio, together with their unique electronic structure, allows CNTs to be used for the sensitive and selective detection of different soil health parameters, such as nutrients, PTEs, and organic contaminants [108]. CNT-based sensors have been used to detect nitrate (NO_3^-) and ammonium (NH_4^+) ions in soil solution, with a detection limit of 0.1 μM [65]. Furthermore, CNT sensors can detect Cu ions in soil with a sensitivity of 0.1 μM and a response time of less than 5 s [198].

Graphene has exceptional electrical, mechanical, and optical properties, making it a very attractive material for sensing applications [128]. Graphene sensors take advantage of changes in electrical conductivity, resistivity, or capacitance when graphene interacts with target analytes [171]. Using this technology, it was possible to detect phosphate ions in soil samples with a sensitivity of 0.2 μM [30] and Zn ions with a detection limit of 0.05 μM [110].

Metal oxide nanoparticles, such as ZnO, SnO_2, and TiO_2, are widely used in sensing applications due to their electrical, optical, and catalytic properties [15]. Sensors using these nanoparticles rely on changes in electrical conductivity or resistance when they interact with analytical targets [195]. For example, sensors with ZnO and SnO_2 are used to detect K ions in a concentration range of 0.1 to 100 mM [67]. Kundu et al. [95] developed an electrochemical nanosensor with CNT-α-Fe_2O_3 nanocomposite to detect NO_3^-, showing high sensitivity and low detection limit. Another example is a sensor with ZnO NPs, which is sensitive in detecting Mn ions in the soil, demonstrating selectivity in relation to other metal ions [29].

Table 1 Nanomaterials used in sensors for detecting chemical elements

Nanomaterial used	Operating principle	Elements	Detection limit	References
CNTs	CNTs improve electron transfer and increase sensitivity	NO_3^-	0.1 μM	[63]
		NH_4^+	0.1 μM	[65]
		NO_3^-	0.1 μM	[185]
		Cu	0.1 μM	[198]
CNTs + Graphene Oxide	The combination with CNTs improves the stability and sensitivity of oxide nanosensors	Cu	1.574 x 10^{-14}	[19]
ZnO and SnO_2 nanoparticles	They depend on changes in electrical conductivity when interacting with analytical targets	K	0.1 a 100 mM	[67]
QDs	Detect changes in fluorescence when interacting with target substances	Cu	0.1 μM	[146]
		Zn	0.5 μM	[146]
		Mn	1 μM	[146]
Graphene QDs		Co	0.014 M	[181]
		Cu	0.018 M	[181]

Research on QDs uses semiconductor nanocrystals with dimensions between 2 and 10 nm [7]. Sensors that use QDs detect changes in fluorescence or photoluminescence when the nanocrystals interact with target substances [55]. The adoption of this technique enabled the simultaneous detection of Cu, Zn and Mn ions in the soil, with detection limits of 0.1, 0.5 and 1 μM, respectively [146]. Lei et al. [102] developed a robust, magnetic, and self-accelerated electrochemiluminescence nanosensor using multifunctional cobalt ferrite magnetite ($CoFe_2O_4$) NPs for extremely sensitive and discriminative detection of Cu ion through a click reaction in a quasi-homogeneous system. Vahab and Keshispour [181] developed a graphene QD nanosensor modified with Eriochrome Black T (EBT) to detect Co and Cu, with the detection limit of 0.014 and 0.018 M, respectively.

7 Nanotechnology and Plant Stress

Globally, agricultural production lands are challenged by a lot of environmental stresses that cause a reduction in biomass and productivity [17, 122, 156]. Plants are constantly exposed to these environmental stressors, which are generally categorized as biotic and abiotic stressors where the main biotic stressors include pathogens, insects, and herbivores, while abiotic stressors include exposure to PTEs, soil salinity, drught, and heat, among many other factors [43, 124, 130]. As a result, these stresses induce a cascade of physiological changes in plants, ultimately affecting their overall health and productivity patterns [27].

Eventually, the most severe consequence of the effects of these stressors is oxidative stress, which can be identified by an imbalance in the antioxidant defense system and reactive oxygen species (ROS) in plant sites such as the apoplast, chloroplast, mitochondria, and peroxisomes [52, 164]. DNA damage can be related to ROS accumulation after they affect nucleic acids, enzyme inhibition, and lipid peroxidation, eventually inducing cell injury, disrupting cell organelles, and resulting in cell death [158].

The response of plants to the deleterious effects of ROS involves complex enzymatic and non-enzymatic defense mechanisms, collectively referred as "antioxidant system" [78]. The enzymes of the antioxidant system include ascorbate peroxidase (APX), catalase (CAT), dehydro-ascorbate reductase (DHAR), glutathione reductase (GR), glutathione peroxidase (GPX), GST, monohydro ascorbate reductase (MDAR), peroxide reduction (PRX) and superoxide dismutase (SOD).

Non-enzymatic antioxidants include ascorbic acid (AA), α-tocopherol, carotenoids, flavonoids, glutathione (GSH), and plastoquinone/ubiquinone [99]. Both groups of antioxidants are necessary for ROS homeostasis, and some previous studies suggest that high antioxidant activity is linked to stress tolerance in plants and, therefore, plays a key role in plant stress adaptation [120].

A new wave of advancements is emerging in the agrotechnology sector, aimed at meeting the global food demand sustainably, with nanotechnology playing a pivotal role in the widespread use of nanomaterials. These innovations highlight the potential of nanomaterials to address pressing challenges in agriculture [92, 132, 192]. Nanobiosensors, which encapsulate living organisms, are non-invasive and highly sensitive materials for detecting environmental samples [85, 174]. Nanosensors offer significant potential for continuous monitoring of crop health under stress, improving crop efficiency by optimizing resource use and reducing crop losses from diseases. According to [83], these materials also facilitate the controlled release of antibiotics or nutrients, boosting plant resistance to diseases (Table 2).

Various types of nanoparticles, such as metal oxides (e.g., TiO_2, MgO, and ZnO) and magnetic NPs, are combined with nanobiosensors for targeted monitoring [163]. According to [118] and [159], nanosensors hold great potential for sustainable agriculture due to their high detection capacity and low detection limits. These portable devices are used in agriculture for on-site measurement of plant and soil conditions, including pH levels, soil fertility, toxic chemicals, moisture content, and the presence of plant pathogens and pests [127, 159].

Environmental stresses like waterlogging, salinity, drought, and PTEs are major barriers to agricultural productivity and food security globally [64–75]. Each year, these stresses contribute to 51–82% of agricultural production losses worldwide [84, 105, 136]. These adverse conditions can disrupt the physiology and biochemistry of plants, leading to changes in cellular composition, phytohormonal activity, leaf light absorption capacity, carbohydrate storage, stomatal function, and the production of harmful radicals and catalysts, ultimately reducing agricultural output [12, 68, 69, 145].

Table 2 Nanosensors characterization techniques used during plant stress

Nanosensors	Nanomaterial technique applied	Stress	Target	Concentration	Key finding	References
Single plasmonic NSs	Gold nano-rods (GNRs)	Abiotic stress	Pb^{2+}; Hg^{2+}	5 nM; 1 nM	Position encoded aptamer functionalized GNRs sensing device	[191]
Triboelectric NSs	Ultra-thin MXene	Abiotic stress	Cu^{2+}; Cr^{3+}; Zn^{2+}	10–300 μM	Simple vacuum filtration method	[185]
Colorimetric NSs	Mesoporous silica NPs	Abiotic stress	Hg	60 pM	Hg(II)'s strong affinity for sulfur, it can interact with solid S2 and produce new Hg(S-R)2, which can disrupt the linear dithioacetal connections and cause the release of cargo Hg (II)'s strong affinity for sulfur, it can interact with solid S2 and produce new Hg (S-R)2, which can disrupt the linear dithioacetal connections and cause the release of cargo	[80]
Surface plasmon resonance nanosensor (SPRNS)	Au	Abiotic stress	Afatoxin B1 (*Aspergillus flavus*)	1.04 pg mL^{-1}	The concentration of aflatoxin affected the reflectivity index of the gold electrode surface	[5]
Electrochemical Biosensor	Colloidal Gold Nanoparticles (GNPs)	Biotic stress	*Pseudomonas syringae*	214 pM	Uses differential pulse voltammetry to evaluate electrochemical changes	[98]
Chemiresistivand biosensors	Polypyrrole (PPY) nanoribbons on Au microelectrode array (Au-MEAs)	Biotic stress	Mosaic virus (Cucumber)	10 ng ml^{-1}	Amperometry	[28]

In recent years, significant environmental changes have negatively impacted global agriculture. Climate change, in particular, has altered the abiotic conditions affecting plants, leading to changes in their development, physiology, and productivity [6]. To minimize crop losses due to abiotic stress, it is crucial to identify and address these stressors promptly. Plants possess various adaptive mechanisms to cope with abiotic stress, such as (a) changes in physiological responses, (b) the production of osmoprotectants, (c) antioxidant defense systems, (d) stress-related proteins, (e) gene expression adjustments, (f) detoxification signaling to repair damage, and (g) restoration of cellular homeostasis, among others [91, 117, 162]. Under abiotic stress, plants produce harmful chemical compounds like diatomic oxygen, including H_2O_2, O_2^-, OH–, etc. [32, 61]. Additionally, stressed plants often display altered phenotypes, and monitoring the airborne sounds emitted by these plants can provide insight into their condition [89]. Therefore, researchers can assess plant health in response to various stresses by analyzing phenotypic changes, chemical composition, phytohormones, and so on [12].

Under stress conditions, plants release chemicals to trigger their defense mechanisms, and detecting these stress-related chemicals can help identify different types of stress. [165] developed a nanosensor based on surface-enhanced Raman scattering (SERS) for real-time detection of various stress-related chemicals in living plants. Additionally, Alova et al. [8] created oxygen nanosensors using quartz micropipettes with carbon and platinum-coated tips to measure oxygen level reductions.

Several approaches that combine electrochemical techniques with nanomaterials are being employed to diagnose various forms of abiotic stress (Table 3). This study focuses on modern strategies to alleviate different plant stresses and the recent advancements in nano-inspired biosensors within plant stress biology in recent years.

8 Nanotechnology and Potentially Toxic Elements

Potentially toxic elements are chemical elements with specific characteristics, such as high toxicity and densities greater than 5 g cm^{-3}. Most heavy metals, such as arsenic (As), lead (Pb), mercury (Hg), chromium (Cr), zinc (Zn), cadmium (Cd), copper (Cu), and nickel (Ni), are widely recognized as significant environmental contaminants, compromising both natural ecosystems and human health [14]. Potentially toxic elements accumulate in the food chain due to their resistance to biodegradation [114].

Soil availability and toxicity vary significantly since PTEs are initially adsorbed and undergo slow adsorption reactions, leading to redistribution into distinct chemical forms [25]. The potentially toxic elements distribution in soils is influenced by a series of reactions, including precipitation and dissolution of minerals, ion exchange, adsorption, and desorption, complexation in aqueous media, biological immobilization and mobilization, and absorption by vegetation (Fig. 2). The potentially toxic elements absorption by roots depends on several factors like their anatomy and, ability

Table 3 Principal insights funds of nanosensors under abiotic stress

Abiotic Stress	Nanosensors/nanomaterial technique applied	Key finding	References
Heavy metal	QDS (Quantum dots)	Improved the sensing performance, overcoming the draw backs of traditional methods like slowness, low sensitivity, high cost ,and complicated operation	[185]
	Smartphone-enabled detection platform capable of sensing Hg (II)	An alternative method for (Hg) and (Cd) detection involved employing a lateral flow assay with Upconversion Nanoparticles (UCNPs), Gold Nanoparticles (AuNPs), aptamers, DNAprobes, and antibody probes. Detection limits were achieved at 2.5 x 10^{-11} and 1.6 x 10^{-9} for Hg and Cd, respectively	[58]
Drought	Cerium oxide nanoparticles	Enhanced drought tolerance in sorghum	[40]
	Carbon nanotube + aquaporin	Mitigate the deleterious impacts of drought and facilitating enhanced water uptake	[116]
Salinity	Electrochemical biosensor + multiwall carbon nanotubes (MWNT) and carbon black	Detection limit of 1.99 M for indole-3-acetic acid (IAA) and 3.30 M for salicylic acid (SA) in soybean seedlings exposed to saline stress	[66, 125]
	Carbon nanoparticle-based seed priming	Boosts lettuce seed vigor and seedling growth by suppressing primary root elongation	[22]

to adapt to the environment in which it is located, allowing the plant to survive in potentially contaminated soils [104].

Traditional methods for the PTEs identification in complex matrices encompass several techniques, such as atomic absorption spectroscopy, neutron activation analysis, X-ray fluorescence spectrometry, energy-dispersive X-ray fluorescence, inductively coupled plasma mass spectroscopy, and atomic or optical emission spectrometry [44, 133, 139, 177, 203]. Although these methods often achieve adequate detection sensitivity, they face considerable challenges, such as equipment weight, high costs, and operational complexity, which limit their application to specialized laboratories.

The main attraction of nanomaterials is their chemical and physical singularities, which include size, composition, conductivity, magnetism, mechanical strength, light absorption, and emission properties [197]. The use of nanomaterials represented a advance in detecting traces of PTEs in soil and proving effective in treating pollutants. These materials can be used to identify chemical markers of pollutants, functioning both in capturing the marker and amplifying the signal associated with its detection [190]. The evolution of nanotechnology has provided significant advances in the

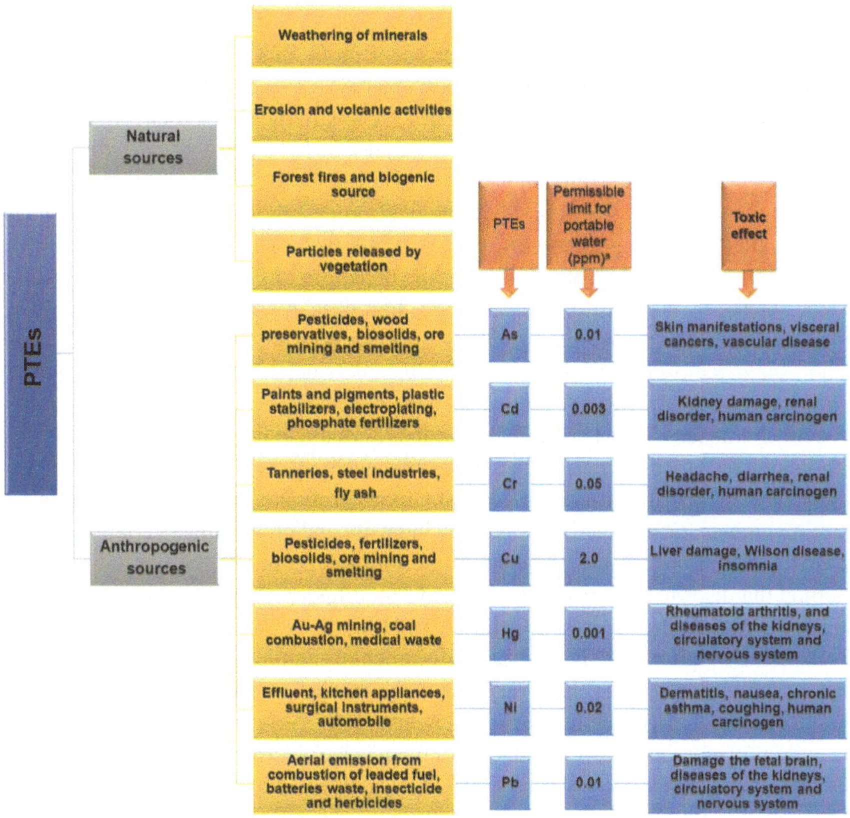

Fig. 2 Potentially toxic elements: sources in the environment [39] and permissible limits by World health organization—WHO[a] and health toxic effects [20]

performance of sensors, especially in sensitivity, detection limits, selectivity, and reproducibility [142].

Nanosystems have proven to be highly effective in identifying traces of heavy metals in soil, offering an innovative solution for environmental monitoring. These systems, which use nanomaterials with specific properties, allow the detection of very low concentrations of contaminants that can harm human health and the ecosystem. These technologies help assess soil quality and support environmental recovery and conservation strategies.

In polluted soil, Fe ions combine with humic substances, while eliminating bound metal ions is a complex procedure requiring a potent chelating agent [169]. Several nanotechnologies have already been tested for the detection of Fe present in soil, aiming to improve the efficiency and accuracy of this process. Carbon nanotubes, single-walled and multi-walled, are fascinating nanomaterials that often play a crucial role as electrodes in electrochemical sensors [30]. In addition, there have been significant advances in the development of CNT-modified electrodes using platinum nanoparticles (Pt NPs), which enable the selective identification of iron ions [11]. Another innovative example is the fluorescent probe that uses highly luminescent carbon QDs to detect Fe ions [189]. Finally, there is also a portable and highly selective colorimetric film, which presents a color change to red when encountering Fe, remaining unchanged in the face of other metal ions [31].

Although copper is a fundamental trace element that plays a key role in several environmental physiological and pathological processes, an excess of Cu ions can cause toxicity in living organisms, posing health hazards [77]. In recent years, the application of fluorescent biosensors for the detection of Cu ions has gained prominence thanks to their simplified structure, low cost, high sensitivity, and reduced operating time [173]. We can highlight several innovations in detecting Cu ions: Fluorescent nanosensors with a core/shell/shell structure composed of CdS/ZnS/CdS doped with Mn were developed for this purpose [196]. An ultrasensitive fluorescent chemosensor was created to identify residual Cu ions in nanofibers and thin films using knife scattering and electrospinning techniques [81]. Furthermore, etched silicon (Si) wafers were used to generate alkyl-terminated porous Si NPs, demonstrating greater stability and fluorescence intensity and presenting a variation in fluorescence depending on the presence of Cu ions [76]. Finally, the Cu ion detection method that uses ultramicroelectrode arrays modified with Au NPs offers improved sensitivity due to the increased surface area [135].

Arsenic is highly toxic in various environmental samples, such as soil, sediments, water, aquatic organisms, and vegetation [3]. In the presence of cationic metals, As ions in the form of arsenate act as chelates [48]. Several effective methods for detecting As have been developed. One of them is the electrochemical method, which uses graphene paste electrodes and AuNPs to identify subnanomolar As concentrations [150]. Another method involves arrays of Au NPs that detect As ions [21]. In addition, electrochemical systems based on flow analysis have been designed to identify arsenite [113]. Finally, an innovative paper-based device that integrates with a smartphone enables the As identification in water and soil samples. This color optical nanosensor uses sucrose-modified Au NPs (Au NPs/SuC) as the sensing element [161].

Mercury ions present in soil result from waste generated during mineral processing and from releases caused by coal burning [186]. The identification and removal of Hg ions in soil are effectively performed through sorption techniques on humic materials and coprecipitation with sulfides. Failure to detect and remove these ions can result in severe kidney damage [154]. Some Hg ion detection techniques include: preparing aqueous solutions of soil samples for analysis through a smartphone equipped with a photometer application and a colorimetric aptamer nanosensor [187], the use of AgNPs functionalized with 2-aminopyrimidine-4,6-diol (APD-AgNPs) obtained from *Areca catechu* leaf extract, where the interaction between APD-AgNPs and Hg ions generates a visible color change, in addition to changes in spectroscopic properties, allowing effective detection in aqueous samples [143], the application of a fluorescent magnetic sensor (S1), which uses imide-Hg-imide complexes formed by the reaction between 1,8-naphthalimide and Hg ions, acting as a selective fluorescent probe to detect Hg ions at trace levels [202], and the synthesis of AgNPs (with an average particle size of 21.81 nm) from *Soymida febrifuga* extract, resulting in the interaction between AgNPs and Hg ions that causes changes in color and SPR band in soil samples [167].

Lead stands out as the most common toxic metal found in soils. When ionic Pb levels exceed 300 ppm, contamination can lead to the accumulation of this metal in plant leaves and roots, which poses a severe risk to human health [149]. Lead detection occurs through DNA hybridization with a complementary 20-base substrate using a Au electrode. In the presence of Pb ions, the catalytic tape breaks the substrate bond, producing fragments that facilitate the transfer of electrons to the electrode, which gives rise to the Faradaic current [188].

9 Nanotechnology and Pesticides

One of the challenges in the agricultural sector is the loss of nutrients through conventional fertilizers. Identifying controlled-release fertilizers (CRFs) has improved the effectiveness of fertilizers through their continuous release. Conventional fertilizers are mostly physically or chemically modified to decrease solubility and mobility [176]. CRFs can be manufactured with unique attributes. This result is mainly achieved by coating or encapsulating conventional fertilizer with a porous material or reducing or controlling water penetration into the fertilizer [148].

Nanotechnology can develop fertilizers to safeguard the environment and enhance fertilizer effectiveness. According to [59], nanofertilizers associated with nanosensors play a crucial role in synchronizing the release of nutrients from fertilizer and absorption by plants. Through this process, the fertilizer concentration for application can be improved, and nutrient leaching into soil and water bodies can be prevented. Nanofertilizers have the following advantages: i) control of the nutrient release profile by encapsulating nanofertilizers or nutrients in nanoporous materials, ii) enhancement of availability by increasing the solubility of nutritional components in water and iii) controlled and slow release of nutrients from nanocarriers without waste can extend the shelf life of nanocarriers by adjusting the release profile according to the growth phases of plants [148].

Kah et al. [82] comprehensively categorized nano fertilizers, dividing them into three main categories that help to better understand their origins and functions. The first category covers nanomaterials that contain micronutrients. The second refers to nanomaterials produced from macronutrients. Finally, the third category encompasses nanomaterials that act as macronutrient carriers.

Among these various categories of nanomaterials, it is important to highlight that those that act as macronutrient carriers stand out due to their great potential to optimize nutrient delivery. The main benefit of using nanocarriers for nutrient delivery instead of materials that are directly composed of nutrients is that these nanocarriers can be made from safe, environmentally friendly materials and are highly compatible with culture media, plants, and other organisms.

Another significant advantage of using nanocarriers in the agricultural sector is the ability to formulate or "adjust" fertilizers to allow the release of nutrients in a controlled manner suited to the needs of plants [59]. There are six types of nanostructured materials as nutrient carriers: nanoclays, hydroxyapatite nanoparticles, mesoporous silica, carbon-based nanomaterials, and polymeric nanoparticles [59]. Regarding the application of nanotechnology to fertilizers, five types of fertilizers stand out: layered double hydroxides (LDH) with controlled release, zeolites, nanoclays, carbon-based nanofertilizers, mesoporous silica, and polymeric nanoparticles [148]. Each of these types has unique characteristics that can be exploited to improve the efficiency of nutrient delivery to plants. Controlled-release LDH has a structure composed of multiple layers, making it possible to intercalate anionic materials in the spaces between them. This characteristic facilitates the controlled release of anions [51]. Layered double hydroxides are fascinating materials that enable the controlled release of substances, allowing elements such as N and P to be intercalated in their interlayer space through an ion exchange process. This mechanism occurs mainly between LDH and salts such as nitrates or phosphates [175], thus forming a structure that can efficiently deliver N and P. The porous structures of the zeolite can accommodate cations by adsorption and ion exchange [115]. A nitrogen-containing zeolite provides nitrogen to the plant throughout its life cycle and reduces N waste. A comparison between N leakage from urea-based fertilizers and those coated with zeolite reveals a reduction of approximately 65% in N waste when using urea with a zeolite coating [42]. A zeolite containing K can be an excellent option as a potassium fertilizer for various plant species [200], highlighting the efficiency and environmental benefits of using zeolites in agricultural fertilization.

Nanoclay has proven to be a promising material for immobilizing enzymes that catalyze organic P, increasing the availability of this nutrient for plants and optimizing its absorption. The immobilization of enzymes also contributes to their increased stability [49]. Nanoclays can be classified into anionic and cationic [62]. The anion exchange capacity of these materials makes them highly effective as carriers of nitrate [175], phosphate [23], and borate [166]. Nanoclays that stand out as effective nutrient carriers generally have two main characteristics. First, they can protect nutrient molecules through physical barriers provided by their structure [166]. Second, they can intercalate nutrients into the nanoclay layers through ion exchange [23] or through non-electrostatic interactions, such as the formation of hydrogen bonds [93]. Due to these characteristics, nanoclays have the potential to retain nutrients for prolonged periods, promoting healthy plant growth, improving N use efficiency [23], balancing nutrient supply [166], and reducing environmental contamination [141].

There are few studies on the use of carbon-based nanofertilizers [16] conducted a detailed study where they showed that CNFs loaded with Cu NPs demonstrated a significant capacity to release Cu in a more controlled and slow manner into water. Kumar et al. [94] conducted an in-depth study on an innovative combination of polymer film, specifically PVAc-starch, and carbon nanofibers, which act as a delivery vehicle for Cu–Zn NPs. The researchers concluded that this polymeric formulation is crucial in protecting the Cu–Zn NPs, preventing them from being rapidly released into the soil.

Mesoporous silica NPs offer a range of significant advantages for their use. Being a natural substance, silica poses no environmental risks. In addition to enabling the controlled release of fertilizers, silica protects these nutrients from premature degradation. When absorbed by plant roots, nanosilica forms a delicate film around the cell walls, providing an effective defense against stresses. Its ordered and stable porous structure and highly functional surface allow for precise control of fertilizer loading and release [148]. Silica NPs play a crucial role in protecting urea, which is often vulnerable to premature degradation due to its chemical properties, making it highly susceptible to photodegradation. In addition, urea also hydrolyzes efficiently in soil, which can result in significant N losses. Thus, mesoporous silica NPs stand out as promising candidates for the controlled release of urea-based fertilizers, as they present properties that allow a more efficient and targeted release over time, increasing the effectiveness of nutrient use by plants [148]. Fertilizers with coatings made of polymers are designed to detach gradually and in a controlled manner, releasing essential nutrients for plants over time. As the polymer layers degrade in the soil, they allow for more efficient absorption of nutrients. Biodegradable polymers emerge as the most suitable and sustainable option to meet this specific need. The time required for the release of the compound will be strongly influenced by the various polymer layers surrounding it, which act as barriers and regulators of the release rate [100]. Polymeric NPs used as fertilizer delivery vehicles play an essential role in modern agriculture since, in order to be effective, they need to be not only biodegradable, allowing safe decomposition in the environment, but also ensure the safety and health of the soil and cultivated plants [86].

Chitosan is a cationic polymer, which allows it to interact effectively with negatively charged materials. In addition, its size adjustment capacity represents a significant advantage. In agriculture, the incorporation of fertilizers that combine N, P, and K can be carried out in a way that promotes their controlled release into the soil, which results in a more efficient and sustained supply of nutrients to plants during their growth cycle [148].

Chitosan, a natural and biodegradable biopolymer extracted from chitin, provides sorbent and bactericidal characteristics, giving it great potential as an innovative and highly promising material for the efficient transport of agrochemicals. This attribute makes it a sustainable option and an effective solution to improve the delivery and efficacy of chemicals used in agriculture [86].

Incorporating fertilizers into chitosan NPs is a reasonably straightforward process, which can be carried out by dissolving the particles in a specific solution followed by magnetic stirring [33]. Chitosan NPs loaded with NPK nutrients, known as chitosan-NPK NPs, have been developed through a method involving methacrylic acid polymerization in an existing chitosan solution. This process is followed by the additional incorporation of NPK nutrients, which enhances the efficiency of the nanoparticles in agricultural applications [1]. Polydopamine is a highly versatile organic polymer with the impressive ability to form an ultrathin, film-like coating with a thickness in the nanometer range on a wide variety of materials [100]. The thickness of polydopamine layers can be carefully manipulated and adjusted by optimizing four very important key factors: the concentration of the dopamine monomer used, the pH of the solution in which the deposition occurs, the deposition time, which is crucial for proper layer formation, and the number of cycles allowed for deposition, which can significantly influence the density and uniformity of the layer [137].

The time required for polydopamine to become biodegradable can be effectively modified by adjusting the dopamine content present in each coating layer [193]. Furthermore, it is quite advantageous to regulate the release of the materials that are trapped within it [134]. This controllability in the release makes polydopamine play an important role as a vehicle for the controlled release of fertilizers [79]. Therefore, manipulating these characteristics can result in more effective and practical applications in agriculture and sustainability.

10 Conclusions and Future Perspectives

Nanosensors represent an innovation in agriculture, playing a fundamental role in crop protection. Given the current global conditions, society must adopt sustainable practices to protect crops. Accurate fertilizer use, through analysis of soil nutritional status, can significantly reduce environmental degradation, especially with N fertilizers. Real-time monitoring of soil moisture levels can optimize irrigation systems and promote water conservation. Analysis of plant morphogenetic changes in response to climate changes, such as rising temperatures and CO_2 concentrations, as well as pathogen invasion, can provide valuable insights for developing more efficient agricultural production systems. Analysis of nutrient and pesticide residues in the environment is essential to protect both the ecosystem and human health, preventing the occurrence of unknown risks. Although numerous studies have demonstrated the great potential of nanotechnology and nanosensors in agriculture, their practical application in the field is still little used. There are several factors to be considered when advancing the development of this technology in future research. The systems developed should be designed to be versatile, not limited to a single region, and should be suitable for different climatic conditions.

Furthermore, developing devices that can perform multiple detections simultaneously is essential rather than being restricted to identifying only a single nutrient. Choosing environmentally friendly chemicals for use in soil sensing devices is essential. Using nanosensors in controlled-release fertilizers presents an exciting prospect, and research in this area must receive broad support and encouragement. In some sensors, nanotechnology is the exclusive basis. In contrast, in others, different types of nanoparticles and other nanostructures are used to enhance their sensing abilities, taking advantage of their unique characteristics. There is a growing interest in fabricating different sensors capable of detecting pathogens, pH, temperature, moisture, and soil nutrients. The literature indicates that nanotechnology has predominantly focused on laboratory environments.

Most studies on nanosensors in agriculture have been conducted under controlled conditions. However, only a few sensors have been able to provide data after their field application. Therefore, we emphasize the importance of implementing them in field conditions, considering a variety of environments. This critical analysis will be essential to ensure the sustainable adoption of these strategies in agriculture in the future. For large-scale implementation, acquiring more profound knowledge in the field is essential. The increasing adoption of nanosensors in agriculture is intrinsically linked to the principles of sustainable development in response to population growth and climate change. Furthermore, the expanded adoption of advanced nanosensors will significantly increase the effectiveness of the agricultural system as a whole, providing various environmental, economic, and social benefits.

Competing Interests The authors declare that they have no competing interest.

Declaration of Generative AI and AI-Assisted Technologies in the Writing Process During the preparation of this work the book chapter author(s) have not taken any assistance from AI, and the book chapter author(s) reviewed and edited the content as needed and take(s) full responsibility for the content of the publication. All authors read and approved the final version of the manuscript.

References

1. Abdel-Aziz HM, Hasaneen MN, Omer AM (2016) Nano chitosan-NPK fertilizer enhances the growth and productivity of wheat plants grown in sandy soil. Span J Agric Res 14:6141–8205
2. Adla S, Rai NK, Karumanchi SH, Tripathi S, Disse M, Pande S (2020) Laboratory calibration and performance evaluation of low-cost capacitive and very low-cost resistive soil moisture sensors. Sensors 20:363
3. Adriano DC (2001) Trace elements in terrestrial environments, 2nd edn. Springer, New York, NY
4. Agegnehu G, Amede T, Erkossa T, Yirga C, Henry C, Tyler R, Nosworthy MG, Beyene S, Sileshi GW (2021) Extent and management of acid soils for sustainable crop production system in the tropical agroecosystems: a review. Acta Agric Scand B Soil Plant Sci 71:852–869
5. Akgönüllü S, Yavuz H, Denizli A (2020) SPR nanosensor based on molecularly imprinted polymer film with gold nanoparticles for sensitive detection of aflatoxin B1. Talanta 219:121219. https://doi.org/10.1016/j.talanta.2020.121219
6. Al-Khayri JM, Rashmi R, Ulhas RS, Sudheer WN, Banadka A, Nagalla P, Aldaej MI, Rezk AA-S, Shehata WF, Almaghasla MI (2023) The role of nanoparticles in response of plants to abiotic stress at physiological, biochemical, and molecular levels. Plants 12:292
7. Alivisatos AP (1996) Semiconductor clusters, nanocrystals, and quantum dots. Science 271:933–937
8. Alova A, Erofeev A, Gorelkin P, Bibikova T, Korchev Y, Majouga A, Bulychev A (2020) Prolonged oxygen depletion in microwounded cells of *Chara corallina* detected with novel oxygen nanosensors. J Exp Bot 71:386–398
9. Amarante AM, Oliveira GS, Bueno CC, Cunha RA, Ierich JCM, Freitas LCG, Franca EF, Oliveira Junior ON, Leite FL (2014) Modeling the coverage of an AFM tip by enzymes and its application in nanobiosensors. J Mol Graph Model 53:100–104
10. Amézketa E (1999) Soil aggregate stability: a review. J Sustain Agric 14:83–151
11. Anguiano DI, García MG, Ruíz C, Torres J, Alonso-Lemus I, Alvarez-Contreras L, Verde-Gómez Y, Bustos E (2012) Electrochemical detection of iron in a lixiviant solution of polluted soil using a modified glassy carbon electrode. Int J Electrochem 2012:1–6
12. Ansari MH, Lavhale S, Kalunke RM, Srivastava PL, Pandit V, Gade S, Yadav S, Laux P, Luch A, Gemmati D, Zamboni P, Singh AV (2020) Recent advances in plant nanobionics and nanobiosensors for toxicology applications. Curr Nanosci 16:27–41

13. Antonacci A, Arduini F, Moscone D, Palleschi G, Scognamiglio V (2018) Nanostructured (bio)sensors for smart agriculture. TrAC Trends Anal Chem 98:95–103
14. Antoun H, Kloepper J (2001) Rhizobacteria promotora do crescimento de plantas (PGPR). Encycl Genet 1477–1480
15. Arafat MM, Dinan B, Akbar SA, Haseeb ASMA (2012) Gas sensors based on one dimensional nanostructured metal-oxides: a review. Sensors 12:7207–7258
16. Ashfaq M, Verma N, Khan S (2017) Carbon nanofibers as a micronutrient carrier in plants: efficient translocation and controlled release of Cu nanoparticles. Environ Sci Nano 4:138–148
17. Aslam A, Mahmood A, Ur-Rehman H, Li C, Liang X, Shao J, Negm S, Moustafa M, Aamer M, Hassan MU (2023) Plant adaptation to flooding stress under changing climate conditions: ongoing breakthroughs and future challenges. Plants 12:3824
18. Auzanneau F (2013) Wire troubleshooting and diagnosis: review and perspectives. Prog Electromagn Res B 49:253–279
19. Bao Z, Zhong H, Li X, Zhang A, Liu Y, Chen P, Cheng Z, Qian H (2021) Core-shell Au@Ag nanoparticles on carboxylated graphene for simultaneous electrochemical sensing of iodide and nitrite. Sens Actuators B: Chem 345:130319. https://doi.org/10.1016/j.snb.2021.130319
20. Barakat M, Kumar R (2015) Synthesis and characterization of porous magnetic silica composite for the removal of heavy metals from aqueous solution. J Ind Eng Chem 23:93–99. https://doi.org/10.1016/j.jiec.2014.07.046
21. Baron R, Sljukic B, Salter C, Crossley A, Compton RG (2007) Electrochemical detection of arsenic on a gold nanoparticle array. Russ J Phys Chem A 81:1443–1447
22. Baz H, Creech M, Chen J, Gong H, Bradford K, Huo H (2020) Water-soluble carbon nanoparticles improve seed germination and post-germination growth of lettuce under salinity stress. Plants 10:1192
23. Benício LPF, Silva RA, Lopes JA, Eulário D, Santos RMM, Aquino LA, Vergutz L, Novais RF, Costa LM, Pinto FG, Tronto J (2015) Layered double hydroxides: nanomaterials for applications in agriculture. Rev Bras Ciência Solo 39(1):1–13
24. Bouyoucos GJ, Mick AH (1940) An electrical resistance method for the continuous measurement of soil moisture under field conditions. Michigan Agric Exp Stn Tech Bull 172
25. Buekers J, Van Laer L, Amery F, Van Buggenhout S, Maes A, Smolders E (2007) Role of soil constituents in fixation of soluble Zn, Cu, Ni and Cd added to soils. Eur J Soil Sci 58:1514–1524
26. Chajanovsky I, Cohen S, Shtenberg G, Suckeveriene RY (2021) Development and characterization of integrated nano-sensors for organic residues and pH field detection. Sensors 21:175842
27. Chakraborti S, Bera K, Sadhukhan S, Dutta P (2022) Bio-priming of seeds: plant stress management and its underlying cellular, biochemical and molecular mechanisms. Plant Stress 3:100052
28. Chartuprayoon N, Zhang M, Bosze, W, Choa Y, Myung N V (2015) One-dimensional nanostructures based bio-detection. Biosens Bioelectron 63:432–443. https://doi.org/10.1016/j.bios.2014.07.043
29. Chen Y, Li J, Wang H (2021) Zinc oxide nanoparticle-based sensor for manganese ion detection in soil with high selectivity. Sens Actuators B Chem 329:129205
30. Chen L, Wang X, Liu H (2020) Graphene-based sensor for phosphate ion detection in soil samples. Sens Actuators B Chem 310:127869
31. Choodum A, Sriprom W, Wongniramaikul W (2019) Portable and selective colorimetric film and digital image colorimetry for detection of iron. Spectrochim Acta A Mol Biomol Spectrosc 208:40–47
32. Choudhury S, Panda P, Sahoo L, Panda SK (2013) Reactive oxygen species signaling in plants under abiotic stress. Plant Signal Behav 8:e23681
33. Corradini E, De Moura MR, Mattoso LHC (2010) A preliminary study of the incorporation of NPK fertilizer into chitosan nanoparticles. Express Polym Lett 4(8):1–3

34. Davidson EA (1991) Fluxes of nitrous oxide and nitric oxide from terrestrial ecosystems. In: Rogers JE, Whitman WB (eds) Microbial production and consumption of greenhouse gases: methane, nitrogen oxide, and halomethanes. ASM Press, Washington, DC, pp 219–235
35. Delgadillo-Vargas O, Garcia-Ruiz R, Forero-Álvares J (2016) Fertilising techniques and nutrient balances in the agriculture industrialization transition: the case of sugarcane in the Cauca river valley (Colombia), 1943–2010. Agric Ecosyst Environ 218:150–162
36. Dias PC, Roque W, Ferreira EC, Dias JAS (2013) A high sensitivity single-probe heat pulse soil moisture sensor based on a single NPN junction transistor. Comput Electron Agric 96:139–147
37. Van Dijk M, Meijerink GW (2014) A review of global food security scenario and assessment studies: results, gaps and research priorities. Glob Food Sec 3:227–238
38. Dimri A, Pathak N, Sharma S (2020) Nanosensors for root zone parameters influencing plant growth. In: Nanomaterials for agriculture and forestry applications. Elsevier, pp 387–406
39. Dixit R, Malaviya D, Pandiyan K, Singh UB, Sahu A, Shukla R, Singh BP, Rai JP, Sharma PK, Lade H (2015) Bioremediation of heavy metals from soil and aquatic environment: an overview of principles and criteria of fundamental processes. Sustain For 7:2189–2212
40. Djanaguiraman M, Nair R, Giraldo JP, Prasad PVV (2018) Cerium oxide nanoparticles decrease drought-induced oxidative damage in sorghum leading to higher photosynthesis and grain yield. ACS Omega 3:14406–14416
41. Du D, Chen W, Zhang W, Liu D, Li H, Lin Y (2010) Covalent coupling of organophosphorus hydrolase loaded quantum dots to carbon nanotube/Au nanocomposite for enhanced detection of methyl parathion. Biosens Bioelectron 25:1370–1375
42. Dubey A, Mailapalli DR (2019) Zeolite coated urea fertilizer using different binders: fabrication, material properties and nitrogen release studies. Environ Technol Innov 16:100452
43. Dumanović J, Nepovimova E, Natić M, Kuča K, Jaćević V (2021) The significance of reactive oxygen species and antioxidant defense system in plants: a concise overview. Front Plant Sci 11:552969
44. El-Araby EH, Abd El-Wahab M, Diab HM, El-Desouky TM, Mohsen M (2011) Avaliação da deposição atmosférica de metais pesados em aerossóis do norte do Egito usando análise de ativação de nêutrons e plasma indutivamente acoplado por emissão óptica. Appl Radiat Isot 69:1506–1511
45. FAO (2023) Standard operating procedure for soil moisture content by gravimetric method. FAO, Rome
46. Fageria NK (2009) The use of nutrients in crop plants. CRC Press, Boca Raton
47. Feldman AF, Koster RD, Cawse-Nicholson K, Crow WT, Holmes TRH, Poulter B (2024) Soil moisture profiles of ecosystem water use revealed with ECOSTRESS. Geophys Res Lett 51:e2024GL108326
48. Fischwasser K (1990) Environmental inorganic chemistry—properties, processes, and estimation methods. Edited by Bodek I, Lyman W, Reehl WF, Rosenblatt DH. Int Rev Der Gesamten Hydrobiol Und Hydrogr 75:681–681
49. Floody MC, Theng BKG, Reyes P, Mora ML (2009) Natural nanoclays: applications and future trends—a Chilean perspective. Clay Miner 44(2):161–176
50. Flynn KD, Wyatt BM, McInnes KJ (2021) Novel cosmic ray neutron sensor accurately captures field-scale soil moisture trends under heterogeneous soil textures. Water 13:3038
51. Forano C, Hibino T, Leroux F, Taviot-Guého C (2006) Layered double hydroxides. Dev Clay Sci 1:1021–1095
52. Forrester SJ, Kikuchi DS, Hernandes MS, Xu Q, Griendling KK (2018) Reactive oxygen species in metabolic and inflammatory signaling. Circ Res 122:877–902
53. Ganeshkumar R, Sopiha KV, Wu P, Cheah CW, Zhao R (2016) Ferroelectric $KNbO_3$ nanofibers: synthesis, characterization and their application as a humidity nanosensor. Nanotechnology 27
54. García Sánchez F, Navas Díaz A, Ramos Peinado MC, Belledone C (2003) Free and sol–gel immobilized alkaline phosphatase-based biosensor for the determination of pesticides and inorganic compounds. Anal Chim Acta 484:45–51

55. Geszke-Moritz M, Moritz M (2013) Quantum dots as versatile probes in medical sciences: synthesis, modification and properties. Mater Sci Eng C 33:1008–1021
56. Ghidan AY, Al Antary TM (2020) Applications of nanotechnology in agriculture. In: Applications of nanobiotechnology. IntechOpen
57. Grimalt S, Dehouck P (2016) Review of analytical methods for the determination of pesticide residues in grapes. J Chromatogr A 1433:1–23
58. Guo Z, Kang Y, Liang S, Zhang J (2020) Detection of Hg(II) in adsorption experiment by a lateral flow biosensor based on streptavidin-biotinylated DNA probes modified gold nanoparticles and smartphone reader. Environ Pollut 266:115389
59. Guo H, White JC, Wang Z, Xing B (2018) Nano-enabled fertilizers to control the release and use efficiency of nutrients. Curr Opin Environ Sci Health 6:77–83. https://doi.org/10.1016/j.coesh.2018.07.009
60. Haddaoui M, Raouafi N (2015) Chlortoluron-induced enzymatic activity inhibition in tyrosinase/ZnO NPs/SPCE biosensor for the detection of ppb levels of herbicide. Sens Actuators, B Chem 219:171–178
61. Hasanuzzaman M, Bhuyan MHMB, Zulfiqar F, Raza A, Mohsin SM, Mahmud JA, Fujita M, Fotopoulos V (2020) Reactive oxygen species and antioxidant defense in plants under abiotic stress: revisiting the crucial role of a universal defense regulator. Antioxidants 9:681
62. Hayles J, Johnson L, Worthley C, Losic D (2017) 5—Nanopesticides: a review of current research and perspectives. In: Grumezescu AM (eds) New pesticides and soil sensors. Academic Press, pp 193–225
63. He L, Cui B, Liu J, Song Y, Wang M, Peng D, Zhang Z (2018) Novel electrochemical biosensor based on core-shell nanostructured composite of hollow carbon spheres and polyaniline for sensitively detecting malathion. Sens Actuators B Chem 258:813–821
64. Hou D, Jia X, Wang L, McGrath SP, Zhu G, Hu Q, Zhao J, Bank MS, Nriagu J (2025) Global soil pollution by toxic metals threatens agriculture and human health. Science 388:316–321. https://doi.org/adr5214
65. Hu H, Trejo M, Nicho ME, Saniger JM, García-Valenzuela A (2010) Adsorption kinetics of optochemical NH_3 gas sensing with semiconductor polyaniline films. Sens Actuators B Chem 142:1–9
66. Hu Y, Wang X, Wang C, Hou P, Dong H, Luo B, Li A (2020) A multifunctional ratiometric electrochemical sensor for combined determination of indole-3-acetic acid and salicylic acid. RSC Adv 10:3115–3121
67. Huang X, Dong F, Xiang D, Zhang X, Zhang L (2019) Ultrasensitive and selective nonenzymatic glucose detection based on Au@Pt core-shell nanoparticles decorated graphene. Sens Actuators B Chem 279:227–233
68. Husen A (2022) Environmental pollution and medicinal plants. CRC Press 6000 Broken Sound Parkway NW, Suite 300, Boca Raton, FL 33487–2742. https://doi.org/10.1201/9781003178866
69. Husen A (2023) Plants and their interaction to environmental pollution (Damage detection, adaptation, tolerance, physiological and molecular responses). Elsevier Inc. https://doi.org/10.1016/C2020-0-03319-0
70. Husen A (2021) The harsh environment and resilient plants: an overview. In: Husen A (ed) Harsh environment and plant resilience. Springer International Publishing AG, Cham, pp 1–23 https://doi.org/10.1007/978-3-030-65912-7_1
71. Husen A (2021) Plant performance under environmental stress (*hormones, biostimulants and sustainable plant growth management*). Springer Nature Switzerland AG, Cham, Switzerland, https://doi.org/10.1007/978-3-030-78521-5
72. Husen A (2024) Biostimulants in plant protection and performance. Elsevier Inc., Cambridge. https://doi.org/10.1016/C2022-0-02207-8
73. Husen A (2024) Essential minerals in plant-soil system (coordination, signalling, and interaction under adverse situations). Elsevier Inc., Cambridge. https://doi.org/10.1016/C2022-0-02437-5

74. Husen A (2025) Agricultural crop improvement: plant and soil relationships. Taylor & Francis Group, Broken Sound Pkwy. https://doi.org/10.1201/9781032630366
75. Husen A (2025) Role of osmolytes in changing environment. Elsevier Inc., Cambridge
76. Hwang J, Hwang MP, Choi M, Seo Y, Jo Y (2016) Sensitive detection of copper ions via ion-responsive fluorescence quenching of engineered porous silicon nanoparticles. Sci Rep 6:35565
77. Jaishankar M, Tseten T, Anbalagan N, Mathew BB, Beeregowda KN (2014) Toxicity, mechanism and health effects of some heavy metals. Interdiscip Toxicol 7:60–72. https://doi.org/10.2478/intox-2014-0009
78. Jena AB, Samal RR, Bhol NK, Duttaroy AK (2023) Cellular red-ox system in health and disease: the latest update. Biomed Pharmacother 162:114606
79. Jia X, Ma Z-Y, Zhang G-X, Hu J-M, Liu Z-Y, Wang H-Y, Zhou F (2013) Polydopamine film coated controlled-release multielement compound fertilizer based on mussel-inspired chemistry. J Agric Food Chem 61(12):2919–2924
80. Jimenez-Falcao S, Villalonga A, Parra-Nieto J, Llopis-Lorente A, Martinez-Ruiz P, Martinez-Mañez R, Villalonga R (2020) Dithioacetal-mechanized mesoporous nanosensor for Hg(II) determination. Micropor Mesopor Mat 297:110054. https://doi.org/10.1016/j.micromeso.2020.110054
81. Kacmaz S, Ertekin K, Mercan D, Oter O, Cetinkaya E, Celik E (2015) An ultra sensitive fluorescent nanosensor for detection of ionic copper. Spectrochim Acta A Mol Biomol Spectrosc 135:551–559
82. Kah M, Kookana RS, Gogos A, Bucheli TD (2018) A critical evaluation of nanopesticides and nanofertilizers against their conventional analogues. Nat Nanotechnol 13:677–684
83. Kah M, Tufenkji N, White JC (2019) Nano-enabled strategies to enhance crop nutrition and protection. Nat Nanotechnol 14:532–540
84. Kajla M, Yadav VK, Khokhar J, Singh S, Chhokar RS, Meena RP, Sharma RK (2015) Increase in wheat production through management of abiotic stresses: a review. J Appl Nat Sci 7:1070–1080
85. Kalyani N, Goel S, Jaiswal S (2021) On-site sensing of pesticides using point-of-care biosensors: a review. Environ Chem Lett 19:345–354
86. Kashyap PL, Rai P, Sharma S, Chakdar H, Kumar S, Pandiyan K (2016) Nanoscience in food and agriculture 2. Sustain Agric Rev 21:253–276
87. Kaur J, Singh PK (2020) Enzyme based optical biosensors for organophosphate class of pesticide detection. Phys Chem Chem Phys 22:15105–15119
88. Kaushal M, Wani SP (2017) Nanosensors: frontiers in precision agriculture. Nanotechnology: an agricultural paradigm. Springer, Singapore, pp 279–291
89. Khait I, Lewin-Epstein O, Sharon R, Saban K, Goldstein R, Anikster Y, Zeron Y, Agassy C, Nizan S, Sharabi G, Perelman R, Boonman A, Sade N, Yovel Y, Hadany L (2023) Sounds emitted by plants under stress are airborne and informative. Cell 186:1328–1336
90. Khaled F, Sayed A (2023) Soil pH and its influence on nutrient availability and plant health. Int J Adv Chem Res 5:68–70
91. Kocabay O, Emregul E, Aydin SS, Aras S (2013) Detection of superoxide radicals in tomato plants exposed to salinity, drought, cold, and heavy metal stress using CMC-G-SOD biosensor. Artif Cells Nanomed Biotechnol 41:352–358
92. Konur O (2021) Nanobiosensors in agriculture and foods: a scientometric review. In: Pudake RN, Jain U, Kole C (eds) Biosensors in agriculture: recent trends and future perspectives. Springer International Publishing, Cham, pp 365–384
93. Kottegoda N, Munaweera I, Adassooriya NM, Karunaratne V (2011) A green slow-release fertilizer composition based on urea-modified hydroxyapatite nanoparticles encapsulated wood. Curr Sci 101:73–78
94. Kumar R, Ashfaq M, Verma N (2018) Synthesis of novel PVA–starch formulation-supported Cu–Zn nanoparticle carrying carbon nanofibers as a nanofertilizer: controlled release of micronutrients. J Mater Sci 53:7150–7716

95. Kundu M, Krishnan P, Chobhe KA, Manjaiah KM, Pant RP, Chawla G (2022) Fabrication of electrochemical nanosensor for detection of nitrate content in soil extract. J Soil Sci Plant Nutr 22:2777–2792
96. Kunte GV, Shivashankar SA, Umarji AM (2009) Humidity sensing characteristics of hydrotungstite thin films. Bull Mater Sci 31:835–839
97. Kwak SY, Wong MH, Lew TTS, Bisker G, Lee MA, Kaplan A, Dong J, Lui AT, Koman VB, Sinclair R, Hamann C, Strano MD (2017) Nanosensor technology applied to living plant systems. Annu Rev Anal Chem 10:113–140
98. Lau H, Mat Ludin AF, Rajab NF, Shahar S (2017) Identification of neuroprotective factors associated with successful ageing and risk of cognitive impairment among Malaysia older adults. Curr Gerontol Geriatr Res 2017(1):4218756. https://doi.org/10.1155/2017/4218756
99. Laus MN, De Santis MA, Flagella Z, Soccio M (2022) Changes in antioxidant defence system in durum wheat under hyperosmotic stress: a concise overview. Plants 11:98
100. Lee H, Dellatore SM, Miler WM, Messersmith PB (2007) Mussel-inspired surface chemistry for multifunctional coatings. Science 318(5849):426–430
101. Lee JH, Park JY, Min K, Cha HJ, Choi SS, Yoo YJ (2010) A novel organophosphorus hydrolase-based biosensor using mesoporous carbons and carbon black for the detection of organophosphate nerve agents. Biosens Bioelectron 25:1566–1570
102. Lei YM, Xiao BQ, Liang WB, Chai YQ, Yuan R, Zhuo Y (2018) A robust, magnetic, and self-accelerated electrochemiluminescent nanosensor for ultrasensitive detection of copper ion. Biosens Bioelectron 109:109–115
103. Leone M (2022) Advances in fiber optic sensors for soil moisture monitoring: a review. Results Opt 7:100213
104. Levy DB, Barbarick KA, Siemer EG, Sommers LE (1992) Distribution and partitioning of trace metals in contaminated soils near Leadville, Colorado. J Environ Qual 21:85–195
105. Li Z, Liu Y, Hossain O, Paul R, Yao S, Wu S, Ristaino JB, Zhu Y, Wei Q (2021) Real-time monitoring of plant stresses via chemiresistive profiling of leaf volatiles by a wearable sensor. Matter 4:2553–2570
106. Li XH, Xie Z, Min H, Li C, Liu M, Xian Y, Jin L (2006) Development of quantum dots modified acetylcholinesterase biosensor for the detection of trichlorfon. Electroanalysis 18:2163–2167
107. Li Z, Yu T, Paul R, Fan J, Yang Y, Wei Q (2020) Agricultural nanodiagnostics for plant diseases: recent advances and challenges. Nanoscale Adv 2:3083–3094
108. Lim JH, Sinha SN, Narayan RJ, Jang SH (2021) Recent advances in carbon nanotube-based sensors for environmental monitoring. Sensors 21:7287
109. Lin Z, Hu F, He G, Yang Y, Liao Y, Luo X, Wang XD (2022) Highly photostable and pH-sensitive nanosensors. Chemosensors 10:93
110. Liu X, Zhang W, Hu Y (2022) Graphene-based sensor for zinc ion detection in soil samples. ACS Sens 7:405–413
111. Long B, Tang L, Peng B, Zeng G, Zhou Y, Mo D, Fang S, Ouyang X, Yu J (2019) Voltammetric biosensor based on nitrogen-doped ordered mesoporous carbon for detection of organophosphorus pesticides in vegetables. Curr Anal Chem 15(1):92–100
112. Mahmoud ZH, Al-Salman HNK, Hussein HH, Adhab AH, Al-Majdi K, Rasheed T, Abdul-hussien HA, Sasirekha N, Abd AN, Kianfar E (2024) Organic chemical nanosensors: synthesis, properties, and applications. Braz J Biol 84
113. Majid E, Hrapovic S, Liu Y, Male KB, Luong JHT (2006) Electrochemical determination of arsenite using a gold nanoparticle modified glassy carbon electrode and flow analysis. Anal Chem 78:762–769
114. Malik LA, Bashir A, Qureashi A, Pandith AH (2019) Detection and removal of heavy metal ions: a review. Environ Chem Lett 17:1495–1521
115. Marakatti VS, Halgeri AB (2015) Metal ion-exchanged zeolites as highly active solid acid catalysts for the green synthesis of glycerol carbonate from glycerol. RSC Adv 5(19):14286–14293
116. Martínez-Ballesta MC, Zapata L, Chalbi N, Carvajal M (2016) Multiwalled carbon nanotubes enter broccoli cells enhancing growth and water uptake of plants exposed to salinity. J Nanobiotechnol 14:1–14

117. Mashabela MD, Masamba P, Kappo AP (2023) Applications of metabolomics for the elucidation of abiotic stress tolerance in plants: a special focus on osmotic stress and heavy metal toxicity. Plants 12:269
118. Mathivanan S (2021) Perspectives of nano-materials and nanobiosensors in food safety and agriculture. In: Novel nanomaterials. IntechOpen, pp 197–217
119. Mihelič R, Pečnik J, Glavan M, Pintar M (2021) Impact of sustainable land management practices on soil properties: example of organic and integrated agricultural management. Land 10:1–17
120. Mishra N, Jiang C, Chen L, Paul A, Chatterjee A, Shen G (2023) Achieving abiotic stress tolerance in plants through antioxidative defense mechanisms. Front Plant Sci 14:1110622
121. Mishra A, Kumar J, Melo JS, Sandaka BP (2021) Progressive development in biosensors for detection of dichlorvos pesticide: a review. J Environ Chem Eng 9(2):105067
122. Mitra S, Chakraborty AJ, Tareq AM, Emran TB, Nainu F, Khusro A, Idris AM, Khandaker MU, Osman H, Alhumaydhi FA et al (2022) Impact of heavy metals on the environment and human health: novel therapeutic insights to counter the toxicity. J King Saud Univ Sci 34:101865
123. Mittal D, Kaur G, Singh P, Yadav K, Ali SA (2020) Nanoparticle-based sustainable agriculture and food science: recent advances and future outlook. Front Nanotechnol 2:579954
124. Mohanta TK, Bashir T, Hashem A, Abd Allah EF, Khan AL, Al-Harrasi AS (2018) Early events in plant abiotic stress signaling: interplay between calcium, reactive oxygen species, and phytohormones. J Plant Growth Regul 37:1033–1049
125. Mondal R, Dam P, Chakraborty J, Paret ML, Katı A, Altuntas S, Husen A (2022) Potential of nanobiosensor in sustainable agriculture: the state-of-art. Heliyon e12207
126. Mu T, Liu G, Yang X, Yu Y (2023) Soil-moisture estimation based on multiple-source remote-sensing images. Remote Sens 15:139
127. Nadporozhskaya M, Kovsh N, Paolesse R, Lvova L (2022) Recent advances in chemical sensors for soil analysis: a review. Chemosensors 10(1):35
128. Nag A, Mitra A, Mukhopadhyay SC (2018) Graphene and its sensor-based applications: a review. Sens Actuators A Phys 270:177–194
129. Nandi R, Shrestha D (2024) Assessment of low-cost and higher-end soil moisture sensors across various moisture ranges and soil textures. Sensors 24(18):5886
130. Nawaz M, Sun J, Shabbir S, Khattak WA, Ren G, Nie X, Bo Y, Javed Q, Du D, Sonne C (2023) A review of plant strategies to resist biotic and abiotic environmental stressors. Sci Total Environ 900:165832
131. Neina D (2019) The role of soil pH in plant nutrition and soil remediation. Appl Environ Soil Sci 2019:5794869
132. Neme K, Nafady A, Uddin S, Tola YB (2021) Application of nanotechnology in agriculture, postharvest loss reduction and food processing: food security implication and challenges. Heliyon 7(12):e08539
133. Obiajunwa EI, Pelemo DA, Owolabi SA, Fasasi MK, Johnson-Fatokun FO (2002) Caracterização de poluentes de metais pesados de solos e sedimentos ao redor de um terminal de produção de petróleo bruto usando EDXRF. Nucl Instrum Methods Phys Res Sect B 194:61–64
134. Ochs CJ, Hong T, Such GK, Cui J, Postma A, Caruso F (2011) Dopamine-mediated continuous assembly of biodegradable capsules. Chem Mater 23(13):3141–3143
135. Orozco J, Fernandez-Sanchez C, Jimenez-Jorquera C (2008) Underpotential deposition - anodic stripping voltammetric detection of copper at gold ultramicroelectrode arrays. Electrochim Acta 42:4877–4882
136. Oshunsanya SO, Nwosu NJ, Li Y (2019) Abiotic stress in agricultural crops under climatic conditions. In: Sustainable agriculture, forest and environmental management. Springer, Singapore, pp 71–100
137. Ouyang R, Lei J, Ju H (2008) Surface molecularly imprinted nanowire for protein specific recognition. Chem Commun 44:5761–5763

138. O'Kennedy S (2022) Soil pH and its impact on nutrient availability and crop growth. Int J Geogr 4(2):236–238
139. O'Neil GD, Newton ME, Macpherson JV (2015) Direct identification and analysis of heavy metals in solution (Hg, Cu, Pb, Zn, Ni) by use of in situ electrochemical X-ray fluorescence. Anal Chem 87:4933–4940
140. Panwar A, Kleidon A (2022) Evaluating the response of diurnal variations in surface and air temperature to evaporative conditions across vegetation types in FLUXNET and ERA5. J Clim 35(19):1–67
141. Pereira EI, Minussi FB, Cruz CCT, Bernardi ACC, Ribeiro C (2012) Nanocompósitos extrudados de ureia-montmorilonita: um novo material de liberação lenta. J Agric Food Chem 60:5267–5272
142. Potadar RS, Moholkar DN, Pawar KD (2022) 14—Nanosensors for the detection of heavy trace metals in soil. In: Denizli A, Nguyen TA, Rajendran S, Yasin G, Nadda AK (eds) In micro and nano technologies. Nanosensors for Smart Agriculture, pp 329–353
143. Prasad KS, Shruthi G, Shivamallu C (2018) Functionalized silver nano-sensor for colorimetric detection. Sensors
144. Priya E, Sarkar S, Maji PK (2024) A review on slow-release fertilizer: nutrient release mechanism and agricultural sustainability. J Environ Chem Eng 12(4)
145. Pérez-Llorca M, Pollmann S, Müller M (2023) Ethylene and jasmonates signaling network mediating secondary metabolites under abiotic stress. Int J Mol Sci 24(6):5990
146. Qu F, Zhu L, Yang M (2020) Quantum dot-based sensor for multiplexed detection of metal ions in soil. Anal Chim Acta 1133:66–75
147. Rai V, Acharya S, Dey N (2012) Implications of nanobiosensors in agriculture. J Biomater Nanobiotechnol 3(02):315
148. Rakhimol KR, Thomas S, Kalarikkal N, Jayachandran K. (2021) Nanotechnology in controlled-release fertilizers. In: Controlled release fertilizers for sustainable agriculture. Academic Press, pp 169–181
149. Rosen CJ (2002) Lead in the home garden and urban soil environment. Univ Minnesota Extension Serv
150. Sanghavi BJ, Gadhari NS, Kalambate PK, Karna SP, Srivastava AK (2015) Potentiometric stripping analysis of arsenic using a graphene paste electrode modified with a thiacrown ether and gold nanoparticles. Mikrochim Acta
151. Sashidhar P, Dubey MK, Kochar M (2019) Sensing soil microbes and interactions: how can nanomaterials help? In: Nanotechnology in the life sciences. Springer, pp 213–236
152. Schirrmann M, Gebbers R, Kramer E, Seidel J (2011) Soil pH mapping with an on-the-go sensor. Sensors 11(1):573–598
153. Schwarzenbach RP, Escher BI, Fenner K, Hofstetter TB, Johnson CA, Von Gunten U, Wehrli B (2006) The challenge of micropollutants in aquatic systems. Science
154. Scragg A (2005) Environmental biotechnology, 2nd edn. Oxford University Press
155. Serrano D, Ávila E, Barrios M, Darghan A, Lobo D (2020) Surface soil moisture monitoring with near-ground sensors: performance assessment of a matric potential-based method. Meas J Int Meas Confed 155:107542
156. Shah H, Hellegers P, Siderius C (2021) Climate risk to agriculture: a synthesis to define different types of critical moments. Clim Risk Manag 34:100378
157. Shang Y, Kamrul Hasan M, Ahammed GJ, Li M, Yin H, Zhou J (2019) Applications of nanotechnology in plant growth and crop protection: a review. Molecules
158. Sharma P, Jha AB, Dubey RS, Pessarakli M (2012) Reactive oxygen species, oxidative damage, and antioxidative defense mechanism in plants under stressful conditions. J Bot 2012:217037
159. Sharma P, Pandey V, Sharma MMM, Patra A, Singh B, Mehta S, Husen A (2021) A review on biosensors and nanosensors application in agroecosystems. Nanoscale Res Lett 16:136
160. Sherma J (2015) Review of advances in the thin layer chromatography of pesticides: 2012–2014. J Environ Sci Health B 50:301–316

161. Shrivas K, Patel S, Sinha D, Thakur S, Patle T, Kant T, Dewangan K, Satnami ML, Nirmalkar J, Kumar S (2020) Colorimetric and smartphone-integrated paper device for on-site determination of arsenic (III) using sucrose modified gold nanoparticles as a nanoprobe. Microchim Acta 187(3)
162. Singh A, Chauhan PS (2017) Ecological significance of soil-associated plant growth-promoting biofilm-forming microbes for stress management. Biofilms in plant and soil health. John Wiley & Sons, Hoboken, NJ, USA, pp 291–326
163. Singh RP (2021) Recent trends, prospects, and challenges of nanobiosensors in agriculture. In: Pudake RN, Jain U, Kole C (eds) Biosensors in agriculture: recent trends and future perspectives. Springer International Publishing, Cham, pp 3–13
164. Singh R, Singh S, Parihar P, Mishra RK, Tripathi DK, Singh VP, Chauhan DK, Prasad SM (2016) Reactive oxygen species (ROS): beneficial companions of plants' developmental processes. Front Plant Sci 7:186069
165. Son WK, Choi YS, Han YW, Shin DW, Min K, Shin J, Lee MJ, Son H, Jeong DH, Kwak S-Y (2023) In vivo surface-enhanced Raman scattering nanosensor for the real-time monitoring of multiple stress signalling molecules in plants. Nat Nanotechnol 18(2):205–216
166. Songkhum P, Wuttikhun T, Chanlek N, Khemthong P, Laohhasurayotin K (2018) Controlled release studies of boron and zinc from layered double hydroxides as the micronutrient hosts for agricultural application. Appl Clay Sci 152:311–322
167. Sowmyya T, Lakshmi GV (2018) Soymida febrifuga aqueous root extract maneuvered silver nanoparticles as mercury nanosensor and potential microbicide. World Sci News 114:84–105
168. Steinegger A, Wolfbeis OS, Borisov SM (2020) Optical sensing and imaging of pH values: spectroscopies, materials, and applications. Chem Rev 120(22):12357–12489
169. Stevenson FJ (1995) Humus chemistry: genesis, composition, reactions. J Chem Educ 72
170. Surya SG, Yuvaraja S, Varrla BMS, Palaparthy VS, Salama KN (2020) An in-field integrated capacitive sensor for rapid detection and quantification of soil moisture. Sens Actuator B Chem 321:128542
171. Suvarnaphaet P, Pechprasarn S (2017) Graphene-based materials for biosensors: a review. Sensors 17(10):2161
172. Tang FHM, Lenzen M, McBratney A, Maggi F (2021) Risk of pesticide pollution at the global scale. Nat Geosci 14:206–210
173. Tang L, Li F, Liu M, Nandhakumar R (2011) Single sensor for two metal ions: colorimetric recognition of Cu^{2+} and fluorescent recognition of Hg^{2+}. Spectrochim Acta A Mol Biomol Spectrosc 78:1168–1172
174. Thakur M, Wang B, Verma ML (2022) Development and applications of nanobiosensors for sustainable agricultural and food industries: recent developments, challenges and perspectives. Environ Technol Innov 26:102371
175. Torres-Dorante LO, Lammel J, Kuhlmann WT, Olfs H-W (2008) Capacity, selectivity, and reversibility for nitrate exchange of a layered double-hydroxide (LDH) mineral in simulated soil solutions and in soil. J Plant Nutr Soil Sci 171(5):777–784
176. Trenkel ME (1997) Controlled-release and stabilized fertilizers in agriculture, vol 11. International Fertilizer Industry Association, Paris
177. Tuzen M, Soylak M (2009) Coprecipitação multielementar para separação e enriquecimento de íons de metais pesados para suas determinações espectrométricas de absorção atômica de chama. J Hazard Mater 162:724–729
178. USEPA (2007) Environmental Protection Agency Nanotechnology White Paper. EPA, Ed. Kaushal M, Wani SP (2017) Nanosensors: frontiers in precision agriculture. In: Nanotechnology: an agricultural paradigm. Springer, Singapore, pp 279–291
179. Umasankareswari T, Dharshana V, Saravanadevi K, Dorothy R, Sasilatha T, Nguyen TA, Rajendran S (2022) Soil moisture nanosensors. In: Nanosensors for smart agriculture, pp 185–216
180. Usman M, Farooq M, Wakeel A, Nawaz A, Cheema SA, Rehman H, Ashraf I, Sanaullah M (2020) Nanotechnology in agriculture: current status, challenges and future opportunities. Sci Total Environ 721:137778

181. Vahab L, Keshipour S (2022) Novel nanosensor of cobalt(II) and copper(II) constructed from graphene quantum dots modified with Eriochrome Black T. Sci Rep 12:13179
182. Veloso CAC, Borges AL, Muniz AS, De IA, Veigas JM (1992) Effect of different materials on soil pH. Sci Agric 49(1):123–128
183. Verdian A (2018) Apta-nanosensors for detection and quantitative determination of acetamiprid—a pesticide residue in food and environment. Talanta 176:456–464
184. De Volder MF, Tawfick SH, Baughman RH, Hart AJ (2013) Carbon nanotubes: present and future commercial applications. Science 339:535–539
185. Wang X, Kong L, Zhou S, Ma C, Lin W, Sun X, Kirsanov D, Legin A, Wan H, Wang P (2022) Development of QDs-based nanosensors for heavy metal detection: a review on transducer principles and in-situ detection. Talanta 239:122903
186. Wuana RA, Okieimen FE (2011) Heavy metals in contaminated soils: a review of sources, chemistry, risks and best available strategies for remediation. ISRN Ecol 2011:1–20
187. Xiao W, Xiao M, Fu Q, Yu Q, Yu S, Shen H, Bian H, Tang Y (2016) A portable smart-phone readout device for the detection of mercury contamination based on an aptamer-assay nanosensor. Sensors 16:1–10
188. Xiao Y, Rowe AA, Plaxco KW (2007) Electrochemical detection of parts-per-billion lead via an electrodebound DNAzyme assembly. JACS Commun 262–263
189. Xie Z, Sun X, Jiao J, Xin X (2017) Ionic liquid-functionalized carbon quantum dots as fluorescent probes for sensitive and selective detection of iron ion and ascorbic acid. Colloids Surf A Physicochem Eng Asp
190. Xu X, Daniel WL, Wei W, Mirkin CA (2010) Colorimetric Cu^2 detection using DNA-modified gold-nano-particle aggregates as probes and click chemistry. Small 6:623–626
191. Ye W, Yu M, Wang F, Li Y, Wang C (2022) Multiplexed detection of heavy metal ions by single plasmonic nanosensors. Biosens Bioelectron 196:113688. https://doi.org/10.1016/j.bios.2021.113688
192. Yu S, Tan Z, Lai Y, Li Q, Liu J (2023) Nanoparticulate pollutants in the environment: analytical methods, formation, and transformation. Eco-Environ Health 2(2):61–73
193. Yu B, Wang DA, Ye Q, Zhou F, Liu W (2009) Robust polydopamine nano/microcapsules and their loading and release behavior. Chem Commun 44:6789–6791
194. Yu Z, Zhao G, Liu M, Lei Y, Li M (2010) Fabrication of a novel atrazine biosensor and its subpart-per-trillion levels sensitive performance. Environ Sci Technol 44:7878–7883
195. Zappa D (2021) Metal oxide-based nanomaterials: Synthesis, characterization and functional applications as chemical sensors. J Nanosci Nanotechnol 21(9):4887–4899
196. Zhai X, Gong Y, Yang W, Kang H, Zhang X (2015) Mn-doped CdS/ZnS/CdS QD-based fluorescent nanosensor for rapid, selective, and ultrasensitive detection of copper(II) ion. RSC Adv 1–3
197. Zhang L, Fang M (2010) Nanomaterials in pollution trace detection and environmental improvement. Nano Today 5:128–142
198. Zhang Y, Wang L, Chen H (2019) Carbon nanotube-based sensor for copper ion detection in soil. Biosens Bioelectron 142:111543
199. Zhang M, Zhang X, Zhang L, Zeng L, Liu Y, Wang X, He P, Li S, Liang G, Zhou W, Ai C (2021) The stronger impact of inorganic nitrogen fertilization on soil bacterial community than organic fertilization in short-term condition. Geoderma 382:114752
200. Zhao H, Vance GF, Ganjegunte GK, Urynowicz MA (2008) Use of zeolites for treating natural gas co-produced waters in Wyoming, USA. Desalination 228(1–3):263–276
201. Zhao F, Wu J, Ying Y, She Y, Wang J, Ping J (2018) Carbon nanomaterial-enabled pesticide biosensors: design strategy, biosensing mechanism, and practical application. TrAC Trends Anal Chem 106:62–83
202. Zhao J, Zhu B, Yu H, Yan L, Wei Q, Du B (2013) A reusable naphthalimide-functionalized magnetic fluorescent nanosensor for the simultaneous determination and removal of trace Hg^{2+} in aqueous solution. J Colloid Interface Sci 389:46–52
203. Zhou L, Wang S, Hao Q, Kang L, Kang C, Yang J, Yang W, Jiang J, Huang LQ, Guo L (2018) Bioaccessibilidade e avaliação de risco de metais pesados e análise da especiação de arsênio em Cordyceps sinensis. Chin Med 13:1–8

204. Zhou L, Zhang X, Ma L, Gao J, Jiang Y (2017) Acetylcholinesterase/chitosan-transition metal carbides nanocomposites-based biosensor for the organophosphate pesticides detection. Biochem Eng J 128:243–249. https://doi.org/10.1016/j.bej.2017.10.008

Nanosensors for Detection of Agricultural Soil Microbes

Ishaan Lakhani, Proxima Dhiman, Vikas Dutt, Hina Alim, Ahmad Ali, and Nimisha Patel

Abstract The role of soil microorganisms in sustaining agricultural productivity cannot be overstated. From nutrient cycling to promoting plant growth, these microscopic entities are indispensable for maintaining healthy soil. Yet, accurately detecting and analyzing them remains a challenge with traditional methods, which often require extensive time, labour intensive and may miss non-culturable species. To address these limitations, advances in nanotechnology have introduced nanosensors, which offer real-time, highly sensitive detection capabilities. Nanosensors not only solves the issue of plant disease management or prevention, it can also boost food security, soil health, inform about the amount of nutrients present in soil. It can certainly optimize the crop yield without harming the environment. Despite these benefits, the practical deployment of nanosensors in large-scale agricultural settings presents challenges that need to be addressed. The chapter will examine the various types of nanosensors used for soil microbe detection, their underlying mechanisms, and how they have revolutionized agricultural practices.

Keywords Nanotechnology · Nanosensors · Soil microbes · Agricultural applications

I. Lakhani · P. Dhiman · N. Patel (✉)
Department of Life Sciences, J. C. Bose University of Science and Technology, YMCA, Faridabad, Haryana, India
e-mail: nimishapatel@jcboseust.ac.in

V. Dutt
Department of Life Sciences, School of Basic Science and Research, Sharada University, Greater Noida, Uttar Pradesh, India

H. Alim · A. Ali
Department of Life Sciences, University of Mumbai, Mumbai, India

A. Husen (ed.), *Nanobiosensors for Agricultural and Other Related Sectors*, Smart Nanomaterials Technology, https://doi.org/10.1007/978-981-96-6499-3_6

1 Introduction

Early detection of soil microbes by nano-biosensors is a sustainable approach towards overseeing soil quality, enhancing agriculture output, and reducing environmental dangers linked to pathogens, pollutants, or disparities in microbial populations. Nano-biosensors are a carefully constructed blend of nanoparticle and biological sensors, where a biological recognition molecule is attached to the surface of a signal transducer [1]. This option is superior to other detection tools and methods due to its higher specificity and sensitivity, capable of detecting subtle interactions that may go unnoticed by other methods. Agricultural professionals can utilize the microbial variety in soil to assess soil and plant health, monitor the ecosystem, detect levels of pollutants or pesticides in soil, and identify pathogens early, all aiding in the improvement of the agriculture industry. This chapter discusses the important microorganisms found in soil and the importance of detecting them early and effectively.

Microbial heterogenicity, related plants, and their habitats, all have an impact on microbiome functioning. Compared to single strains, complex microbial inoculum can give plants greater resistance to diseases [2]. It has been demonstrated that soil bacteria belonging to the genus Bacillus can lower aphid populations on Arabidopsis and broccoli plants by either drawing in natural pathogens towards the host plant or triggering their defence responses against the herbivores [3].

The root exudation profile changes during pathogen infection, increasing the secretion of phenolics, according to *Ralstonia solanacearum* a pathogen which is responsible for wilting in tomato, Changes in soil microbial distributions were linked to the exudates of diseased plants thus makes it quite possible to determine the disease, also when Arabidopsis was injected with *Pseudomonas fluorescens WCS417r*, it changed the plant's volatile emission profile in response to an herbivorous. Several advantageous biocontrol agents exhibit clear benefits for disease management and show significant potential for further exploration, these were isolated and examined in various environmental samples and utilized for managing [4]. It is observed that presence of diverse microbes in soil could decide its texture and available nutrients for plant growth.

Proteobacteria, Bacteroidetes, and *Actinobacteria* were found to be more abundant in the soil microbiota of the rhizosphere, making for 73–80% of all readings, as opposed to 46–56% in bulk soil [3]. Similarly, the existence of both filamentous bacterial and fungal species is intricately connected to soil texture and the composition of the surrounding environment [5]. When plants release different metabolites to influence the community of microorganisms in the rhizosphere or attract helpful microbes during times of stress, rhizosphere microbes may be advantageous since they can use these metabolites as carbon source or energy to establish themselves in the rhizosphere.

Soil microbes like *Bacillus, Pseudomonas,* and *Enterobacter* species have the ability to utilize very limited phosphorus in order to increase the growth of crops by using citric, fumaric, gluconic, and malic acids that are organic chelators cause a reduction in soil pH. Most of the *Staphylococcal sp.* present in roots of plants like apple can help in increasing levels of important enzymes, (e.g. foliar enzymes), leading to a decrease in Fe^{3+} levels and promoting faster absorption of iron in situations of iron deficiency [6]. All these microbial activities seen in plant interacting soil, directly or indirectly help in plant growth, thus real time detection of such microbial communities makes it easier to estimate the crop yield.

Some microbial biosensors are very beneficial for analysing pesticides this mechanism includes various sensors like calorimeters, conductometers, colorimeters, transducers, potentiometers, and fluorescent biosensors, which are capable of recognizing electroactive pesticides, while biosensors with luminescence and fluorescence detect light-emitting pesticides in microbial metabolic reactions. One frequently utilized enzyme for this task is β-galactosidase (β-gal) coded by the lacZ gene to identify contaminants through either colorimetric or fluorescent properties. Similarly, the gene lux, responsible for bioluminescence, was successfully cloned in *Vibrio fischeri* and *Photorhabdus luminescens*, its codes for luciferase, an enzyme that triggers light emission, acting as a reporter gene to detect pollutants. In certain microorganisms, the Green fluorescent protein (GFP) which has been utilized to identify pollutants by emitting light easily detectable with a modern potentiometer. Another example is of crtA gene from *Rhodovulum sulfidophilum*, a purple photosynthetic bacterium, was used as reporter gene as it detects pollutants through a colour change in the solution caused by spheroidene build-up. The benefit of utilizing this gene is that it does not require the inclusion of a specific reagent or substrate to produce colour.

Nutrient cycling can decide the amount of organic matter present in soil, and they also impact the climate by contributing in global carbon cycle, hence soil microorganisms are important in extracting organic material in farms such as nitrogen, phosphorus and sulphur. Recent research indicates that there in microbial genomes there are more sequences for nitrogenase than previously thought. Furthermore, soil bacteria and fungi show the ability to degrade the organic compounds like phosphorus through mineralization, resulting in the release of orthophosphate into the soil solution. Soil microbes are essential in the nutrient cycle through the activation of their genes and extracellular enzymes. Bacteria are essential for enzymatic changes in soil ecosystems. For instance, *Azospirillum*, a bacterium that thrives in low oxygen environments, helps with nitrogen fixation alongside grass roots. Nitrifying bacteria like *Nitrosomonas* carry out the process of converting ammonia into nitrite ions, whereas *Nitrobacter* is responsible for oxidizing nitrites into nitrates. *Clostridium pasteurianum*, a microorganism that requires no oxygen, transforms nitrogen from the air into ammonia, which adds nitrogen to the soil. *Alnus rubra* also known as red alder boosts most important prokaryotic process known as nitrification, emphasizing autotrophic nitrifiers as the primary absorber of nitrogen in form of ammonia in soil [7].

Biosensors based on nanomaterials are employed as a quick detection method in contemporary agriculture. *Pseudomonas syringae pv. lachrymans* (Psl) has ability to detect 337 CFU/mL then it is detected by biosensors. A gold electrode modified with an antibody (4-aminothiophenol-glutaraldehyde-anti-Psl) was employed as the biological analyte in this experiment. The detection procedure relies on recombinase polymerase amplification (RPA) isothermally amplifying DNA sequences (the target pathogen) and then electrochemically evaluating the results. *Pseudomonas syringae* was detected calorimetrically once again using DNA-gold nanoparticle linked with thiol as probes. The detection procedure is done by observing the colour changing from red to purple, which verifies the favourable outcome [8]. This significant procedure may encounter multiple obstacles. The high sensitivity of nano sensors is utilized to identify incorrect substances in soil which contains a mixture of organic particles, minerals, metals, pollutants, and microorganisms, potentially affecting the nano-detection process.

At times, the chemical composition of soil, such as its alkalinity or acidity or the presence of pesticides, can change how the bio-sensor reacts to the analyte. The performance of the nano-sensor setup may diminish over time due to the excessive buildup of microbial population and alterations in organic matter levels. Likewise, fluctuations in temperature and weather can disrupt the production of nanoparticles. Despite its sensitivity, this process cannot identify very small amounts of microbial strains that may be harmful or beneficial to soil health. Using signal amplification techniques can greatly enhance the accuracy of detecting low levels of biomarkers [9]

Nanosensors, unlike traditional sensors, have the ability to detect small disturbances at the nanoscale, which can lead to increased levels of noise or inaccurate measurements. This sensitivity process requires the creation of specific algorithms for processing signal that are able to differentiate genuine structural changes from data noise, which may be caused by slight environmental fluctuations or material irregularities [10].

2 Principle of Nanosensors

Biosensors encompass a wide range of devices, such as physical tools, genetically engineered cells, and methods for both qualitative and quantitative analysis. They function as a signal detector by utilizing biological receptors to detect biochemical or physiological changes. Transforming biological processes into measurable signals can be conveniently achieved with a signal transformation system, allowing for tracking and evaluation [11].

The sensitive nature, specificity of analysing as well as stability of the biosensors is improved by incorporating nanomaterials adorned with polymers and biometric components. The entire system utilizes a transducer to convert the interaction among the bio-sensitive elements and target analyte into specific signals. Bio-detection is achievable through the processes of Bio-catalysis and bio-affinity. Bio-catalysis utilizes enzymes or biological organisms to catalyse chemical reactions, whereas bio-affinity-based sensors employ bio-receptors to attach to the target analyte. Biosensors could be categorized into electrochemical, optical, and calorimetric types, depending on various transduction pathways.

The main components of biological recognition elements are enzymes, peptides, antibodies, and nucleic acids, as well as biotic components like cell antibodies, recombinant proteins, peptide nucleic acids, and aptamers, along with other substances including artificial catalysts, ligands, and molecularly imprinted polymers [12].

2.1 Nanomaterials Used

These are materials that have structures or components measured in nanometer, which lies between 1 and 100 nm. Nanostructured materials exhibit distinctive optical, electronic, magnetic, mechanical, thermal, and chemical properties due to a higher concentration of atoms residing on their surfaces. Nanomaterials exhibit a very high surface area to volume ratio, when organized or controlled on a nanoscale. With a significantly larger number of atoms located on the surfaces of materials. Various kinds of NMs such as nanoparticles (NPs), nanotubes, nanowires, nanocomposites, and quantum dots have distinct features ideal for particular uses. Nanomaterials can be categorized according to three main factors: size, composition, and structure [13].

Nanomaterials important for biosensing: Various nanoparticles have been created, engineered, and are employed to improve the overall functionality of biosensors. Different types of nanomaterials (NMs) are utilized in creating biosensors, showcasing their unique sizes and qualities. Precise handling and integration of these nanomaterials in biosensor configurations lead to unmatched sensitivity, specificity, and efficiency in identifying target analytes.

Quantum dots: These fall within the category of nanomaterials with zero dimensions, typically with a diameter between 2.0 and 10.0 nm. These nanomaterials display varied colours based on their size. example: Ni-doped CdTe is type of QD comes under Fluorescence type sensor and is sensed by analyte like Pyrazinamide [14].

Nanowires: Nanowires are long structures that are usually 1 to 100 nm in diameter and can range in length from nanometers to micrometres. They are classified as one-dimensional nanomaterials [13]. AuPt NWs network with PDa Coating uses electrochemical mechanism to detect pesticides [14]

Metallic nanomaterials: Such as gold nanorods are used for identifying two common and commercially important viruses found in orchids, specifically the Odontoglossum Ringspot and Mosaic viruses.

Carbon nanotubes: CNTs are a type of fullerenes with various agricultural uses. They are able to enter plant cells, impact plant's growth and regulation, work as a medium for biosensors, and aid in smart agricultural practices.

Liposomes: Lipid-based vesicles, are commonly utilized in different fields like biosensor development, medicine, environmental surveillance and many more. Nano-biosensors based on liposomes are being used to detect extremely small amounts of organophosphorus pesticides like dichlorvos and paraoxon.

Piezoelectric biosensors: They are utilized for examining the harmful bacteria strain, Staphylococcal enterotoxin B, found in fruit juices specifically made from apple [11]

2.2 Mechanism of Sensing

Different techniques are frequently employed in biosensing, such as Localized Surface Plasmon Resonance (LSPR). Metal nanoparticles such as gold and silver demonstrate LSPR, in which conduction electrons vibrate when exposed to light. Change in the local dielectric environment can impact the frequency of this oscillation, which can be utilized to detect analytes by monitoring shifts in the LSPR peak. The fluorescence intensity changes when analytes bind or interact with quantum dots, allowing for sensitive detection (Fig. 1).

One example is when single-walled carbon nanotubes (SWCNTs) interact with graphene quantum dots (GQDs), causing fluorescence to increase at lower concentrations and decrease at higher concentrations. Another method of Surface-enhanced Raman Scattering constructs greatly boosts the Raman signal, allowing for the identification of minute analyte levels. The same phenomenon can be observed in Electrochemical biosensing where conductive nanomaterials are used such as carbon nanotubes (CNTs) as well as graphene undergo alterations in electrical conductivity upon contact with analytes and these alterations are quantifiable. Metal nanoparticles have catalytic characteristics that make electrochemical reactions easier, enabling the identification of analytes by observing alterations in current or potential. Responses to binding or interacting with analytes can occasionally be thermal or calorimetric. Analyte presence and concentration can be ascertained by tracking temperature variations or heat flow [13].

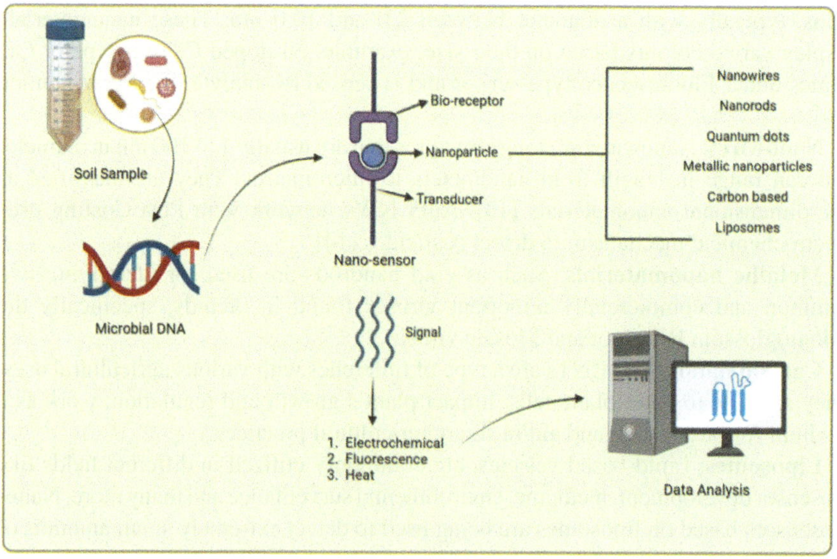

Fig. 1 Schematic representation of the nanosensor mechanism

3 Detection of Soil Microbes

The identification of bacteria has been transformed by molecular approaches based on proteins and nucleic acids, which were previously restricted by the incapacity of traditional culture methods to detect their presence [15]. The quantity of mRNA retrieved from a sample indicates the growing cells due to its short half-life and gene activation signals, while the metagenome comprises all DNA content from samples.

Ribosomal RNA, or rRNA, is a more stable type of RNA that may last for years in dead cells. Because of this property, this molecule is a good fit and the gold standard for determining which bacterial population is most prevalent in a certain setting. For these reasons, the 16S rRNA gene sequence is commonly used for bacterial taxonomy and phylogeny studies, mainly due to its stable functionality ensuring that random sequence changes are a dependable indication of evolutionary time; it is abundant in bacterial cells; it is present in every bacterium, typically found as a group of multiple genes; and it is of sufficient size (1500 base pair gene) for analysis of information. Since the 1960s, protein electrophoresis has been employed in ecological studies to identify genetic diversity in various samples using molecular techniques [16] (Table 1).

According to microbiology, these methods are often employed and refined so that phylogenic positions, identity, and the absence of microorganisms from culture medium from diverse environmental samples can be examined [17]. Ever since DNA sequencers were considerably more inexpensive in the early 1990s, accessible, and technologically advanced, these methods have grown in popularity.

These tests are far more sensitive, affordable, reliable, and appropriate for diseases or microorganisms that cannot be cultured or when culturing presents a risk. These tests have the potential to uncover and characterize new uncultivable species, genera, and gene families in addition to detecting the existence of previously identified uncultivable bacteria. Additional cutting-edge techniques for identifying the existence of

Table 1 Comparison of conventional and nanosensors based methods

Parameter	Conventional methods	Nanosensors
Sensitivity	Moderate	High (detects low concentrations of microbes)
Time required	Hours to days (e.g., culturing)	Minutes to hours
Specificity	May require additional steps	High (functionalized nanosensors)
Portability	Not portable	Compact and field-deployable devices
Cost	Low (but time-intensive)	Higher initial cost, but cost-effective long-term
Multiplexing capability	Limited (one target at a time)	High (simultaneous detection of multiple targets)
Environmental impact	Minimal	Requires regulation of nanomaterial disposal

bacteria that are challenging to cultivate include DNA microarrays, fluorescent in situ hybridization (FISH), metaproteomics, and metagenomics.

3.1 Conventional Methods

3.1.1 Flowcytometry

Because of its great precision and resilience, flow cytometry (FCM) is a useful tool for characterizing cell populations in a short amount of time. Consequently, by enabling the sorting of several facets of biodiversity in bacteria, FCM may enhance the capacities of soil research. But in order to apply FCM, bacteria must first be dispersed, separated, and purified from intricate soil matrices. Soil analysis and management techniques require intensive analysis as well.

The following phenotypic diversity of soil bacteria has been discovered using FCM: (a) the physiology or viability of soil bacteria specially the one which is native of a particular area; (b) variations in native soil bacteria their cell volumes and genetic contents when cultured; (c) the use of biosensors and biotechnology tools for detecting quorum sensing in bacteria, genotoxins, and metal bioa-vailability in polluted soils; and (d) Examining variation in genes of specific soil bacteria using techniques like 16S rRNA gene probing in conjunction with fluorescent cell sorting [18].

3.1.2 Epifluorescence Microscopy

Epifluorescence microscopy offers an alternate method for detecting, visualizing, and estimating the number of microbial cells [19]. By detecting both culturable and nonculturable methods and providing highly accurate quantitative estimates of microorganisms from environmental samples, this method has solved the prior huge count plate anomaly [20]. With the help of certain fluorescent dyes: SYBR green, ETBR and DAPI. In several investigations of various soils and sludge samples, such as sand filters in water treatment plants providing drinking water [21], medical equipment [19], pharmaceuticals, and food quality control [22], it has been utilized to determine the microbial count.

The main benefits of this method include direct visual feedback, accurate cell counts (biovolume), and the ability to distinguish between live and dead, culturable and nonculturable microorganisms in environmental samples [19]. Its disadvantages include long processing periods and low sample throughput [23].

3.1.3 16 s rRNA Gene Analysis

For many years, the B1500 bp 16S rRNA gene sequence—which consists of nine variable sections scattered across a highly conserved sequence—has been employed as a tool for bacterial investigation [24]. The use of 16S rRNA in phylogenetics was first introduced by Woese et al. [25]. Its sequencing has served as a major resource for identifying and detecting germs from unidentified origins [26].

The 16S rRNA sequencing of various species, whether closely or distantly related, serve as a dependable molecular clock because they contain sequences which are conserved in certain regions and perform similar functions [27]. The amount of conserved sequences varies widely, higher-level taxonomy corresponds to highly conserved sections, whereas lower-level taxonomy corresponds to less conserved parts, including genus and species [28]. The most interesting possible applications of 16S rRNA sequencing is identifying species and genera for introducing isolates which do not match any known biochemically existing profiles, for populations only provides "appropriate" or "lower probability" identifications using commercial techniques, sometimes for taxa that are seldom related to the infectious illnesses in humans.

The 16S rRNA sequence analysing the similarity is used to investigate highest and least similarity with previously known species [29], as well as cut-off values that prove originality at a given level [30]. In order for 16S rRNA gene sequencing to be effectively used for microbiological identification, two conditions must be met: accurate nucleotide sequences need to be included in databases, and each sequence must be correctly "labelled". Even though 16S rRNA is a molecular marker of bacteria that can detect bacterial populations with high resolution and coverage, the overall representation of the bacteria or their communities in a sample is not optimal [31].

3.1.4 Quantitative Real Time- PCR

Q-PCR is a real time technique used to detect microbes by amplifying and measuring the amount of DNA/RNA present. Despite having already transformed nucleic acid detection, traditional PCR is not as popular for quantitative analysis [32]. However, by offering sufficient information for microbial identification and quantification in a brief amount of time, the qPCR has also surpassed a commonly used culture plate approach. It also uses fluorescent dyes different in specificity as compared to epifluorescence technique.

The amount of intensity of fluorescent dye after each cycle increases proportional to the initial amount of nucleic acid present. All things considered, a number of factors significantly impact the detection and quantification, such as the qualitative specificity of the targets, the detection and quantification limit, and the effectiveness, accuracy, and precision of qPCR. There is diversity for the many operational parameters, which has made it a difficult process because there are no universally accepted standard processes.

3.2 Nanosensor Based Method

Sensor-based analytical methods have a significant influence on a number of industries, including agriculture, food, healthcare, and environmental monitoring. Since they are now a crucial instrument for detecting environmental pollutants and microorganisms, biosensors are becoming more and more popular. Biosensors have remarkable qualities in a small package, including resilience, sensitivity, selectivity, and affordability (Fig. 2).

3.2.1 Electrochemical Detection

Because of their intrinsic accuracy [33] and usefulness for point-of-care testing [34]. He most often used microbial sensors are electrochemical ones. Researchers find it interesting to working with different microbial nanosensors because of their ability to recognize a wide range of analytes, such as glucose [35], phenolic chemicals, bacteria, viruses, fungi, and heavy metals [36].

Electrochemical biosensors utilize bioreceptors attached to a transducer surface to interact with analytes and generate various patterns in the electrochemical signal. It enables a high signal-to-volume ratio and label-free detection. This particular kind of biosensor have potentiometric, impedimetric, amperometric, and capacitive

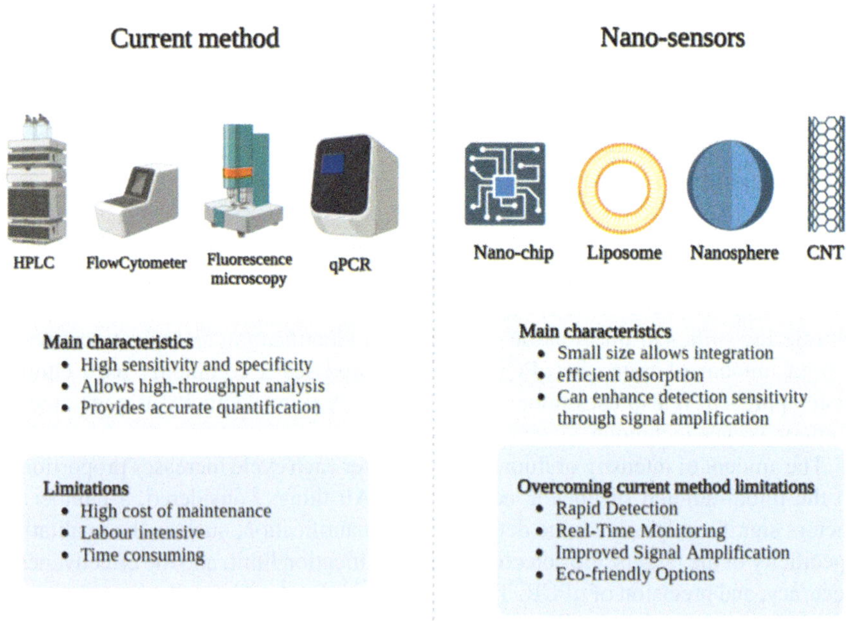

Fig. 2 Difference between conventional methods and nanosensor based methods

transducers which are used in converting biological signal to a chemical signal which is further measured by different methods, based on its operational principle [37]. Electrochemical nanosensors typically have a data processing unit and a transducer element for detection placed on top of a working electrode. The signal produced by a chemical interaction between the components of the solution, proportionate to the analyte concentration, is picked up by the transducer and further expressed electrically.

One may classify the signal transduction method across the transducer into four types: conductometric, amperometric, potentiometric, and voltametric nanosensors [38]. Microorganisms' metabolic function is assessed through the use of electric conductivity (EC) and redox potential, directly is connected to a count of viable microbes.

3.2.2 Optical Detection

Since the soil microbiome has an impact on human health in addition to plant development, researchers have been paying more and more attention to its detection. Traditional approaches for detecting the soil microbiome lacked real-time applications since the majority of them rely on culture, and the soil is home to a wide variety of microbial species not being cultured in a lab [39]. Colorimetric, surface plasmon resonance, and nanosensors using fluorescence are some of the techniques used to identify soil microbiomes. Colorimetric nanosensors monitor and compare analytes in coloured liquids using small components. The effects are also immediately visible to the unaided eye. This approach of microorganism detection is rapid and efficient despite its simplicity [40].

Surface improved. Raman spectroscopy is a regularly used method for detecting microorganisms. This technology takes advantage of nanomaterials' unique plasmon enhanced scattering and excitation capabilities to increase its sensitivity. This approach shows promise for multiplex detection [41]. Using nanomaterials, fluorescence-based nano sensing for microbial detection exploits an optical phenomenon in which energy absorption causes longer-wavelength fluorescent photons to emit shorter-wavelength photons [42].

3.2.3 Microfluidics

By combining sample preparation and reactions, separation and detection on a single chip, known as microfluidic chips—can be referred to as lab-on-a-chip —replace conventional biochemistry techniques [43]. Terry pioneered the concept of microfluidic chips. The scientist created a tiny gas analysis apparatus on a silicon wafer using photolithography and chemically induced etching techniques [44].

Bacterial detection is important in many fields, including industrial production, agriculture, food safety, and the management and prevention of foodborne infections. Bacterial detection is therefore extremely important. The traditional methods for detecting bacteria are intricate, ineffective, and need a large amount of sample preparation.

With its unmatched sensitivity and selectivity above traditional methods, this technology combines a variety of biochemical units that are compatible with a micro setup [45]. These sensors need certain sequences to identify microorganisms since they have different surface traits and properties. Target microorganism identification has lately benefited from the use of DNA microarray sensor technology, which is extremely sensitive and fast [46].

4 Applications

4.1 Plant Disease Prevention and Management

Phytopathology is the field of science that studies pathogens. Initially, pathogen identification relied on eye inspection, which might take days [47]. Novel biosensing techniques enable early identification of infections, resulting in disease control and reduced production losses. Identifying pathogens accurately can help design effective illness treatment options [48]. Nanobiosensors provide for faster detection than traditional approaches. This innovative diagnostic process is sometimes referred to as nano diagnostics. Nanobiosensors were utilized to identify contaminated microorganisms. Gold nanoparticle-based nanobiosensors are to diagnose agroterrorism agents. Gold nanoparticle-based sensors were created to detect plant diseases. DNA is the biological component used in the majority of nanobiosensors. Pathogen population genetics, gene transfer across hosts, and plant-pathogen interactions may all be tracked using nanobiosensors.

Rapid and precise DNA sequencing is made possible by the nanopore sensor technology, [49] created a nanopore technology to track DNA transport and detect pathogens using genome analysis. According to Tomar, real-time PCR are useful for identifying dangerous microbes. High-sensitivity nanobiosensors can enhance agricultural food safety (Fig. 3).

A biosensor for *Xanthomonas axonopodis*, the pathogen that causes tomato and pepper disease, was created [50] utilizing silica nanoparticles immobilized with pathogen antibodies.

It is possible to develop nanobiosensors using calcium phosphate and fluorogenic substrate nanocomposites. In cell-based activities such as signal transmission, membrane formation, and cell division, phosphorus is an essential element [51]. By using the nanobiosensors based on quantum dots fluorescence resonance transfer of energy, witches' broom disease-causing bacterium, *Phytoplasma aurantifolia*, was identified in lemon plants.

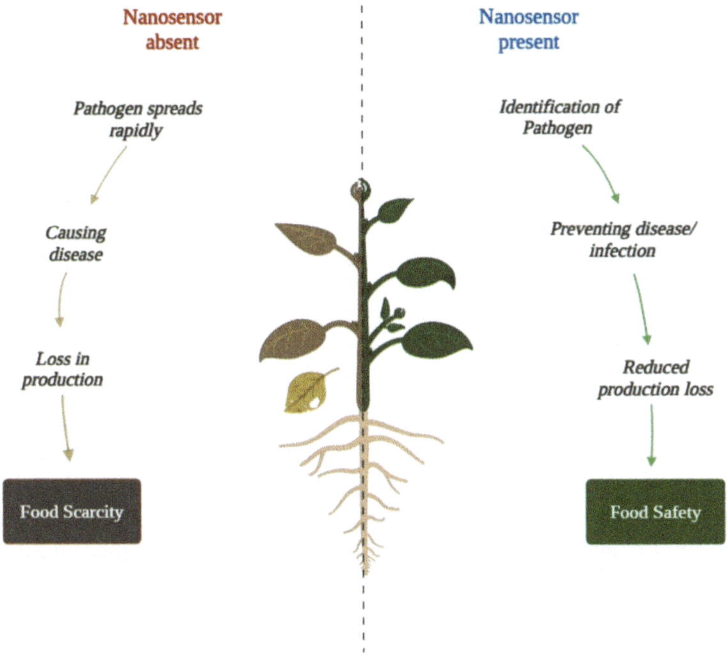

Fig. 3 Application of nanosensor in pathogen/pesticide detection

4.2 Soil and Crop Health Monitoring

Due to the dynamic nature of soil health, before beginning any agricultural projects, it is important to conduct monitoring of various soil indicators at site. Nanoparticle-based measurements or sensor systems are important for agricultural precisions. Soil can contain a variety of compounds providing nutrition in various forms, each of which emits a distinct signal. For example, high levels of nitrate indicate imbalance of microbes and ammonia signs presence of soil-feeding insects.

The development of plants and the health of the soil are greatly influenced by the many microorganisms that are present in soil. Ammonia, nitrate, and nitrite are all sources of nitrogen [52, 53]. To detect nitrites in soil samples, scientists have developed flexible lipid-based nanobiosensors, having a detection threshold of 2.1 μg/L, this double layer of lipid molecules, connected by hydrophobic forces, can be duplicated using the majority of elements necessary for maintaining thermodynamic balance [54]. Moisture content present in soil is considered as another soil measure used to detect and maintain crop productivity. Advanced plant science research and agriculture might be facilitated by wearable sensors and other smart nanobiosensors, which are now at the forefront of identifying hazardous substances and plants. On the various plant surfaces, such as the leaves, stems etc., the sensors in a wearable detecting system will be positioned.

Because of its affordability, adaptability, lightweight design, and biocompatibility, wearable sensors have become more and more popular in the diagnostics field [55]. They created a flexible strain sensor to monitor plant growth, leaf surface temperature, and humidity levels. A biosensor that can be worn by plants was created for the purpose of detecting pesticides such as methyl parathion, using laser-induced graphene technology for electrochemical sensing. Intelligent leaf sensors for plants, such as graphene sensors on tape, are able to monitor the movement of water in real-time [56, 57].

4.3 Crop Optimization

Numerous elements, including biological, microbiological, chemical, biochemical, and mechanical ones, might affect a crop's nutritional content. In particular, bacteria, fungus, mites, insects, and other pollutants are examples of biological and microbiological influences. Various chemical and biological processes, such as the Maillard reaction, lipid oxidation, and mechanical degradation, can take place during post-harvest storage. Germination, sprouting, and senescence that results in spoiling are a few other mechanical aspects.

Glutamate dehydrogenase and glutamate oxidase are enzymes used as biosensors to measure toxin levels like monosodium glutamate, to prevent neurotoxicity and diseases impacting the human kidney, liver, and brain [58]. The advancement of multiplex sensing in nanobiosensors enables the simultaneous detection of many analytes. A nanobiosensor has been created by Zhang et al. to identify and quantify different sweeteners like glucose, sucrose, cyclamate, and saccharin.

The broad variety of applications for nanobiosensors in the agricultural industry is demonstrated by the use of crop supply nanobiosensors to detect genetically modified organisms.

4.4 Other Applications

In addition to the previously described applications, nanobiosensors may be used to track the amount of signalling molecules produced by plants [59]. For example, Strigolactones, a tiny molecule biosensor, were employed in discovering the signalling hormone that halts growth of plant shoots. Phytoestrogens occur naturally in numerous plants, including soybeans, apples, and cabbages.

They are produced in plants as a defensive mechanism against fungal damage and are known as dietary estrogens. Similarly, [60] created a fluorescent activation sensor using Yoshimu lactone green for the identification of strigolactone compounds. Real-time sensing technique could also be combined with automation and enhanced management of natural resource. Water is a key natural resource needed for future sustainable agriculture, and its efficient use might be accomplished by using sensor technology to automate irrigation systems [61].

Thus far, a FRET probe was created for the identification of phytoestrogens like genistein and daizein. and resveratrol using fluorescence signals and an estrogen binding domain because catecholamines and dopamine are essential for plant synthesis, growth, and development. Nanobiosensors were utilized to assess dopamine levels [59]

5 Conclusion

Nanotechnology is a developing innovative area that has the ability to revolutionize agriculture utilizing contemporary technologies. Nanotechnology has several uses in agriculture, including pesticide delivery, biopesticide balancing, delayed release of fertilizers, and genetic material transfer for crop amendment. Nanosensors can assess plant nutrition, moisture, and physiological condition, allowing for rapid remedial treatments. Although there have been reports of nanobiosensors in medical diagnosis and applications. The high cost of manufacturing, automated testing, assessment, and validation of field trials remains a key hurdle for miniaturizing prototypes for production in industry. This may explain why there are fewer commercial nanobiosensors available. Using waste biomass to create innovative nanomaterials can be a cost-effective solution. Future research should focus on the versatility of nanobiosensors. Creating nanomaterials for biosensing based on bioassays can facilitate the commercialization of portable nanobiosensors. Nanobiosensor technology can be connected to GPS systems in order to improve precision farming and enhance smart agriculture practices. Farmers can enhance decision-making on irrigation, fertilization, insect control, and harvesting with decreased use of natural resources. The article concludes that nanobiosensors that too customized to provide great specificity with sensitivity will become a reality in the near future.

Competing Interests: The authors declare that they have no competing interest.

Declaration of Generative AI and AI-Assisted Technologies in the Writing Process During the preparation of this work the book chapter author(s) have not taken any assistance from AI, and the book chapter author(s) reviewed and edited the content as needed and take(s) full responsibility for the content of the publication. All authors read and approved the final version of the manuscript.

References

1. Sreejith S, Ajayan J, Radhika JM, Uma Reddy NV, Manikandan M (2024) Recent advances in nano biosensors: an overview. Measurement 236:115073
2. Finkel OM, Castrillo G, Paredes SH, González IS, Dangl JL (2017) Understanding and exploiting plant beneficial microbes. Curr Opin Plant Biol 38:155–163

3. Sanchez-Mahecha O, Klink S, Rothballer M, Sturm S, Weisser WW, Zytynska S, Heinen R (2024) Microbe-induced plant resistance against insect pests depends on timing of inoculation but is consistent across climatic conditions. Funct Ecol 38:848–860
4. Wei D, Zhu D, Zhang Y, Yang Z, Hu Y, Song C, Yang W, Chang X (2024) Pseudomonas chlororaphis IRHB3 assemblies' beneficial microbes and activates JA-mediated resistance to promote nutrient utilization and inhibit pathogen attack. Front Microbiol 15:1328863
5. Purohit HJ, Pandit P, Pal R, Warke R, Warke GM (2024) Soil microbiome: An intrinsic driver for climate smart agriculture. J Agri Food Res 101433
6. Zhuang Y, Wang H, Tan F, Wu B, Liu L, Qin H, Yang Z, He M (2024) Rhizosphere metabolic cross-talk from plant-soil-microbe tapping into agricultural sustainability: current advance and perspectives. Plant Physio Biochem 210:108619
7. Bayu T (2024) Systematic review on the role of microbial activities on nutrient cycling and transformation implication for soil fertility and crop productivity. Cold Spring Harbor Laboratory, p 610905
8. Mondal R, Dam P, Chakraborty J, Paret ML, Katı A, Altuntas S, Sarkar R, Ghorai S, Gangopadhyay D, Mandal AK, Husen A (2022) Potential of nanobiosensor in sustainable agriculture: the state-of-art. Heliyon 8(12)
9. Kim C, Kang MS, Raja IS, Joung YK, Han D (2024) Advancements in nanobiosensor technologies for in-vitro diagnostics to point of care testing. Heliyon e40306
10. Han D, Hosamo H, Ying C, Nie R (2023) A comprehensive review and analysis of nanosensors for structural health monitoring in bridge maintenance: innovations, challenges, and future perspectives. Appl Sci 13:11149
11. Bharti A, Jain U, Chauhan N (2024) From lab to field: nano-biosensors for real-time plant nutrient tracking. Plant Nano Biol 9:100079
12. Wang X, Zhou J, Wang H (2024) Bioreceptors as the key components for electrochemical biosensing in medicine. Cell Rep Phys Sci 5:101801
13. Darwish MA, Abd-Elaziem W, Elsheikh A, Zayed AA (2024) Advancements in nanomaterials for nanosensors: a comprehensive review. Nanoscale Adv 6:4015–4046
14. Malik S, Singh J, Goyat R, Saharan Y, Chaudhry V, Umar A, Ibrahim AA, Akbar S, Ameen S, Baskoutas S (2023) Nanomaterials-based biosensor and their applications: a review. Heliyon 9:e19929
15. Hirsch PR, Mauchline TH, Clark IM (2010) Culture-independent molecular techniques for soil microbial ecology. Soil Biol Biochem 42:878–887
16. Baker AJ, Tavares ES, Elbourne RF (2009) Countering criticisms of single mitochondrial DNA gene barcoding in birds. Mol Ecol Resour 9:257–268
17. Zhang Z, Willson RC, Fox GE (2002) Identification of characteristic oligonucleotides in the bacterial 16s ribosomal RNA sequence dataset. Bioinformatics 18:44–250
18. Gougoulias C, Shaw LJ (2012) Evaluation of the environmental specificity of fluorescence in situ hybridization (FISH) using fluorescence-activated cell sorting (FACS) of probe (PSE1284)-positive cells extracted from rhizosphere soil. Syst Appl Microbiol 35(8):533–540
19. Daims H, Wagner M (2007) Quantification of uncultured microorganisms by fluorescence microscopy and digital image analysis. Appl Microbiol Biotechnol 2:237–248
20. Cragg BA, Parkes RJ (2014) Bacterial and archaeal direct counts: a faster method of enumeration, for enrichment cultures and aqueous environmental samples. J Microbiol Methods 98:35–40
21. Frossard A, Hammes F, Gessner MO (2016) Flow cytometric assessment of bacterial abundance in soils, sediments and sludge. Front Microbiol 27379043
22. Nemati M, Hamidi A, Dizaj SM, Javaherzadeh V, Lotfipour F (2016) An overview on novel microbial determination methods in pharmaceutical and food quality control. Adv Pharma Bull. 6:301
23. Pernthaler J, Pernthaler A, Amann R (2003) Automated enumeration of groups of marine picoplankton after fluorescence in situ hybridization. Appl Environ Microbiol 69:2631–2637
24. Johnson JS, Spakowicz DJ, Hong BY, Petersen LM, Demkowicz P, Chen L (2019) Evaluation of 16s rRNA gene sequencing for species and strain-level microbiome analysis. Nat Commun 10:1–11

25. Woese CR, Kandler O, Wheelis ML (1990) Towards a natural system of organisms: proposal for the domains archaea, bacteria, and eucarya. Proc Natl Acad Sci USA 87:4576–4579
26. Lu J, Salzberg SL (2020) Ultrafast and accurate 16s rRNA microbial community analysis using kraken. Microbiome 8:1–11
27. Tsukuda M, Kitahara K, Miyazaki K (2017) Comparative RNA function analysis reveals high functional similarity between distantly related bacterial 16S rRNA. Sci Rep 7:1–8
28. Yang B, Wang Y, Qian PY (2016) Sensitivity and correlation of hypervariable regions in 16s rRNA genes in phylogenetic analysis. BMC Bioinform 17:1–8
29. Rossi-Tamisier M, Benamar S, Raoult D, Fournier PE (2015) Cautionary tale of using 16s rRNA gene sequence similarity values in identification of human-associated bacterial species. Int J Syst Evol Microbiol 65:1929–1934
30. Pulschen AA, Bendia AG, Fricker AD, Pellizari VH, Galante D, Rodrigues F (2017) Isolation of uncultured bacteria from antarctica using long incubation periods and low nutritional media. Front Microbiol 8:1346
31. Caporaso JG, Lauber CL, Walters WA, Berg-Lyons D, Huntley J, Fierer N (2012) Ultra-high-throughput microbial community analysis on the illumina Hiseq and Miseq platforms. ISME J 8:1621–1624
32. Zhang T, Fang HH (2006) Applications of real-time polymerase chain reaction for quantification of microorganisms in environmental samples. Appl Microbiol Biotechnol 70:281–289
33. D'souza S (2001) Microbial biosensors. Biosens Bioelectron 16:337–353
34. DeBusschere BD, Kovacs GT (2001) Portable cell-based biosensor system using integrated cmos cell cartridges. Biosens Bioelectron 16:543–556
35. Odaci D, Kayahan SK, Timur S, Toppare L (2008) Use of a thiophene-based conducting polymer in microbial biosensing. Electrochim Acta 53:4104–4108
36. Tag K, Riedel K, Bauer HJ, Hanke G, Baronian KH, Kunze G (2007) Amperometric detection of Cu21 by yeast biosensors using flow injection analysis (FIA). Sens Actuators B Chem 122:403–409
37. Malhotra BD, Ali MA, (2018) Nanomaterials in biosensors: fundamentals and applications. In: Nanomaterials for biosensors, pp 1–74
38. Su L, Jia W, Hou C, Lei Y (2011) Microbial biosensors: a review. Biosens Bioelectron 26:1788–1799
39. Sashidhar P, Dubey M, Kochar M (2019) Sensing soil microbes and interactions: how can nanomaterials help. 213–236
40. Chen B, Wang F (2020) Emerging frontiers of upconversion nanoparticles. Trend Chem 2(5):427–439
41. Zong C, Mengxi X, Xu LU, Wei T, Ma X, Zheng XS (2018) Surface-enhanced Raman spectroscopy for bioanalysis: reliability and challenges. Chem Rev 118:4946–4980
42. Hegyi J, Hegyi V (2016) New developments in fluorescence diagnostics. In: Hamblin MR, Avci P, Gupta GK (eds) Imaging in dermatology. Academic Press, Boston, pp 89–94
43. Xu K, Zhu P, Colon T, Huh C, Balhoff M (2017) A microfluidic investigation of the synergistic effect of nanoparticles and surfactants in macro-emulsion-based enhanced oil recovery. SPE J 22:459–469
44. Terry SC, Jerman JH, Angell JB (1979) A gas chromatographic air analyser fabricated on a silicon wafer. IEEE Trans Electron Devices 26:1880–1886
45. Gencturk E, Mutlu S, Ulgen KO (2017) Advances in microfluidic devices made from thermoplastics used in cell biology and analyses. Biomicrofluidics 11:051502
46. Zhou W, Le J, Chen Y, Cai Y, Hong Z, Chai Y (2019) Recent advances in microfluidic devices for bacteria and fungus research. TrAC, Trends Anal Chem 112:175–195
47. Khiyami MA, Almoammar H, Yasser M, Awad M, Mousa A, Alghuthaymi Kamel A, Abd-Elsalam A (2014) Plant pathogen nanodiagnostic techniques: forthcoming changes. Biotechnol Biotechnol Equip 28:775–785
48. Ali Q, Ahmar S, Sohail MA, Kamran M, Ali M, Saleem MH, Rizwan M, Ahmed AM, Mora-Poblete F, Junior ATD, Mubeen M, Ali S (2021) Research advances and applications of biosensing technology for the diagnosis of pathogens in sustainable agriculture. Environ Sci Pollut Res Int 8:9002–9019

49. Ozsolak F (2012) Third-generation sequencing techniques and applications to drug discovery. Expert Opin Drug Discov 7:231–243
50. James A, Franco F, Merca EM, Rodriguez S, Johny F, Balidon-Veronica P (2019) DNA-based electrochemical nanobiosensor for the detection of Phytophthora palmivora (Butler) Butler, causing black pod rot in cacao (Theobroma cacao L.) pods. Physiol Mol Plant Pathol 107:14–20
51. Caon T, Martelli SM, Fakhouri FM (2017) New trends in the food industry: application of nanosensors in food packaging. Nanobiosensors 773–804
52. Ji R, Brune A (2006) Nitrogen mineralization, ammonia accumulation, and emission of gaseous NH3 by soil-feeding termites. Biogeochemistry 78:267–283
53. Nettleton WD, Peterson FF (2011) Landform, soil, and plant relationships to nitrate accumulation, Central Nevada. Geoderma 160:265–270
54. Siontorou CG, Georgopoulos KN (2016) A biosensor platform for soil management: the case of nitrites. J Clean Prod 111:133–142
55. Nassar JM, Khan SM, Villalva DR, Nour MM, Almuslem AS, Hussain MM (2018) Compliant plant wearables for localized microclimate and plant growth monitoring. Npj Flex Electron 2:1–12
56. Oren S, Ceylan H, Schnable PS, Dong L (2017) High-resolution patterning and transferring of graphene-based nanomaterials onto tape toward roll-to-roll production of tape-based wearable sensors. Adv Mater Technol. 2:1700223
57. Zhao F, He J, Li X, Bai Y, Ying Y, Ping J (2020) Smart plant-wearable biosensor for in-situ pesticide analysis. Biosens Bioelectron 170:112636
58. Arugula MA, Simonian AL (2016) Biosensors for detection of genetically modified organisms in food and feed. In: Genetically modified organisms in food. Elsevier, Amsterdam, the Netherlands, pp 97–110
59. Kwak SY, Wong MH, Lew TTS, Bisker G, Lee MA, Kaplan A, Dong J, Liu AT, Koman VB, Sinclair R, Hamann C, Strano MS (2017) Nanosensor technology applied to living plant systems. Annu Rev Anal Chem 10:113–140
60. Tsuchiya Y, Yoshimura M, Sato Y, Kuwata K, Shigeo T (2015) Probing strigolactone receptors in striga hermonthica with fluorescence. Science 349:864–868
61. Ramnani PN, Saucedo M, Mulchandani A (2016) Carbon nanomaterial-based electrochemical biosensors for label-free sensing of environmental pollutants. Chemosphere 143:85–98

Nanosensors and Their Role in Bioremediation

Shrejal Tiwari and Anant

Abstract Environmental pollution poses a threat of human health and ecosystem. Bioremediation is a sustainable approach which utilizes microorganism to biodegrade pollutant into less toxic substances. However, monitoring and assessing bioremediation process is challenging. Nanosensors are promising devices as they are highly specific and selective engineered to detect and quantify various substances such as heavy metals, organic compounds, and microbial toxins present in environment. Furthermore, integration with biological systems enables the tracking of bioremediation processes enabling precise resulting in paced cleanup and improved outcomes. The integration between nanotechnology and bioremediation provides a sustainable solution for environmental challenges. This abstract highlight the potential of nanosensors to revolutionize bioremediation strategies, encouraging innovative approaches to environmental restoration.

Keywords Nanosensors · Nanotechnology · Bioremediation · Environment · Restoration

1 Introduction

Globalisation and human growth were aided by economic and scientific advancements in the second half of the eighteenth century. This led to the rapid exploitation of natural resources for product manufacturing, which in turn resulted in inadequate waste disposal in the environment. Increased air pollution, contaminated water and soil, and biomagnification have long-term negative effects on both human and animal health. Pollutants that have contaminated the environment include heavy metals, azo

S. Tiwari (✉)
Department of Floriculture and Landscaping, Rani Laxmi Bai Central Agricultural University, Jhansi, India
e-mail: shrejaltiwari@gmail.com

Anant
Department of Genetics and Plant Breeding, Rani Laxmi Bai Central Agricultural University, Jhansi, India

A. Husen (ed.), *Nanobiosensors for Agricultural and Other Related Sectors*, Smart Nanomaterials Technology, https://doi.org/10.1007/978-981-96-6499-3_7

dyes, steroids, endocrine disrupting compounds, pesticides, antibiotics, and chlorinated phenols. These xenobiotics may react to produce dangerous molecules when they are released [1]. The ecosystem, climate, and human health are all at risk due to the sharp rise in their numbers and longevity in recent years. And one of the most pressing issues facing the world now is environmental contamination. Effective management techniques must be used for environmental protection in order to overcome these problems. Technological developments in recent years have increased the effectiveness of pollutant removal [2]. With the use of bioaugmentation (inoculation of inactive oil-degrading bacteria) and bio stimulation (growth of native microorganisms), bioremediation is one such efficient technology that can significantly reduce the impact of pollution in the environment. It offers an effective method for eliminating pollutants from a variety of environments and permits management that is internationally applicable and relatively scalable [3, 4]. Although bioremediation is a sustainable process, it has advantages and disadvantages.

On the other hand, futuristic technologies like nanotechnology are used in a variety of sectors, including biodegradation, cosmetics, medicine, food production, and agriculture. These developments in nanotechnology and the use of nanomaterials can provide more practical ways to push bioremediation much beyond its existing limits [5]. Thus, encouraging the application of bioremediation in conjunction with nanotechnology provides an economical, environmentally friendly method of eliminating harmful pollutants. Nano bioremediation enables precise control and management of materials at the atomic and molecular levels. Being extremely small particles with high surface area to volume ratio, nanoparticles provide better coverage [6, 7]. Open up extensive possibilities for their application in the remediation of petroleum hydrocarbons, pesticides, and metal-contaminated sites. With the disruption of ecosystem and increase of contaminants at alarming rate there is need for monitoring, detecting, identifying, and remediating even trace amounts of contaminants present in air, water, and soil. Several methods available for monitoring of environment including gas chromatography, HPLC, mass spectroscopy further use of number of sensors (Ultrasonic sensors, Pressure sensors, Temperature sensors, Infrared sensors, Radar sensors, Soil sensors, Gas sensors, Rainfall sensors, Light sensors, and Wind speed and Direction sensors) these technologies are sensitive, specific and reproducible but have number of drawbacks. The traditional technologies are expensive, time consuming, requires skilled and knowledgeable person to handle it, moreover, data loss and misinterpretation will be the common problem. All of these issues are addressed by nano bioremediation, which also provides a system that can monitor the toxins using nanosensors.

Nano sensors are devices engineered at nanoscale to recognize certain molecules, biological entities, or to monitor environmental conditions and can detect at a much smaller level. They are highly specific, portable, and economical. When compared to conventionally constructed sensors that have fewer detecting parts overall, nano sensors can take use of these extraordinary properties and potentially improve upon them. In addition to this nano sensor are not only used to detect at nanoscale but finds their application at macroscale where they can be used in larger devices that can

Table 1 Sensing techniques and their benefits

Sensing techniques	Benefit	Used
Optical	Flexible detection and offers less scattering	Metal NP Silica NP
Fluorescence	In vivo detection	AuNPs
Mechanical	Low sampling volumes but affected by viscous fluids	Cantilevers or resonators
Colorimetric	Easy comprehensive signals High probe concentration required	AuNPs, AgNPs
Electrochemical	Development of chemical sensors and biosensors	Carbon nanotubes, graphene quantum dots

take advantages of nanotechnology and measure the activates at nanoscale. Furthermore, Nano sensors readily interact with substance such as protein at nanoscale where macro scale is impossible to use. Multiple method including top-down lithography, bottom-up assembly, and molecular self-assembly are used for designing of Nano sensors. Classification of Nano sensors is based on material used to build Nano sensors. Biosensors are integrated self-contained device used for analysis of data through signal transduction moiety. These sensors identify compounds such as glasses, heavy metal ions, carbohydrates, amino acids and various molecules linked to diseases while physical sensors use physical properties (force, pressure) as a signal. When we talk about monitoring of environment chemical sensors are mostly used to detect the presence of pollutant in air, medicine development. Chemical Nano sensors are also employed in diagnostic application. Different kind of nano material are used for making of nanosensors including, thin films, metal oxide nano particles, polymers, and biomaterials. These nanomaterials provide ample advantages over bulky machines due to their small size. Below enlisted Table 1 the different sensing technique used in nanosensors for detection of pollutants.

2 Strategies for Nano Bioremediation.

I. Nanoencapsulation: It includes microorganisms which are encapsulated, enzymes or other within nanoparticles. This phenomenon protects active agent from various environmental conditions which can be very harsh for them and enhance their mobility with stability. It also permits targeted delivery to contaminated locations [8]. Also, nanoencapsulation can stop GMOs from entering the environment.
II. Nano scale Sorbents: Nanoparticles can efficiently absorb and immobilise pollutants if we use them as sorbents. They can be functionalised with particular chemical groups to target particular pollutants e. g. Metal nanoparticles can absorb heavy metals but carbon nanotubes can absorb organic contaminants.

III. Enhanced Biodegradations: Nanoparticles can be employed to improve their biodegradation by making contaminants more bio available to microorganisms [9]. By supplying oxygen to anaerobic condition for instance nZVI can encourage the development of aerobic microbes and speed up the breakdown of organic pollutants.
IV. Bioinspired Nanomaterial: Nanomaterials which are bio inspired are created by imitating natural phenomenon or systems. For instance, scientists have created nanomaterial that effectively degrade pollutant by immidating the structure and function of enzymes. For the cleanup of pollutants bioinspired nanomaterials provide enhanced catalytic activity and selectively by utilising the special characteristics of biological systems at the Nanoscale.
V. Nanoparticle based remediation: Carbon nanotube, titanium dioxide and zero valent iron are example of nanoparticles that can be employed as career to transport remedial material to the contaminated sites. These nanoparticles can interact with contaminants and aid in their breakdown because of their large surface area and strong reactivity. To improve their efficacy in removing pollutants they can be functionalized with particular enzymes or microbes.
VI. Nanoscale Reactive Barriers: In order to intercept and treat pollutants nano particles are injected into the subsurface to create nanoscale reactive barriers. These barriers may be made of carbon-based nanomaterials, metal oxides or zero valent metals. The contaminants encounter the reactive nanomaterials as they pass through the barrier which cause them to change or degrade.
VII. Nano enabled Sensors: Highly sensitive and selective sensors for real time pollutant detection and monitoring can be created using nanotechnology. Nano sensors can help with the evaluation and clean-up process by quickly and accurately providing information regarding pollution concentrations.
VIII. Biological Agents: The integration of biological agents such as microbes and enzymes further enhance the efficiency of environmental clean-up. These biological entities can be natural occurring for generally engineered to have specific pollutant degrading capabilities.

    (a) Bio remediation:

        This technique uses enzymes and microbes to break down or change contaminants into non-toxic forms. For instance, hydrocarbons in oil spills can be metabolised by certain micro-organisms [10].

    (b) Phyto remediation:

        Pollutants from soil and water can be absorbed and accumulated by plants. The effectiveness of phytoremediation can be increased by using nanoparticles to improve the uptake and transport of pollutants by the plants [11].

IX. Photocatalysis: Through photo catalytic process some nano particles like titanium oxide can use light to break down contaminants. When exposed to ultraviolet radiation these nanoparticles generate reactive oxygen species which can break down organic substances into harmless byproducts.

X. Nanotoxicology: Researching the possible toxicity and environmental effects of nanoparticles is essential in addition to their usefulness. To guarantee the safe and responsible use of nanoparticles in bio remediation nanotoxicology studies their behaviour destiny and biological impacts.

3 Mechanism of Action for Nanosensors in Bioremediation

Mechanism of Nanosensors-pollutant interaction involve several steps for detection and quantification of pollutants which covered in Fig. 2 which are (i) recognition of biological element like enzyme (react with specific pollutant and alter their structure), antibody (bind to specific pollutant with high specificity), nucleic acid (hybridize with the complementary DNA/RNA sequence of pollutants), or a whole cell, specifically designed to interact with target pollutant followed by (ii) transduction including mechanism like electrochemical (electrical properties are measured like current and voltage), optical (detect change in light properties like fluorescence and absorbance), mass sensitive (change in mass is analysed through quartz crystal microbalance and thermal (alteration in temperature are measured) followed by (iii) signal processing involves amplification (enhance signal strength for its detection), filtering (remove irrelevant data to isolate signal of interest) and conversion (transform signal into readable format) and (iv) output (provide real time information about presence and concentration of pollutant. e.g. Detection of heavy metals in water (Fig. 1).

a. Mechanism of adsorption

By adsorption pollutant adhere to Nanosensors surface for which several key factors are required like surface area (nanomaterial have so many sites for pollutant molecule to adhere because they have a very high surface area to volume ratio), surface chemistry (certain chemical groups increase binding affinity for target molecule), physical adsorption (include weak Vander Waal

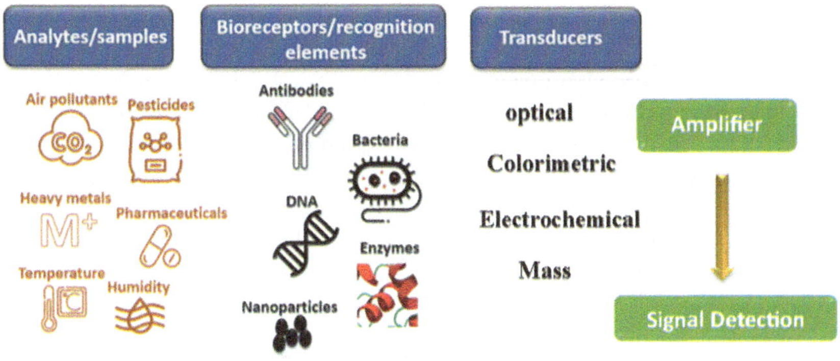

Fig. 1 Working of nanosensors in bioremediation

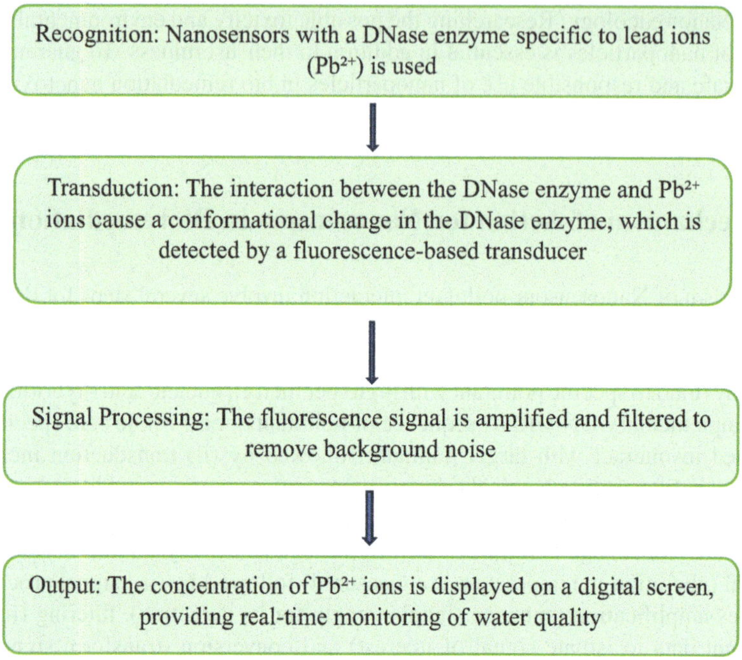

Fig. 2 Stepwise flowchart for mechanism of nanosensors-pollutant interaction

forces allowing binding of pollutant reversibly) and chemical adsorption (more stable with irreversible binding due to stronger covalent or ionic bond).

b. Mechanism of detection

Detection mechanism converts Nanosensors-pollutant interaction into a measurable signal. Primary detection mechanism includes (i) electrochemical detection (change in electrical properties of Nanosensors are measured eg with gold nanoparticles heavy metals like lead can be detected by biosensors through measuring changes in electrical conductivity) (ii) optical mechanism includes change in optical properties like absorbance, fluorescence, surface plasmon resonance for example when quantum dots attach to particular contaminants, they can release fluorescence, and it is possible to detect how intense this fluorescence is, (iii) mass sensitive detection use quartz crystal microbalance to detect change in mass for example when pollutants attach to a nano-biosensor's surface, the mass shift can be detected by a piezoelectric crystal, (iv) thermal detection measure change in temperature for example When a pollutant binds, a nanosensors can identify exothermic or endothermic events that result in a detectable temperature change (Fig. 3).

c. Surface interaction mechanism

It includes mechanisms like (i) physical adsorption/physisorption (occurs at lower temperature which have weak van der Waals forces between pollutant and nanomaterial surface and is reversible for example Carbon nanotubes' vast

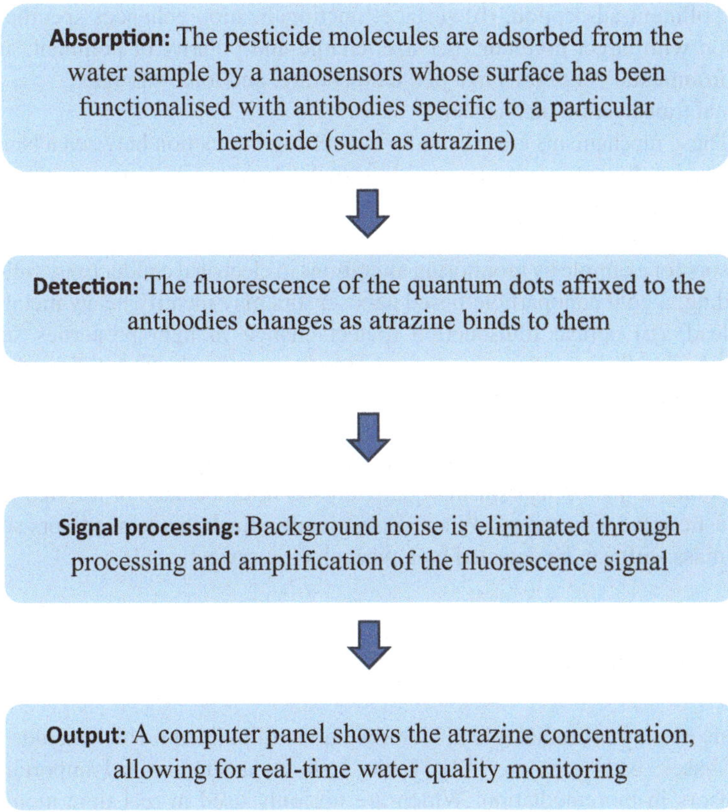

Fig. 3 Stepwise flowchart depicting the working mechanism of nanosensors

surface area and π-π interactions allow them to absorb organic contaminants like benzene), (ii) Chemical Adsorption/Chemisorption (occurs at higher temperature and is irreversible which involves formation of bonds like covalent and ionic between pollutant and nanomaterial surface for example Ionic interactions allow metal oxide nanoparticles to bind heavy metals like lead), (iii) electrostatic interactions (influenced by pH and environmental ionic strength and involve interaction between charged pollutant molecule nanomaterial surface which is oppositely charged for example Ammonium ions (NH^4+) and other positively charged contaminants interact with negatively charged graphene oxide), (iv) hydrogen bonding (involves the creation of hydrogen bonds between functional groups on the surface of the nanomaterial and polar pollutant molecules for example Phenolic chemicals and hydroxyl groups on functionalised silica nanoparticles form hydrogen bonds). There are certain factors which affect adsorption like (a) higher surface area and porosity provide more active sites for pollutant adsorption, (b) surface functionalization enhances specific interaction with target pollutant, (c) size, charge and polarity of pollutant, and (d) environmental conditions like pH, temperature, and ionic strength.

d. Signal transduction mechanisms

   These mechanisms are crucial for converting interaction between a Nanosensors and pollutant into a measurable signal. These mechanisms are (i) electrochemical transduction (measure change in current through amperometry sensors, in voltage through potentiometric sensors and impedance through impedimetric sensors for example by monitoring alterations in electrical conductivity following binding, a gold nanoparticle-based nanosensors may identify heavy metals such as lead, (ii) optical transduction (detect change in light properties such as absorbance, fluorescence or surface plasmon resonance through different sensors for example when quantum dots attach to particular contaminants, they can release fluorescence, and it is possible to detect how intense this fluorescence is) and (iii) mass sensitive transduction (detects change in mass when a pollutant binds to nanosensors using quartz crystal microbalance or through surface acoustic wave for example When contaminants attach to a nanosensors surface, the mass shift can be detected by a piezoelectric crystal.

4 Nanosensors Based Bioremediation

Nanosensors plays a crucial role in bioremediation of pollutant which include heavy metals, gases, pathogens etc [12]. Table 2 shows a summarized importance of nanosensors in bioremediation. Which are not only used in real time monitoring and detection but can also contribute to enhanced sensitivity, mapping of pollutant location and data collection and assessment.

4.1 Detection of Heavy Metals in Soil and Water

Heavy metal pollution poses a high level of warning to the environment as it can significantly affect the ecosystem, human health, and aquatic species. While it naturally occurs at low levels in the environment but anthropogenic activities like fossil fuel burning, mining, and agricultural practices also contributed significantly. Heavy metals such as lead, mercury, cadmium, chromium, arsenic, and nickel become pollutants when accumulate at very large levels in the environment [13]. To address this issue continuous monitoring of water quality and research is very important. Among various methods, electrochemical and colorimetric detection is a powerful and widely used method for quantifying and monitoring heavy metals in different environmental samples, including water, soil, and air [14, 15]. For the detection of heavy metals electrochemical and colorimetric nanosensors are mostly used (Fig. 4).

Table 2 Importance of nanosensors in bioremediation

Role	Importance
Real-time monitoring	Optical, electrochemical and magnetic nanosensors provide real time monitoring of environmental parameters like pH, temperature, oxygen levels, and pollutant concentrations. Enabling the real time, early detection of contamination along with assessment of effective strategies for bioremediations
Early detection	Nanosensors being highly specific and sensitive it can detect the contamination at very low concentration accurately which can be missed by time taking and costly traditional methods, allowing for early intervention and prevention of progressing contamination
Mapping pollutant	Nanosensors with its high spatial resolution is capable of mapping the traces of pollutant distribution in an area furthermore it can provide information regarding pollutant hotspot for commencing bioremediation
Detection of specific pollutant	Being selective, nano sensors can be specifically used for detection of one or other pollutant at a time like development of sensors for detection of heavy metals, organic compounds, and pathogens
Assessing bioavailability of pollutants	Nanosensors are highly specific and can determine the amount of pollutants available for bioremediation by agents. Furthermore, based on the bioavailability microbial strains or plants are chosen for the process
Monitoring activity of microbes	By real time monitoring of key parameters, nanosensors ca be used to optimize process of bioremediation for enhancing the activity of microbes
Toxicity assessment	Various nanosensors like carbon nanotubes have a potential to measure the toxic amount of pollutant and its by product present along with assessing bioremediation strategies and environment safety
Prevention of secondary pollution	Nanosensors are used to assess the bioremediation process as well as monitor byproducts of this remediation. Nanosensors can be used to ensure the environment safety by monitoring byproducts of bioremediation process
Optimization	Nanosensors when integrates with IoT can optimize the process of selection of suitable microbes for the treatment, along with monitoring the activities of microorganism in the process of biodegradation and collection of data from contaminated sites
Data analysis	Nanosensors are capable of using remote sensing and data transmission capabilities for large-scale. It can be used for future prediction of pollution trends for prevention of future contamination

Arsenic (As) is one of the most toxic elements present in soil and water [16]. Arsenic contamination affects biological activities as a teratogen, carcinogen, and mutagen as well as having detrimental effects on the immune system. There are some bioremediation strategies for reducing its effects such as the use of plants that act as accumulators, indicators, and excluders [17]. Another feasible strategy is phosphate fertilization for successful phytoremediation using As hyperaccumulating

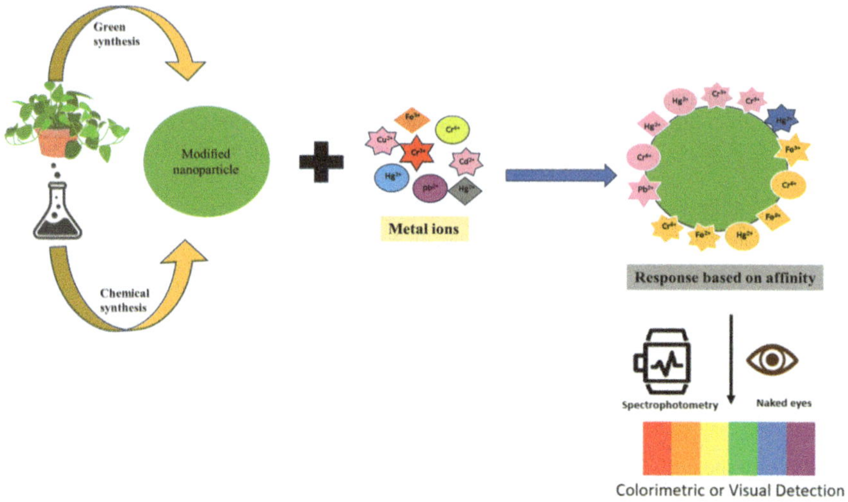

Fig. 4 Detection of heavy metals ions through nanosensors

fern viz. *Pteris vittata.* Arbuscular mycorrhizal fungus formed a stable association with Chinese brake fern and possesses resistance to arsenic toxicity. Grasses such as vetiver have great potential for remediation of arsenic contamination in soil. Mercury is a worldwide problem due to its wide application in industry (paper, textile, chlorine production, etc. Bioremediation of mercury occurs via molecular and transgenic approaches and enhanced phytovolatilization. Certain bacteria are capable of oxidizing, reducing or modifying the metal ions to less toxic species such as mer operon which contains genes that sense mercury (merB), transport mercury (merT) & sequester mercury to periplasmic space(merP). Chromium is the chief heavy metal contaminant found in tannery effluent. Mycorrhizal fungi have the ability to enhance plant tolerance or phytoaccumulation of Cr. Excess fluoride in drinking water causes harmful effects on human health in many parts of India. Using absorption technique especially clays that contain oxides of Fe, Al, and Si are appropriate for removal of fluoride. Nano biosensors can detect even trace amounts of heavy metals which is crucial for monitoring minute concentrations during the bioremediation process. Nanomaterials such as gold, carbon nanotubes, and quantum dots are often integrated into biosensors to increase their sensitivity and selectivity. Nano biosensors allow for continuous, real-time monitoring of heavy metal concentrations in soil, water or air. It is cost effective and portable for on-site heavy metal monitoring [16]. The summarized form of different nanoparticles used for the detection of specific pollutant is mentioned in table 3.

Table 3 Nanosensors used for detection of different pollutants

Nanosensors type	Used nanomaterial	Detected pollutant	Limit of detection	Key references
Electrochemical nano sensors	Fe	Cd(II) Pb(II) Hg(II)	0.036 nM 0.003 nM 0.011 nM	[18]
Electrochemical nano sensors	Fe and Au	Hg(II)	0.5 μg/L	[19]
Electrochemical nano sensors	Ag	Cu(II) Cd(II) Pb(II) Hg(II)	0.4 ppb 2.5 ppb 7.3 ppb 0.7 ppb	[20]
Electrochemical nano sensors	Carbon	Cd(II) Pb(II) As(III)	0.03 μg/L 0.05 μg/L 0.15 μg/L	[21]
Colorimetric (UV–Vis) nanosensors	TiO_2	Cu^{2+}	2.51 nM	[22]
Colorimetric (UV–Vis) nanosensors	AgCl NPs	Fe^{3+} Cu^{2+} Cr^{6+}	1.69 ppb 3.18 ppb 5.05 ppb	[23]
Colorimetric (UV–Vis) nanosensors	AuNPs	Pb^{2+}	1.07 μM	[24]
Optical	Nanohybrid CdSe QDs	Cd (II)	25 nM	[25]
Colorimetric nanosensors	Mesoporous silica nanoparticles (MSN)	Hg (II)	60 pM	[26]
Electrochemical sensor	Nano sheets of Fc-NH_2-UiO-66, and thermally reduced graphene oxide (trGNO)	Cd(II) Pb(II) Cu(II)	8.5 nM 0.6 nM 0.8 nM	[25]
Magnetic-fluorescent based nanosensors	Carboxymethyl chitosan-functionalized magnetic-fluorescent nanocomposites	Hg (II)	9.1×10^{-8} mol/L	[27]

(continued)

Table 3 (continued)

Nanosensors type	Used nanomaterial	Detected pollutant	Limit of detection	Key references
Fluorescent-nanosensors	3-aminopropyl-triethoxysilane coated Yb_2O_3	Imazapyr	0.2 ppm	[28]
Surface plasmon resonance (SPR) based affinity sensor	Atrazine imprinted nanoparticles	Atrazine	0.7134 ng/mL	[29]
Fluorescence sensor	Copper (II) oxide carbon nanotubes (MWCNTs)	Glyphosate	0.67 ppb	[30]
Electrochemical luminescence sensor	Luminol-gold nanoparticles-L-cysteine-Cu(II) composites	Glyphosate	0.5 nM	[31]
Electrochemical sensor	CuO-TiO2 hybrid nanocomposites	Methyl parathion	1.21 ppb	[32]
Electrochemical nanosensors	Chitosan-iron oxide	Malathion	0.01 ng/ml	[33]
Optical nanosensors	Silver nano dendrites	Dimethoate	0.002 ppm	[34]
Electrochemical	AuNPs	*Pseudomonas aeruginosa*	60 CFU mL^{-1}	[35]
Optical	COF-BA Au@Ag NPs	*Staphylococcus aureus*	10 CFU mL^{-1}	[25]
Optical	AuNCs	*Escherichia coli*	89 CFU mL^{-1}	[36]
Colorimetric	Fe_3O_4/Au-PEI NPs	*Escherichia coli*	0.52 CFU mL^{-1}	[37]
Electrochemical	MPDA/MnO_2	*Staphylococcus aureus*	3 CFU mL^{-1}	[38]

4.2 Detection of Pesticide

In order to increase production and productivity, pesticides are often used chemicals in agriculture to prevent and maintain invasive species, weeds, insects, fungi, etc. Hap hazardous use of pesticides in agricultural fields leads to the contamination of groundwater, biomagnification (ex- organophosphates imparts severe health risk), thereby affecting off targets like decrease in thickness of eggshells of birds and humans (mutagenicity, neurotoxicity, carcinogenicity to genotoxicity) further leading to increase in water BOD and increased chances of algal bloom affecting marine life [39]. There are several methods used for detection of pesticides residues like high-performance liquid chromatography, colorimetric assays, enzyme-linked immune sorbent assay, liquid/gas chromatography-mass spectrometry, electrophoresis, and fluorometric assay procedures but these methods are nevertheless costly and time taking. Hence detection by these methods is not suitable for real time detection. Various identification elements for several pesticides are present in optical nanosensors which function in conjunction with transducers to generate signals that cause pesticide particles to adhere to sensors. Four type optical probes are used to base on signal output. fluorescence (FL), colorimetric (CL), surface-enhanced Raman scattering (SERS), and surface plasmon resonance (SPR) optical sensors. Immunochromatographic strip (ICTS) nanosensors is another preferred sensor used in agroecosystem for pesticide detection [40]. Furthermore, electrochemical properties of nanoparticles make electrochemical nanosensors as an effective tool for pesticide detection. Recently the electrochemical biosensors used for detection of enzyme cholinesterase appeared as propitious devices is used for detection of special class of pesticide residues including carbamates and organophosphates [41].

The integration of nanotechnology, using materials like nanoparticles, nanocomposites, and nanotubes, provides superior surface areas and unique chemical properties, enabling high sensitivity and selectivity for residual pesticide detection [42]. Enzyme-based electrochemical sensors are widely applied, but their limitations, includes enzyme instability and interference from environmental impurities, prompted the creation of non-enzymatic electrochemical sensors that make use of nanomaterials. Numerous investigations show how well nanosensors work; for example, optical sensors based on silver nanodendrite and upconverting nanoparticles identified dimethoate and metribuzin at 0.002 ppm and 6.8×10^{a} M, respectively [43]. Similarly, chitosan-iron oxide nanocomposite-based electrochemical apt sensors demonstrated an impressive sensitivity of 0.001 ng/mL, whereas CuO nanoparticle-3D graphene nanocomposite electrochemical nanosensors detected malathion at 0.01 nM [44]. These developments highlight the promise of nanosensors as practical, sensitive, and reasonably priced instruments for detecting pesticide residue in agroecosystems.

4.3 Monitoring Microbial Activity

Monitoring microbial communities is essential in understanding which species of microorganisms are actively involved in the degradation of pollutants In order to break down or purify environmental toxins including hydrocarbons, heavy metals, and organic pollutants, bioremediation depends on the activity of microorganisms and species of microorganisms used for specific purpose is shown in Table 4. Experts in bioremediation can assess the process' efficacy by monitoring the expansion, composition, and alterations of microbial communities [45]. Optical nanosensors, such as fluorescence-based or surface plasmon resonance (SPR)-based sensors, can monitor microbial metabolites (e.g., CO_2, organic acids) and enzyme activity indicative of pollutant degradation [46]. For instance, it is possible to functionalise nanomaterials such as gold nanoparticles or quantum dots to detect certain biomolecules, enabling real-time monitoring of microbial activity. Redox reactions or electrochemical signals generated by microbial metabolic activities are detected by electrochemical nanosensors that use nanomaterials such as metal nanoparticles, graphene, and carbon nanotubes. These sensors work especially well for monitoring microbial respiration, organic pollutant degradation, and heavy metal reduction. Furthermore, nanosensors integrated with microbial biosensors (e.g., whole-cell biosensors) can identify the presence and bioavailability of pollutants while simultaneously assessing microbial health and activity. This integration enhances the precision and reliability of bioremediation monitoring [47].

Table 4 Microorganisms used in bioremediation

Organisms	Contaminants	Key references
Pseudomonas spp.	PCBs, halogenated hydrocarbons, alkylbenzene sulphonates, organophosphates, benzene, anthracene, 2,4-D, DDT and 2,4,5-tri chlorophenoxyacetic Acid etc	[48]
Clostridium Trichoderma	Malathion	
Flavobacterium spp.	Organophosphates, PCBs (polychlorinated biphenyls)	
Euglena gracilis	2,6-dinitrotoluene, Phenanthrene, di-nbutylphthalate, Toxaphenepetroleum	
Yeast (Saccharomyces)	DDT	
Trametes hirsute	trichlorophenol,	

4.4 Nanosensors for Waste Water Management

A healthy economy and the multifaceted development of society depend on access to clean and safe water. Every year, illnesses spread by drinking water tainted with harmful bacteria claim the lives of millions of people [49]. Although various methods for wastewater treatment have been explored in the last few decades but their use is restrained by many limitations including use of chemicals, formation of disinfection by-products (DBPs), time consumption and cost [50]. The extremely low concentrations of pollutants and the complexity and diversity of wastewater matrices make monitoring water quality difficult on both large and small scales. Furthermore, The rise of biological contaminants, such as antibiotic-resistant pathogens and their antibiotic-resistance genes (ARGs), in wastewater poses a serious challenge. Methicillin-resistant *S. aureus* (MRSA), a well-known antibiotic-resistant pathogen, harbors the *mecA* gene responsible for resistance. Chemicals known as organophosphorus (OP), which are often employed as pesticides, are increasingly found in rivers, industrial effluent, and agricultural runoff. Even at low levels, these compounds can be neurotoxic. To detect the OP pesticide diazinon, they developed aptamer-based optical nanosensors that use multi-walled carbon nanotubes (MWCNTs) and reduced graphene oxide quantum dots (rGQDs) [16]. These nanosensors are a viable tool for real-time water monitoring since they demonstrated great selectivity and accuracy for diazinon detection in tap water, agricultural runoff, and rivers. Similar to this, triclosan (TCS), an antibiotic substance used in personal hygiene products, builds up in sediments and wastewater, endangering human health and causing long-term damage to aquatic life [14]. To detect TCS in wastewater, Atar and his colleagues [51] created a molecularly imprinted surface plasmon resonance (SPR)-based nanosensor [51]. Excellent performance from the nanosensor showed that it may be used to identify new pollutants in intricate water matrices [52]. Because caffeine is used so widely in beverages, medications, and energy drinks, it is also a common pharmaceutical and personal care product (PPCP) found in wastewater. Using silver nanoparticles (AgNPS) doped in molecularly imprinted polymers (MIP), [53] created a nanosensor to identify caffeine in wastewater. The sensor's efficacy in wastewater surveillance was demonstrated by its detection limit of 100 ng L^{-1}, which was much lower than the caffeine levels found in different rivers. Hexavalent chromium (Cr (VI)), another important pollutant, comes from the steel, leather tanning, and painting industries. In contrast to its trivalent sibling, Cr (III), an important trace element, Cr (VI) is very poisonous, mutagenic, and carcinogenic [54]. In order to provide a sensitive and affordable solution for chromium monitoring in wastewater, [55] created a carbon dot-based nanosensor that is sensitive to Cr (VI) with a detection limit of 2.3 nM at pH 6 in ambient water samples. These developments show how nanosensors can revolutionise wastewater treatment.

4.5 Detection of Organic Pollutants

Using various microbes to detect organic contaminants, including hydrocarbons, pesticides, and industrial chemicals. In the presence of dangerous substances, microorganisms may grow and survive in very hot and oxygenated environments as well as below-freezing temperatures [56]. A technique for cleaning up oil-contaminated places using microorganisms has been developed by TERI (The Energy Resources Institute), located in New Delhi. Toxic contaminants can be eliminated by the association of fungus, bacteria, and plants. More than fifty years ago, research on the breakdown of pesticides in agricultural soils began. Endosulfan degradation is linked to a variety of fungi, including *Aspergillus niger*, *A. terreus*, *Fusarium ventricosum*, and *Trichoderma harzianum*, as well as algae, including Anabaena sp. and Cholorococcum sp. Certain kinds of organic pollutants, including pesticides, dyes, and polycyclic aromatic hydrocarbons (PAH), can be detected using nanobiosensors. Organophosphate and carbamate pesticides can be detected and their degradation tracked by sensors with enzyme-linked systems [39]. They are often employed to track the deterioration of organic contaminants in water bodies, including sewage effluents, agricultural runoff, and industrial wastewater.

5 Conclusion

Nanosensors enhance bioremediation by providing accurate, real-time data on pollutant levels, bioavailability, and environmental conditions. Their integration with biological and technological systems ensures efficient, targeted, and sustainable cleanup of polluted environments. Overall, nanosensors provide a powerful, non-invasive, and highly sensitive approach for monitoring microbial processes, ensuring efficient assessment and optimization of bioremediation strategies in polluted environments.

Competing Interests The authors declare that they have no competing interest.

Declaration of Generative AI and AI-Assisted Technologies in the Writing Process During the preparation of this work the book chapter author(s) have not taken any assistance from AI, and the book chapter author(s) reviewed and edited the content as needed and take(s) full responsibility for the content of the publication. All authors read and approved the final version of the manuscript.

References

1. Thompson LA, Darwish WS (2019) Environmental chemical contaminants in food: review of a global problem. J Toxicol 2345283

2. Bhatt P, Bhandari G, Bhatt K, Maithani D, Mishra S, Gangola S, Bhatt R, Huang Y, Chen S (2021) Plasmid-mediated catabolism for the removal of xenobiotics from the environment. J Hazard Mater 420:126618
3. Bhatt P, Bhatt K, Sharma A, Zhang W, Mishra S, Chen S (2021) Biotechnological basis of microbial consortia for the removal of pesticides from the environment. Crit Rev Biotechnol 41(3):317–338
4. Mishra S, Pang S, Zhang W, Lin Z, Bhatt P, Chen S (2021) Insights into the microbial degradation and biochemical mechanisms of carbamates. Chemosphere 279:130500
5. Vázquez-Núñez E, Molina-Guerrero CE, Peña-Castro JM, Fernández- Luqueño F, de la Rosa-Álvarez M (2020) Use of nanotechnology for the bioremediation of contaminants: a review. Processes 8(7):826
6. Chaudhary P, Ahamad L, Chaudhary A, Kumar G, Chen WJ, Chen S (2023) Nanoparticle-mediated bioremediation as a powerful weapon in the removal of environmental pollutants. J Environ Chem Eng 11:109591. https://doi.org/10.1016/j.jece.2023.109591
7. Fernández-Rodríguez P, de Mendoza JH, López-Colón JL, López-Peinado AJ, Martín-Aranda RM (2019) Nanotechnology: concepts of nanotoxicity. In: Nanocatalysis. CRC Press, pp 3–21
8. Mansor M, Xu J (2020) Benefits at the nanoscale: a review of nanoparticle-enabled processes favouring microbial growth and functionality. Environ Microbiol 22(9):3633–3649. https://doi.org/10.1111/1462-2920.15174
9. Okonkwo CC, Edoziuno F, Orakwe LC (2020) Environmental nanoremediation in Nigeria: a review of its potentials. Alger J Eng Technol 3:043–057
10. Periyasamy V, Dharmalingam KKS, Peraman M, Manickam R, Vasantharaj K (2024) Recent strategies for natural bioremediation of emerging pollutants: development of a green and sustainable environment. Bioremed Emerg Contamin Water 2:1–19
11. Shang Y, Hasan MK, Ahammed GJ, Li M, Yin H, Zhou J (2019) Applications of nanotechnology in plant growth and crop protection: a review. Molecules 24(14):2558. https://doi.org/10.3390/molecules24142558
12. Nano-Bioremediation: Fundamentals and Applications (2021) Elsevier, Netherlands
13. Devarajan T, Arun KS, Jeyabalan S, Charles OA, Saher I, Abdel T (2021) Nanosensors for detection and evaluation of organic compounds in soil. Elsevier, pp 205–219. https://doi.org/10.1016/B978-0-12-822891-3.00010-4
14. Javaid M, Abid H, Singh R, Rab S, Rajiv S (2021) Exploring the potential of nanosensors: a brief overview. Sens Int 2:100130. https://doi.org/10.1016/j.sintl.2021.100130
15. GadelHak Y, Hafez SH, Mohamed HF, Abdel-Hady EE, Mahmoud R (2023) Nanomaterials-modified disposable electrodes and portable electrochemical systems for heavy metals detection in wastewater streams: a review. Microchem J 193:109043
16. Dei J, Bandyopadhyay R, Behera BK (2024) Nano-biosensor applications for water quality monitoring. In: Behera BK (eds) Current trends in fisheries biotechnology. Springer, Singapore. https://doi.org/10.1007/978-981-97-3165-7_5
17. Shroti K, Singh MK (2023) Nano-technological bioremediation: revolutionizing environmental. Int J Res Appl Sci Eng Technol 11(XI)
18. Huang P, Xiong Y, Ge Y, Wen Y, Zeng X, Zhang J, Wang P, Wang Z, Chen S (2023) Magnetic Fe_3O_4 nanoparticles decorated phosphorus-doped biochar-attapulgite/bismuth film electrode for smartphone-operated wireless portable sensing of ultra-trace multiple heavy metal ions. Microchim Acta 190(3):94
19. Butmee P, Mala J, Damphathik C, Kunpatee K, Tumcharern G, Kerr M, Mehmeti E, Raber G, Kalcher K, Samphao A (2021) A portable selective electrochemical sensor amplified with Fe3O4@ Au-cysteamine-thymine acetic acid as conductive mediator for determination of mercuric ion. Talanta 221:121669
20. Naseri M, Mohammadniaei M, Ghosh K, Sarkar S, Sankar R, Mukherjee S, Pal S, Ansari Dezfouli E, Halder A, Qiao J, Bhattacharyya N (2023) A robust electrochemical sensor based on butterfly-shaped silver nanostructure for concurrent quantification of heavy metals in water samples. Electroanalysis 35(2):e202200114

21. Bu L, Xie Q, Ming H (2020) Simultaneous sensitive analysis of Cd (II), Pb (II) and As (III) using a dual-channel anodic stripping voltammetry approach. New J Chem 44(15):5739–5745
22. Ghasemi Z, Mohammadi A (2020) Sensitive and selective colorimetric detection of Cu (II) in water samples by thiazolylazo pyrimidine-functionalized TiO_2 nanoparticles. Spectrochim Acta Part A Mol Biomol Spectrosc 239:118554
23. Aygun A, Sahin G, Tiri RNE, Tekeli Y, Sen F (2023) Colorimetric sensor based on biogenic nanomaterials for high sensitive detection of hydrogen peroxide and multi-metals. Chemosphere 339:139702
24. Li X, Zhao CX, Lin L (2022) Plasma-based instant synthesis of functionalized gold nanoparticles for colorimetric detection of leadions. Chem Eng Sci 260:117849
25. Wang J, Jiang C, Wang X et al (2016) Fabrication of an "ion-imprinting" dual-emission quantum dot nanohybrid for selective fluorescence turn-on and ratiometric detection of cadmium ions. Analyst 141:5886–5892
26. Jimenez-Falcao S, Villalonga A, Parra-Nieto J et al (2020) Dithioacetal-mechanized mesoporous nanosensor for Hg (II) determination. Microporous Mesoporous Mater 297:110054
27. Yang CH, Ding YL, Qian J (2018) Design of magnetic-fluorescent based nanosensor for highly sensitive determination and removal of HG^{2+}. Ceram Int 44(8):9746–9752
28. Kumar S, Sachdeva S, Chaudhary S, Chaudhary GR (2020) Assessing the potential application of bio-compatibly tuned nanosensor of Yb_2O_3 for selective detection of imazapyr in real samples. Colloids Surf A Physicochem Eng Asp 593:124612
29. Yılmaz E, Özgür E, Bereli N et al (2017) Plastic antibody based surface plasmon resonance nanosensors for selective atrazine detection. Mater Sci Eng C 73:603–610
30. Chang Y-C, Lin Y-S, Xiao G-T et al (2016) A highly selective and sensitive nanosensor for the detection of glyphosate. Talanta 161:94–98. https://doi.org/10.1016/j.talanta.2016.08.029
31. Liu H, Chen P, Liu Z et al (2020) Electrochemical luminescence sensor based on double suppression for highly sensitive detection of glyphosate. Sens Actuators B Chem 304:127364
32. Tian X, Liu L, Li Y et al (2018) Nonenzymatic electrochemical sensor based on CuO–TiO_2 for sensitive and selective detection of methyl parathion pesticide in ground water. Sens Actuators B Chem 256:135–142
33. Prabhakar N, Thakur H, Bharti A, Kaur N (2016) Chitosan-iron oxide nanocomposite based electrochemical aptasensor for determination of malathion. Anal Chim Acta 939:108–116
34. Pham TB, Bui H, Pham VH, Do TC (2020) Surface-enhanced Raman spectroscopy based on Silver nano-dendrites on microsphere end-shape optical fibre for pesticide residue detection. Optik (Stuttg) 219:165172. https://doi.org/10.1016/j.ijleo.2020.165172
35. Das R, Dhiman A, Kapil A, Bansal V, Sharma TK (2019) Aptamer-mediated colorimetric and electrochemical detection of Pseudomonas aeruginosa utilizing peroxidase-mimic activity of gold NanoZyme. Anal Bioanal Chem 411:1229–1238
36. Yan R, Shou Z, Chen J, Wu H, Zhao Y, Qiu L, Jiang P, Mou XZ, Wang J, Li YQ (2018) On–off–on gold nanocluster-based fluorescent probe for rapid Escherichia coli differentiation, detection and bactericide screening. ACS Sustain Chem Eng 6(4):4504–4509
37. Ilhan H, Guven B, Dogan U, Torul H, Evran S, Çetin D, Suludere Z, Saglam N, Boyaci IH, Tamer U (2019) The coupling of immunomagnetic enrichment of bacteria with paper-based platform. Talanta 201:245–252
38. Cui A, Hou Y, Zhang J, Mu X, Wang H, Sun Y, Xu H, Shan G (2023) Dual-mode sensing platform based on aptamer-tunable catalytic activity of mesoporous polydopamine/MnO2 nanozymes for detecting S. aureus. Sens Actuators B Chem 393:134218
39. Rabisa Z, Ayesha T, Sumaira Y, Syeda ZB, Faiza L, Yusra F, Kaneez F, AI Midrarullah ZB Sadia 2022 Application of nanosensors for pesticide detection Elsevier 259–302 https://doi.org/10.1016/B978-0-12-824554-5.00006-9
40. Sharma P, Pandey V, Sharma MMM et al (2021) A review on biosensors and nanosensors application in agroecosystems. Nanoscale Res Lett 16:136. https://doi.org/10.1186/s11671-021-03593-0
41. Ammar A, Zaheer A, Rizwana M, Khurram S, Zahid HS, Muhammad ZA, Hameed A, Muhammad B (2022) Nanobioremediation of insecticides and herbicides. Elsevier, pp 655–674. https://doi.org/10.1016/B978-0-12-823970-4.00023-3

42. Anwar A, Minhaz A, Khan NA et al (2018) Synthesis of gold nanoparticles stabilized by a pyrazinium thioacetate ligand: a new colorimetric nanosensor for detection of heavy metal Pd(II). Sens Actuators B Chem 257:875–881. https://doi.org/10.1016/j.snb.2017.11.040
43. Saleh SM, Alminderej FM, Ali R, Abdallah OI (2020) Optical sensor film for metribuzin pesticide detection. Spectrochim Acta Part A Mol Biomol Spectrosc 229:117971
44. Xie Y, Yu Y, Lu L et al (2018) CuO nanoparticles decorated 3D graphene nanocomposite as non-enzymatic electrochemical sensing platform for malathion detection. J Electroanal Chem 812:82–89
45. Loeffler FE, Edwards EA (2006) Harnessing microbial activities for environmental cleanup. Curr Opin Biotechnol 17:274–284
46. Rodríguez-Lorenzo L, de La Rica R, Álvarez-Puebla RA, Liz-Marzán LM, Stevens MM (2012) Plasmonic nanosensors with inverse sensitivity by means of enzyme-guided crystal growth. Nat Mater 11(7):604–607
47. Gui Q, Lawson T, Shan S, Yan L, Liu Y (2017) The application of whole cell-based biosensors for use in environmental analysis and in medical diagnostics. Sensors 17:1623
48. Dhir B (2013) Phytoremediation: role of aquatic plants in environmental clean-up. Springer, New Delhi
49. Jain K, Patel AS, Pardhi VP, Flora SJS (2021) Nanotechnology in wastewater management: a new paradigm towards wastewater treatment. Molecules 26(6):1797
50. Nasseri B, Soleimani N, Rabiee N, Kalbasi A, Karimi M, Hamblin MR (2018) Point-of-care microfluidic devices for pathogen detection. Biosens Bioelectron 117:112–128
51. Atar N, Eren T, Yola ML, Wang S (2015) A sensitive molecular imprinted surface plasmon resonance nanosensor for selective determination of trace triclosan in wastewater. Sens Actuators B 216:638–644
52. Kumar A, Bisht BS, Joshi VD, Dhewa T (2011) Review on bioremediation of polluted environment: a management tool. Int J Environ Sci 1(6):1079–1093
53. Hu R, Tang R, Xu J, Lu F (2018) Chemical nanosensors based on molecularly-imprinted polymers doped with silver nanoparticles for the rapid detection of caffeine in wastewater. Anal Chim Acta 1034:176–183
54. Westerhoff P, Alvarez P, Li Q, Gardea-Torresdey J, Zimmerman J (2016) Overcoming implementation barriers for nanotechnology in drinking water treatment. Environ Sci Nano 3(6):1241–1253
55. Zhang S, Jin L, Liu J, Wang Q, Jiao L (2020) A label-free yellow-emissive carbon dot-based nanosensor for sensitive and selective ratiometric detection of chromium (VI) in environmental water samples. Mater Chem Phys 248:122912
56. Zhuang X, Chen J, Shim H, Bai Z (2007) New advances in plant growth-promoting rhizobacteria for bioremediation. Environ Int 33(3):406–413
57. Dhahi TS, Yousif D, AK, Tayfour OE, Mubarakali A, Alqahtani AS, Tayfour AAE, Elobaid, ME, Adam T, Gopinath, SCB (2024) Advances in nano sensors for monitoring and optimal performance enhancement in photovoltaic cells. iScience. 27(4):109347. https://doi.org/10.1016/j.isci.2024.109347. PMID: 38550986; PMCID: PMC10972835

Nanosensors for Heavy Metal Pollution Detection in Agricultural System

Shefali Singh, Smita Rai, Hiba Khan, Mariyam Shakeel, Archana Vimal, Reena Vishvakarma, Poonam Sharma, and Swati Sharma

Abstract Heavy metal contamination of soil is a prevalent problem driven by the surface treatment, mining industry and negligent dumping of waste in landfills. Nanosensors with enhanced sensitivity, simplicity and selectivity accessibility of use are utilized extensively in environmental monitoring for heavy metal detection. A vital tool for monitoring environmental pollutants, the presence of various toxins and bio hazardous microbes in organic matter, biosensors are real-time sensors that can physically monitor soil health, soil moisture content and temperature as well as sense the microenvironment and nutrient status of soil. These nanosensors enable prior information, accurate, quick and fast that helps mitigate and predict crop losses in agro ecosystems. This chapter covers the recent development in the nanosensors field to detect the heavy metals in soil and explores the different types of biosensors that are used to identify and treat the heavy metal pollution of soil. Certain heavy metal elements are required in trace amounts for physiochemical processes in plants but can cause pollution at higher concentrations. Pollution of metal elements is the main reason to increased accumulation in plants through absorption by roots. Nanosensors including electrochemical, optical and biosensors along can be integrated with latest and advanced technologies as AI and machine learning for effective monitoring. Around the world, agriculture is the primary and main source of food, revenue and employment for mankind. Utilizing biosensors or nanosensors has been shown to accelerate agricultural productivity and make way for precision agriculture.

Keywords Nanosensors · Nano materials · Agriculture system · Heavy metal

S. Singh · S. Rai · H. Khan · M. Shakeel · S. Sharma (✉)
Department of Biosciences, Integral University, Lucknow, U.P., India
e-mail: sw_sh@rediffmail.com

A. Vimal · R. Vishvakarma · P. Sharma
Department of Bioengineering, Integral University, Lucknow, U.P., India

A. Husen (ed.), *Nanobiosensors for Agricultural and Other Related Sectors*, Smart Nanomaterials Technology, https://doi.org/10.1007/978-981-96-6499-3_8

1 Introduction

Heavy metal pollution is now-a-days has reached threatening levels in all the realms of earth as air, water, soil, and concerned ecosystems. Heavy metals used in various industrial or agricultural activities become a major source of pollution also industrial and municipal wastes, Landfill leachates, natural phenomena including weathering, rock abrasion and Landfill leachates add to it. Moreover, they can leak into the environment by other sources cars, coal mines, thermal power plants, sewage, agricultural fertilizers, pesticides, and inappropriate trash disposal. Heavy metals can be highly poisonous/toxic even minute concentrations. Their exposure can damage a number of organs, including liver, stomach, kidneys, skin, lungs and the reproductive as well as neurological system [1]. Ions of heavy metals can even cause cancer. Furthermore, they can bio accumulate and reach hazardous levels in biological systems and cause environmental problem worldwide. These factors offer a substantial threat to both human and environmental health. As a result, the most crucial strategy is appropriately and timely control of the pollution sources beforehand, by getting rid of the heavy metals that have contaminated the soil and water [2]. For this we need to constantly detect their levels specifically in areas where they can possibly bio accumulate.

Traditionally heavy metal detection was a costly affair due to complex and expensive equipment as well as it was time-consuming due to requirement of arduous sample preparation like for atomic absorption spectroscopy [3]. This field has gained impact within last 20 years but from 1990's the potential of using nanotechnology in detecting various environmental pollutants, including heavy metals was being explored. These ideas laid the groundwork for their future applications in soil (Fig. 1). After 2000 AD, with the advancement in nanomaterials significant studies were conducted in this field. Nanosensors which exhibited unique optical and electronic properties were leveraged for sensing applications. Early prototypes of nanosensors were specifically designed to enhance sensitivity and selectivity. The time of 2010 to 2015 AD was marked by the studies in optimization and miniaturization of nanosensors improving their performance, sensitivity, selectivity, and stability for in-situ soil testing. This lead to the development of portable and user-friendly devices.

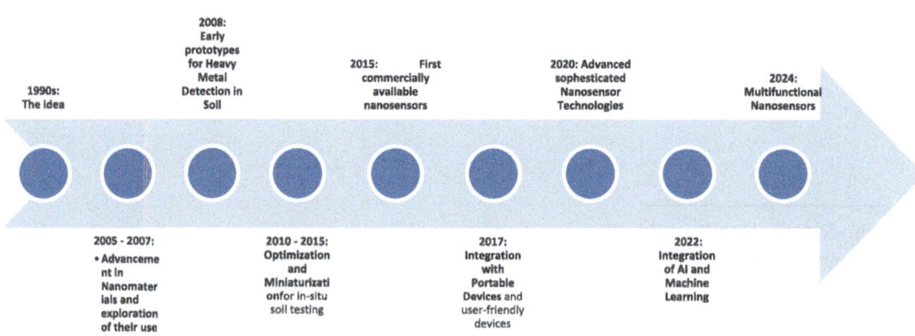

Fig. 1 Development of nanosensors for heavy metal detection in soil

Recent advances in electrochemical and colorimetric techniques made their detection sensitive and affordable such as microelectrode arrays and biosensors. With commercialization of nanosensors, they could be used in environmental monitoring and remediation projects. Their integration with portable devices lead to the development of user-friendly devices which allowed for real-time monitoring of heavy metal contamination. Techniques such as surface-enhanced Raman spectroscopy (SERS) were employed to enhance detection capabilities. At the nanoscale materials acquire unique characteristics making them sensitive and focused for detection of specific heavy metal ions [4]. Advances in nanotechnology led to the development of highly specific, accurate and more sophisticated nanosensors. Electrochemical and colorimetric nano sensors are comparatively affordable for on-demand, in situ field detection of heavy metal ions [3]. Furthermore, nanocomposite formation and covalent functionalization permitted multifold miniaturization of nanosensors which lead to concept of lab-on-chip (LOC) [5]. With advent of artificial intelligence (AI) and machine learning (ML) algorithms and its integration with nanosensor data, predictive insights and models could be generated for forecasting.

Multifunctional Nanosensors which can detect heavymetals along with other soil parameters, such as pH and nutrient levels are also currently being developed. Even on-site screening of heavy metal contamination in food and the environment can now be exceptional sensitivity, allowing the detection of harmful metals at ultra-low levels by carbon nanomaterial-based electrochemical sensors. According to the fusion of nanotechnology with biosensors has greatly improved the sensitivity and selectivity of heavy metal detection in environmental samples, yielding fast and accurate findings. This tremendously improved our monitoring of the environment and public health [6] facilitating a quick as well as selective investigation of a variety of samples including agricultural and consumable items [6].

2 Nanosensors and Their Types

Nanosensors are advanced devices that operate at the nanoscale (1–100 nm) and are crucial in various fields, including environmental monitoring, healthcare and agriculture. Their unique properties are used, which enhance sensitivity, response time and selectivity [7]. They can be mechanical Nanosensors based on microelectromechanical or nanoelectromechanical systems, Chemical Nanosensors, operating by undergoing a chemical reaction or interaction with the analyte, leading to a change in signal or Biological nanosensors utilizing receptors such as DNA./antibodies/enzymes that interact specifically with the target analyte. They can also be optical or electrical based, recently the technologies have also been developed for making Carbon nanotubes, quantum dots, peptide or DNA-based fluorescent nanosensors, Plasmon coupling, Plasmonic enhancing–/quenching–MRI or Photoacoustic-based nanosensors or Multimodal nanosensors. Nanomaterials have some characteristics like, detection sensitivity high stability, high carrier capacity, electrical conductivity and high thermal ability to change color. Many sensing systems can detect an extensive range of clinically significant molecules in a variety of sensing media, from buffers to more complex environment. These molecules include cancer antigens, pharmaceuticals, viruses, contaminants, toxins, whole cells, narcotic drugs and nucleic acids [8] and henceforth can be optimized for use in agricultural systems also [9]. In food and agriculture sector Nano sensors have direct application in detection of soil and food contaminants such as preservatives, antibiotics, heavy metal ions, toxins, microbial load, and pathogens along with the rapid monitoring of environmental factors as temperature or humidity.

2.1 Materials Used for Nanosensors

Metal/metal oxide nanoparticles, nanoclusters, graphene, carbon nanotubes (CNTs), quantum dots (QDs), nanocomposites, nano rods and nanowires are some of the nanomaterials that are involved in the production of Nano biosensors which can be used for detection of heavy metals as given in Table 1.

2.1.1 Gold and Silver Nanoparticles Based Nanosensors

There are some studies as reviewed by Potadar et al. [18] which focus on the recent advancements in the field of nanosensors for detection of some metals like iron, arsenic copper, mercury, platinum, lead, rhodium and palladium. Due to their unique size and shape dependent optical properties, gold nanoparticles-based colorimetric sensors (Ag/Au-NPs-CSns) allow potential prospects for the development of efficient sensors [19]. An efficient and sensitive aptamer base colorimetric test for Hg2 + has been demonstrated by Qi et al. [20] by detection of visual and spectral changes in cationic gold nanoparticles (AuNPs). For colorimetric detection of heavy metal ions (Hg2 + and Pb2 +) silver nanoparticles are exposed to ambient sunlight irradiation. Unique features of silver (Ag) nanoparticles which makes them a good choice for use in the field of colorimetric sensors is that they show distinct optical properties and high excitation coefficient in the visible spectrum [21].

2.1.2 Quantum Dots (QDs) Based Nanosensors

Due to superior optical properties, interface effects and quantum size effects, QD nanomaterials have created a lot of interest for fluorescence chemical sensors. QD based sensors have great potential uses in agriculture, food safety and environmental monitoring (Shiet al. 2023). QDs are classified as semiconductor nanocrystals with radii that are close to or less than Bohr radius of the excitation. They have adjustable and distinct photophysical properties as their broad excitation spectrum, tunable emission wavelengths, negligible photobleaching and narrow emission spectrum [22]. Fluorescence QD nanoparticle probes for detecting heavy-metal ions have the

Table 1 Nanomaterials used in heavy metal detecting nanosensors and their mechanism

Type of nanosensor	Mechanism	Applications	Advantages	Property	Key reference
Gold nanoparticles (AuNPs)	Optical and electrochemical properties	Bioaccumulation of Pb, Hg, As in plants and animals	High sensitivity and specificity, ease of functionalization	Colorimetric sensors that change color upon binding with heavy metals	[10]
Quantum dots (QDs)	Fluorescence quenching and enhancement	Tracking bioaccumulation of Cd, Pb, Hg in biological tissues	High fluorescence quantum yield, tunable emission spectra	Fluorescent sensors that exhibit a change in fluorescence intensity	[11]
Carbon nanotubes (CNTs)	Changes in electrical conductivity	Detection of Zn, Cu, Pb bioaccumulation in soil and plants	High surface area, excellent electrical conductivity	Electrochemical sensors that detect changes in current due to metal ion adsorption on CNTs	[12]
Graphene-based	Changes in electrical resistance or electrochemical properties	Monitoring bioaccumulation of Pb, Hg, As in aquatic systems	High conductivity, large surface area, flexibility	Field-effect transistors (FETs) with graphene channels functionalized to detect specific heavy metals	[13]
Magnetic nanoparticles (MNPs)	Magnetic separation with optical/ electrochemical detection	Bioaccumulation of Cd, Pb, As in plants and animals	Easy separation from samples, high surface area	Sensors that use MNPs for pre-concentration of metals followed by electrochemical detection	[14]
Silica nanoparticles	Optical properties and surface modifications	Detection of Hg, Pb, Cd bioaccumulation in biological systems	High stability, ease of surface functionalization	Fluorescent silica nanoparticles that change fluorescence in the presence of heavy metals	[15]

(continued)

Table 1 (continued)

Type of nanosensor	Mechanism	Applications	Advantages	Property	Key reference
Zinc oxide nanoparticles (ZnO NPs)	Photocatalytic and electrochemical properties	Assessment of Pb, Cd bioaccumulation in plants and soil organisms	High sensitivity, stability under various conditions	Electrochemical sensors utilizing ZnO NPs for detecting trace levels of heavy metals	[16]
Surface-enhanced Raman spectroscopy (SERS)	Enhancement of Raman scattering signals	Detection of Hg, Pb, As bioaccumulation in various organisms	High sensitivity, ability to detect very low concentration	SERS-active substrates with nanostructures that enhance the Raman signals of adsorbed heavy metal ions	[17]

ability to solve issues in the field where huge equipment is either slow or inaccurate due to their low cost, high sensitivity, simple equipment and high specificity. Because of their strong and size tunable luminescence properties, carbon based fluorescent nanomaterials like Graphene quantum dots (GQDs) and carbon quantum dots can be considered as a great option [23].

2.1.3 Magnetic Nanoparticles (MNPs) Based Nanosensors

MNPs can be used in extensive agricultural activities and are synthesized at a low cost. In recent decades, there has been lot of interest in using nanoparticles to treat soil for the removal of toxic metal ions. Due to the extraordinary magnetic capabilities, Magnetic nanoparticles (MNPs) have shown great potential in the heavy metals detection [24]. MNPs have their magnetic core which focus on smart treatment delivery systems in plants and enables them to use the nanoparticles to distribute the nanoparticles to desired location [25]. MNMs have high electrical conductivity, low toxicity, simple production and facile functionalization for managing dangerous environmental contaminants [26]. Inorganic materials like metal or non-metal elementary substance, metal oxide or metal sulphide and silica are used to surface functionalization of MNPs [27]. Although these impacts also depend on the type and concentration of metal oxide, It was discovered that metal oxide have less impact on macroscopic properties of soil and less harmful to soil bacterial communities [28]. Instead of using real fields that contain heavy metals, the majority of the studies were conducted in hydroponic or laboratory conditions scale. Iron-based MNPs have been shown in multiple studies to translocation of heavy metals and decrease the absorption to plant aerial parts. Additionally, the MNPs can treat the soil that has been contaminated by metals [29].

2.2 Nanobiosensors

Biosensors generally involve enzymes or cells based receptor systems integrated with optical and electrical nano based or wireless systems (Fig. 2). Biosensors based on silver (Ag), quantum dots (QDs), carbon nanotubes (CNTs), graphenes, gold (Au) and silica nanocomposites modified nanostructures with large surface to volume ratio has been synthesized in order to improve electrochemical parameters.

Adsorption, catalysis, and fluorescence quenching mechanisms are the sensing methods used by biosensors in agriculture. These mechanisms, which involve molecular identification and transduction, are essential for improving agriculture activities [30]. The functionalization of attapulgite with fluorescent polymers facilitates the efficient adsorption and detection of Cu(II) ions for ion detection. This enables for a maximum adsorption capacity of 41.67 mg/g, which is essential for measuring the health of the soil and the levels of contamination [31]. Biosensors can optimize fertilization techniques, measure essential nutrients in soil and improve crop yields

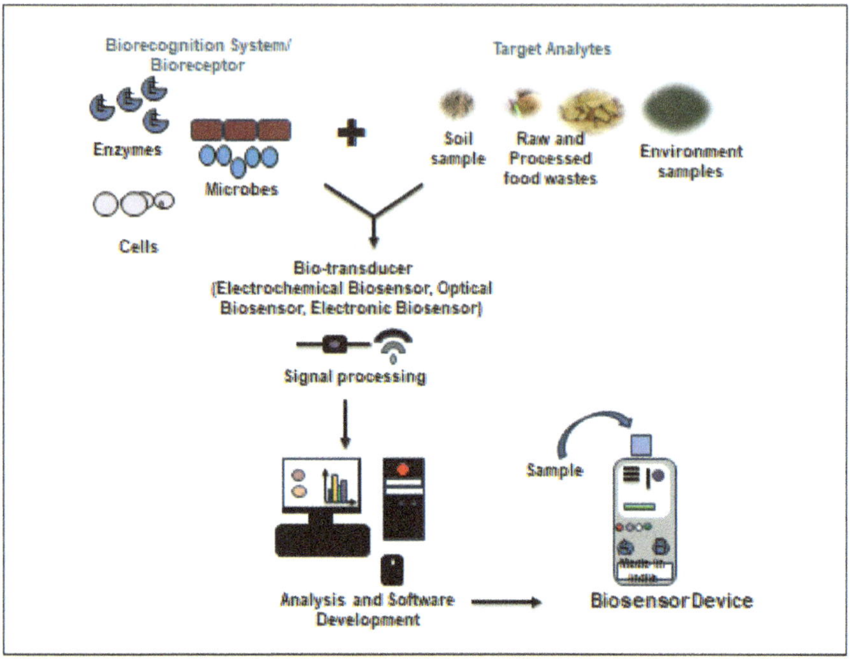

Fig. 2 Pictorial representation of biosensor device development

[32]. Table 2 shows some of the microorganism-based biosensors used for heavy metal detection in soil.

Repurposed biological components are used by genetically encoded plant-based biosensors (GEPBs) to detect and respond to stimuli, allowing in vivo monitoring for plant development, stress management, and environmental monitoring in agriculture. Using techniques like aptamers, electrochemical approaches, fluorescence spectroscopy, nanoparticles, and antibodies, biosensors in agriculture can quickly and accurately detect pesticides with high sensitivity and low detection limits [40].

Plants are composed of cellulose, hemicellulose and lignin [41], they require some metal elements in trace amount to carry out their physicochemical process. These heavy metal elements are present in minimum amount in soil water and air but can result in the accumulation as they can be absorbed through the roots of the plants which can be toxic to the human life. By detecting metabolite changes, field-effect transistor-based biosensors in agriculture can identify plant stress factors. This improves agricultural monitoring capacities by enabling early diagnosis of health issues before they appear as symptoms [42]. The design and construction of biosensors with high specificity and sensitivity is the aim of plant-based biosensor engineering.

Table 2 Microorganism-based biosensors for heavy metal detection

Microorganism-based biosensor	Microorganism used	Target heavy metals	Application	Manufacturer	Key reference
E. coli biosensor	*Escherichia coli*	Pb, Hg, Cd	Detection of heavy metals in soil and water	EnviroLogix	[33]
Bacillus subtilis biosensor	*Bacillus subtilis*	Cd, Pb	Monitoring heavy metal contamination in soil	Agilent technologies	[34]
Pseudomonas putida biosensor	*Pseudomonas putida*	Cr, Cd, Pb	Detection of heavy metals and pollutants in soil	Microbial BioSystems	[35]
Saccharomyces cerevisiae biosensor	*Saccharomyces cerevisiae*	Hg, Pb	Detection of heavy metal ions in water	Bioventures	[36]
Agrobacterium tumefaciens biosensor	*Agrobacterium tumefaciens*	Pb, Cd	Early detection of heavy metals in contaminated sites	Genomatica	[37]
Rhodococcus sp. biosensor	*Rhodococcus sp.*	Pb, Zn	Monitoring heavy metal pollution in soil	Rhodia	[38]
Desulfovibrio desulfuricans biosensor	*Desulfovibrio desulfuricans*	Hg, Cd	Detection of heavy metals in anaerobic conditions	Biotechnology Co	[39]

3 Applications of Nanosensors in Heavy Metal Detection

The sensitivity and accuracy of nanosensors has increased tremendously as nanoscale dimension enables the sensors to detect even in very low amounts of heavy metals in agricultural soils and help maintain human health, food safety as well as the sustainability of the environment. Heavy metal sensing nanosensors have a wide range of applications in agriculture (Fig. 3).

3.1 Soil Contamination Mapping

Soil maps can be made on farm, regional, state or national level based on soil properties like colour, texture, organic matter, nutrient and moisture content or pH using sensors thus guiding in the management of irrigation, fertilization and inputs based

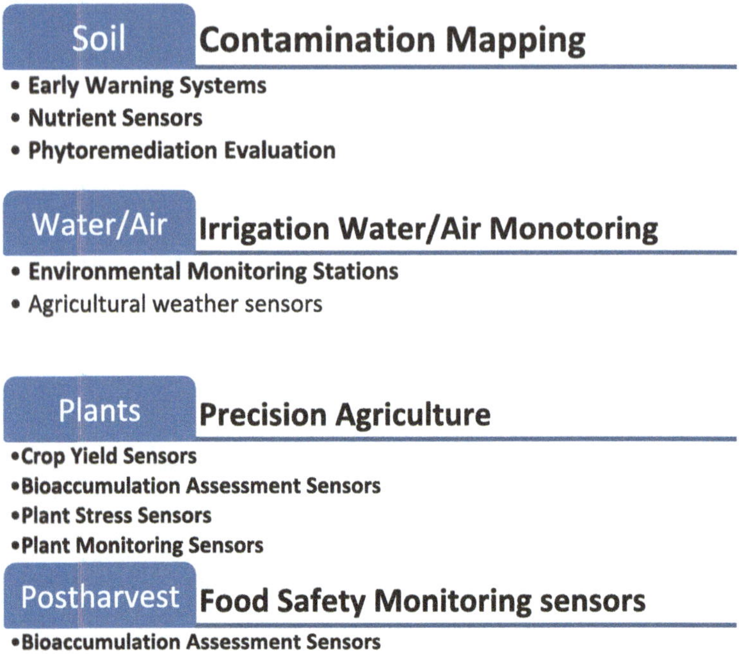

Fig. 3 Applications of nanosensors in heavy metal detection

on variation in soil properties. For map soil texture and organic matter content Penetrating Radar (GPR) and Electromagnetic induction (EMI) are used [43] and nanosensors can be incorporated to make heavemetal related maps. In geographical area, they can be helpful to visualize as well as identify the distribution and intensity of pollutants, particularly heavy metals and visualize in an area. This is called as soil contamination mapping of that area. As heavy metals, such as lead (Pb), mercury (Hg), arsenic (As), cadmium (Cd), and chromium (Cr) are the most insistent and toxic contaminants in soil. Contamination mapping can be a way to provide policy-makers scientific data or information, to support the development of appropriate soil protection.

3.1.1 Early Warning Systems

Nanosensors which offer sensitivity and efficiency in detection of heavy metals, represent an innovative method for early warning systems in soil contamination. These systems provide sensitive, rapid and real-time monitoring of pollutants to solve the heavy metal contamination issue in soil also environmental and health risks. The continual usage of heavy metals in industrial processes and poor waste management are contaminating agricultural soil. Consequently, these contaminations cause modify biological macromolecules and changes in biochemical cycles. Based on the data researchers developed different techniques to detect heavy metals Indian Council of Medical Research (ICMR) recommend toxicity level [44]. One of the most effective analytical method for monitoring heavy metals in different environmental samples, including water, soil, and air is electrochemical detection [45]. Many sensor based techniques have been developed for detection of heavy or hazardous metals. Chemical and optical nano-sensors are the two newest technologies available for detecting toxic heavy metals. There are many nanoparticle based nanosensors which have the Optical and electrochemical properties. Nanoparticles, specifically metal nanoparticles, present several advantages in the electrochemical sensing field. These can be integrated with AI and used for making early warning systems which can predict their future risk levels and alert the system.

3.1.2 Nutrient Sensors

The dynamic balance between physical, biological and chemical characteristics of soil is called as soil health and heavy metal pollution can cause severe imbalance as it directly effects the soil nutrient status. Thus on assessing availability of other nutrients we can indirectly evaluate the effect of heavy metal pollution. The utilization of nano biosensors for monitoring of enzymatic activity and urea detection in soil was developed in laboratory and carried out under carefully monitored conditions. By utilizing the best optimum resource and ultimately enhancing the crop productivity and application of nanobiosensor can support sustainable agriculture [46]. Furthermore, it can help in measuring the various factors such as chlorophyll content, soil pH, protein content, total nutrient uptake (macronutrients and micronutrients) and photosynthetic content by the plants [32]. For evaluation of nutrient level and composition of soil, biosensors are used to assist farmers. To identify the gradient in the soil with regard to particular nutrients like phosphorus, nitrogen and potassium in the soil. Because of this, farmers can use variable rate technology (VRT) to apply fertilizers more effectively focusing deficiency and avoiding toxicity [43]. Plant wearable biosensors can detect the nutrient levels (like nitrogen or phosphorus) in plant tissues.

3.1.3 Phytoremediation Evaluation

Heavy metals and other contaminates can be stabilized, detoxified and eliminated from contaminated soil, air and water using plants in a process known as phytoremediation which is both environmentally and economical beneficial. Plant species, soil type and properties and concentration of heavy metals are the factors that affect phytoremediation efficiency to treat heavy metal contamination. Numerous efforts have been made to identify economical remediation alternatives for heavy metal contaminated soil [47]. Phytoextraction and phytostabilization are the main mechanisms for phytoremediation underground and aboveground parts of plants to stabilize and remove heavy metals in soil [48]. According to [49], the metal accumulation in shoot, leave and root part of the *N. mucronata* plant varied but the concentrations were more than natural soils. According to this study, *N. mucronata* is useful and effective accumulator plant for phytoremediation of soil contaminated by heavy metals. Generally, localized phytoremediation strategy components are detecting the natural vegetation growing on contaminated soil and metal tolerant plants selection that may have potential remediation value [50]. The remediation potential of plant species growing in a saline alkali soil contaminated with heavy metals was identify by Mousavi and Moudi [51]. They found seven plants (*A. santolina, L. arborescens, P. gnaphalodes, P. harmala, Z. eurypteru, A. javanica and P. olivieri*) have good heavy metal stabilization effects while *A. javanica* and *Z. eurypterum* also have relatively high heavy metal extraction abilities and good heavy metal respectively [52].

3.2 *Bioaccumulation Assessment Sensors:*

To be able to identify and analyze heavy metal contamination in soil and to better understand how toxic metals accumulate in ecosystems and crop systems, bioaccumulation assessment sensors are vital tools. These sensors are developed to measure heavy metals such as lead, mercury, cadmium, and arsenic often at micro or nano molar levels [52]. Technologies like optical sensors, biosensors and electrochemical sensors are frequently used. Plant cell based biosensors can detect the heavy metal concentration in plant tissues [31].

3.3 *Precision Agriculture*

Precision agriculture involves integration of biosensors with remote sensing technologies for detection and management of plant need or area specific measures as addition of moisture or nutrients or pesticides. This helps in optimizing the resource management. Unmanned aerial vehicles, harvester-mounted systems integrated with plant based sensors are examples of crop yield sensors. Better crop yield prediction and management, are made by these technologies which facilitate spatial yield mapping easier. This is particularly for horticultural crops, which are still in the early stages of yield assessment. Crop yield sensors provide real time data that optimizes resource management and enhances decision-making in modern agriculture. These sensors can be categorized into proximal and remote sensing technologies. Each has unique advantages for monitoring and yield prediction. Proximal and space borne sensing system provide more accurate yield prediction for many crops. Numerous studies have demonstrated the remote sensing techniques can be integrated with nanosensors and used to estimate heavy metal contamination in crops under controlled laboratory conditions and in field-scale trials, allowing it a potentially efficient and cost-effective to traditional destructive sampling methods for identify heavy metal contamination in soil [53].

3.3.1 Plant Stress Sensors

Plant stress sensors are advance devices made to detect and monitor stress factors that impact plant health such as nutrient deficiencies, drought, environmental contaminants such as heavy metals and disease. These sensors assess physiological responses like photosynthetic activity, temperature variations and stomata conductance changes using a various technologies such as fluorescence imaging, chlorophyll sensors, and thermal imaging. Biosensors that can identify molecular markers such as stress-induced proteins or hormones like abscisic acid which plants release in reaction to adverse conditions, are incorporated into advanced system. While soil-embedded sensors offer real-time data on nutrient levels, soil moisture and toxic metal concentrations and plant stress patterns. By using these sensors, researchers and farmers can detect stress early, enabling timely interventions to prevent yield loss. Agricultural methods can be become more and better and sustainable to agricultural practices by combining plant stress sensors with data analytics platforms and predictive models.

4 Challenges and Future Perspectives

Though this technology can revolutionize the way we deal with heavy metal related issues in agriculture it has its own challenges. The foremost being the toxicity related to nanoparticles used in the sensors which can directly impact environmental and human health. The stability as well as resilience of the sensors is crucial for withstanding the environmental effects over a long period of time. With the advent of Nano biosensors the technology can be further fine-tuned using enzyme, antibodies, DNA and whole cell-based detectors with specific analytical detection of heavy metals. More studies and experiments are needed for making the sensors portable as well as scalable. The process of data collection and interpretation are also needed to be standardized for large scale use. Precision agriculture further requires integration of these sensors to automated systems and signaling devises connected offline or online for analysis. Moreover, the cost of these sensors also needs to be rationalized for pervasive use.

Moreover these nanosensors can be used to detect heavy metal concentration in the irrigation system as well as in air near agricultural fields. They can also detect heavy metal levels in storage or postharvest so that they can be monitored efficiently.

5 Conclusion

Nano sensors have created a new way of precision agriculture. Nano biosenseor can be further used to fine-tune analytical detection of heavy metals using biomolecules as enzyme/antibodies/DNA or whole cell-based detectors. Further studies are needed for optimizing the cost and making the sensors portable and scalable. A for using in large scale integration of technologies and IOT as well as AI/ML mediated data collection and interpretation will be required. Thus integration of these sensors to offline or online linked automated systems or signaling devises. But all of this will increase the cost so attempts should be made to decrease the cost and make the system scalable. Moreover, the sensors should be able to withstand the constant and extreme environmental conditions on a long run. Thus, the development of nanosensors can be a key element for online and offline analysis of heavy metal contamination in soil as well as plants.

Competing Interests The authors declare that they have no competing interest.

Declaration of Generative AI and AI-Assisted Technologies in the Writing Process During the preparation of this work the book chapter author(s) have not taken any assistance from AI, and the book chapter author(s) reviewed and edited the content as needed and take(s) full responsibility for the content of the publication. All authors read and approved the final version of the manuscript.

References

1. Aziz KH, Mustafa FS, Omer KM, Hama S, Hamarawf RF, Rahman KO (2023) Heavy metal pollution in the aquatic environment: efficient and low-cost removal approaches to eliminate their toxicity: a review. RSC Adv 13(26):17595–17610. https://doi.org/10.1039/D3RA00723E
2. Sheydaei M (2024) Investigation of heavy metals pollution and their removal methods: a review. Geomicrobiol J 41(3):213–230. https://doi.org/10.1080/01490451.2024.2318227
3. Fakayode SO, Walgama C, Fernand Narcisse VE, Grant C (2023) Electrochemical and colorimetric nanosensors for detection of heavy metal ions: a review. Sensors 23(22):9080
4. Satpute N, Shrivas K, Dewangan K (2023) Smart nanosensors for pesticides and heavy metals detection. In: Lim K-T, Abd-Elsalam KA (eds) Nanorobotics and nanodiagnostics in integrative biology and biomedicine. Springer International Publishing, pp 433–452. https://doi.org/10.1007/978-3-031-16084-4_18
5. Waheed A, Mansha M, Ullah N (2018) Nanomaterials-based electrochemical detection of heavy metals in water: current status, challenges and future direction. TrAC, Trends Anal Chem 105:37–51
6. Mayegowda SB, Chikkud V, Barua S, Manjula NG (2024) Heavy metal detection by nanotechnology-based sensors. In: Nanotechnology-based sensors for detection of environmental pollution. Elsevier, pp 237–263
7. Adam T, Gopinath SC (2022) Nanosensors: recent perspectives on attainments and future promise of downstream applications. Process Biochem 117:153–173
8. Abdel-Karim R (2024) Nanotechnology-enabled biosensors: a review of fundamentals, materials, applications, challenges, and future scope. Biomed Mater Dev 2(2):759–777
9. Sharma S, Prasad A, Prasad M (2023) Ubiquitination from the perspective of plant pathogens. J Exp Bot 74(15):4367–4376. https://doi.org/10.1093/jxb/erad191
10. Liu J, Lu Y (2004) Colorimetric biosensors based on DNAzyme-assembled gold nanoparticles. J Fluoresc 14:343–354
11. Medintz IL, Uyeda HT, Goldman ER, Mattoussi H (2005) Quantum dot bioconjugates for imaging, labelling and sensing. Nat Mater 4(6):435–446
12. Wang J (2005) Carbon-nanotube based electrochemical biosensors: a review. Electroanalysis 17(1):7–14
13. Novoselov KS, Colombo L, Gellert PR, Schwab MG, Kim K (2012) A roadmap for graphene. Nature 490(7419):192–200

14. Pastucha M, Farka Z, Lacina K, Mikušová Z, Skládal P (2019) Magnetic nanoparticles for smart electrochemical immunoassays: a review on recent developments. Microchim Acta 186(5):312
15. Bolouki A, Rashidi L, Vasheghani-Farahani E, Piravi-Vanak Z (2015) Study of mesoporous silica nanoparticles as nanocarriers for sustained release of curcumin. Int. J. Nanosci. Nanotechnol. 11(3):139–146
16. Djurišić AB, Chen X, Leung YH, Ng AMC (2012) ZnO nanostructures: growth, properties and applications. J Mater Chem 22(14):6526–6535
17. Shanmukh S, Jones L, Driskell J, Zhao Y, Dluhy R, Tripp RA (2006) Rapid and sensitive detection of respiratory virus molecular signatures using a silver nanorod array SERS substrate. Nano Lett 6(11):2630–2636
18. Potadar RS, Moholkar DN, Pawar KD (2022) 14—Nanosensors for the detection of heavy trace metals in soil. In: Denizli A, Nguyen TA, Rajendran S, Yasin G, Nadda AK (eds) Nanosensors for smart agriculture. Elsevier, pp 329–353. https://doi.org/10.1016/B978-0-12-824554-5.00001-X
19. Ali S, Ahmad M, Hamayun M, Alrefaei A (2023) Rhizobacteria containing ACC deaminase enhance plant growth and tolerance under abiotic stress conditions. J Adv Nut Sci Tech 3:72–77. https://doi.org/10.15228/ANST.2023.v03.i03-4.p08
20. Qi Y, Ma J, Chen X, Xiu FR, Chen Y, Lu Y (2020) Practical aptamer-based assay of heavy metal mercury ion in contaminated environmental samples: convenience and sensitivity. Anal Bioanal Chem 412:439–448
21. Tamilselvan S, Soniya RM, Vasantharaja R, Kannan M, Supriya S, Batvari BP, Ramesh T, Govindaraju K (2022) Silver nanoparticles based spectroscopic sensing of eight metal ions in aqueous solutions. Environ Res 212:113585. https://doi.org/10.1016/j.envres.2022.113585
22. Zou L, Gu Z, Sun M (2015) Review of the application of quantum dots in the heavy-metal detection. Toxicol Environ Chem 97(3–4):477–490. https://doi.org/10.1080/02772248.2015.1050201
23. Chini MK, Kumar V, Javed A, Satapathi S (2019) Graphene quantum dots and carbon nano dots for the FRET based detection of heavy metal ions. Nano-Struct Nano-Objects 19:100347. https://doi.org/10.1016/j.nanoso.2019.100347
24. Emeji IC, Okoye-Chine CG, Garcia-Rodriguez O, Igberase E, Osifo PO (2023) Magnetic nanomaterials for heavy metals detection. In: Aigbe UO, Ukhurebor KE, Onyancha RB (eds) Magnetic nanomaterials: synthesis, characterization and applications. Springer International Publishing, pp 87–96. https://doi.org/10.1007/978-3-031-36088-6_4
25. Spanos A, Athanasiou K, Ioannou A, Fotopoulos V, Krasia-Christoforou T (2021) Functionalized magnetic nanomaterials in agricultural applications. Nanomaterials 11(11):Article 11. https://doi.org/10.3390/nano11113106
26. Nouri M (2024) 9—Magnetic nanotechnology-based sensors for environmental contaminants' detection. In: Policarpo Tonelli FM, Roy A, Ozturk M, Murthy HCA (eds) Nanotechnology-based sensors for detection of environmental pollution. Elsevier, pp 155–174. https://doi.org/10.1016/B978-0-443-14118-8.00009-7
27. Zhu N, Ji H, Yu P, Niu J, Farooq MU, Akram MW, Udego IO, Li H, Niu X (2018) Surface modification of magnetic iron oxide nanoparticles. Nanomaterials 8(10):810
28. Ben-Moshe T, Frenk S, Dror I, Minz D, Berkowitz B (2013) Effects of metal oxide nanoparticles on soil properties. Chemosphere 90(2):640–646
29. Le Wee J, Law MC, Chan YS, Choy SY, Tiong ANT (2022) The potential of Fe-based magnetic nanomaterials for the agriculture sector. ChemistrySelect 7(17):e202104603. https://doi.org/10.1002/slct.202104603
30. Kundu M, Krishnan P, Kotnala RK, Sumana G (2019) Recent developments in biosensors to combat agricultural challenges and their future prospects. Trends Food Sci Technol 88:157–178
31. Wang X, Cheng M, Yang Q, Wei H, Xia A, Wang L, Ben Y, Zhou Q, Yang Z, Huang X (2019) A living plant cell-based biosensor for real-time monitoring invisible damage of plant cells under heavy metal stress. Sci Total Environ 697:134097
32. Bharti A, Jain U, Chauhan N (2024) From lab to field: nano-biosensors for real-time plant nutrient tracking. Plant Nano Biol 100079

33. Souiri M, Gammoudi I, Ouada HB, Mora L, Jouenne T, Jaffrezic-Renault N, Dejous C, Othmane A, Duncan AC (2009) Escherichia coli-functionalized magnetic nanobeads as an ultrasensitive biosensor for heavy metals. Proc Chem 1(1):1027–1030
34. Verma N, Singh M (2006) A Bacillus sphaericus based biosensor for monitoring nickel ions in industrial effluents and foods. J Anal Methods Chem 2006(1):083427
35. Raghunandan ME, Sharma JS, Pradhan B (2015) A review on the effect of rubber membrane in triaxial tests. Arab J Geosci 8(5):3195–3206
36. Gayda G, Gonchar M, Sibirny A (2017) Yeast-based systems for environmental control. In: Sibirny AA (ed) Biotechnology of yeasts and filamentous fungi. Springer International Publishing, pp 373–390
37. Bai S, Han X, Feng D (2023) Shoot-root signal circuit: phytoremediation of heavy metal contaminated soil. Front Plant Sci 14:1139744
38. Zhao Z-M, Liu Z-H, Zhang T, Meng R, Gong Z, Li Y, Hu J, Ragauskas AJ, Li B-Z, Yuan Y-J (2024) Unleashing the capacity of Rhodococcus for converting lignin into lipids. Biotechnol Adv 70:108274
39. Rensing C, Maier RM (2003) Issues underlying use of biosensors to measure metal bioavailability. Ecotoxicol Environ Saf 56(1):140–147
40. Kalyani N, Goel S, Jaiswal S (2021) On-site sensing of pesticides using point-of-care biosensors: a review. Environ Chem Lett 19(1):345–354. https://doi.org/10.1007/s10311-020-01070-1
41. Saxena A, Hussain A, Parveen F, Ashfaque M (2023) Current status of metabolic engineering of microorganisms for bioethanol production by effective utilization of pentose sugars of lignocellulosic biomass. Microbiol Res 276:127478. https://doi.org/10.1016/j.micres.2023.127478
42. Elli G, Hamed S, Petrelli M, Ibba P, Ciocca M, Lugli P, Petti L (2022) Field-effect transistor-based biosensors for environmental and agricultural monitoring. Sensors 22(11):4178. https://doi.org/10.3390/s22114178
43. Mandviwala MS (2024) Agri J. World 4(11)
44. Shrivastava P, Jain VK, Nagpal S (2022) Nanoparticle intervention for heavy metal detection: a review. Environ Nanotechnol Monit Manag 17:100667. https://doi.org/10.1016/j.enmm.2022.100667
45. GadelHak Y, Hafez SH, Mohamed HF, Abdel-Hady EE, Mahmoud R (2023) Nanomaterials-modified disposable electrodes and portable electrochemical systems for heavy metals detection in wastewater streams: a review. Microchem J 109043
46. Mandal N, Adhikary S, Rakshit R (2020) Nanobiosensors: recent developments in soil health assessment. In: Rakshit A, Ghosh S, Chakraborty S, Philip V, Datta A (eds) Soil analysis: recent trends and applications. Springer, pp 285–304
47. Chatterjee S, Chetia M, Singh L, Chattopadhyay B, Datta S, Mukhopadhyay SK (2011) A study on the phytoaccumulation of waste elements in wetland plants of a Ramsar site in India. Environ Monit Assess 178:361–371
48. Shi J, Qian W, Jin Z, Zhou Z, Wang X, Yang X (2023) Evaluation of soil heavy metals pollution and the phytoremediation potential of copper-nickel mine tailings ponds. PLoS ONE 18(3):e0277159
49. Chehregani A, Noori M, Yazdi HL (2009) Phytoremediation of heavy-metal-polluted soils: screening for new accumulator plants in Angouran mine (Iran) and evaluation of removal ability. Ecotoxicol Environ Saf 72(5):1349–1353
50. Matanzas N, Afif E, Díaz TE, Gallego JLR (2021) Screening of pioneer metallophyte plant species with phytoremediation potential at a severely contaminated Hg and As mining site. Environments 8(7):63
51. Mousavi Kouhi SM, Moudi M (2020) Assessment of phytoremediation potential of native plant species naturally growing in a heavy metal-polluted saline–sodic soil. Environ Sci Pollut Res 27(9):10027–10038
52. Sharma J, Chattree A, Dan S, Imran M (2022) Heavy metal detection in soil and its treatment (bioremediation) with nanomaterials. In: Malik JA (ed) Advances in bioremediation and

phytoremediation for sustainable soil management: principles, monitoring and remediation. Springer International Publishing, pp 249–259. https://doi.org/10.1007/978-3-030-89984-4_16

53. Liu M, Wang T, Skidmore AK, Liu X (2018) Heavy metal-induced stress in rice crops detected using multi-temporal Sentinel-2 satellite images. Sci Total Environ 637–638:18–29

Gold Nanomaterials and Composites Based Nanosensors for Agricultural Sectors

Aisha Kamal, Ayeesha Khatoon, Nida Sultan, and Sazia Siddiqui

Abstract Gold nanomaterials and composite-based nanosensors have surfaced as pivotal tools of transformation in agricultural sectors, offering advanced solutions for pathogen detection, nutrient monitoring, and abiotic stress management. The nanosensors exploit the special properties inherent in gold nanomaterials, such as high surface-to-volume ratios, exceptional conductivity, and biocompatibility, to provide enhanced sensitivity and selectivity in detecting various agricultural analytes. They enable real-time, on-site monitoring of contaminants, pathogens, and nutrient levels, addressing critical challenges in sustainable agriculture and food security. Recent developments highlight their potential in identifying heavy metals, pesticides, and plant stress markers with unprecedented precision through techniques such as surface plasmon resonance (SPR), electrochemical sensing, and surface-enhanced Raman scattering (SERS). However, despite these advancements, several challenges remain, including high production costs, environmental concerns, and the need for integration with existing agricultural systems. Ensuring the stability and durability of nanosensors in harsh field conditions, along with the development of eco-friendly synthesis methods, is crucial for their practical application. Moreover, addressing regulatory and consumer acceptance issues is essential for the widespread adoption of these technologies. Future research should focus on enhancing cost-effectiveness, scalability, and field validation to bridge the gap between laboratory prototypes and real-world applications. By overcoming these challenges, gold nanomaterials and composite-based nanosensors can play a pivotal role in revolutionizing modern agriculture, contributing to increased productivity, resource efficiency, and environmental sustainability.

Keywords Gold nanoparticles · Gold nanomaterial based composites · Plant protection · Graphene · Au-NMs

A. Kamal (✉)
Department of Bioengineering, Integral University, Lucknow, India
e-mail: aishakamal04@gmail.com

A. Khatoon · N. Sultan · S. Siddiqui
Department of Bioscience, Integral University, Lucknow, India

A. Husen (ed.), *Nanobiosensors for Agricultural and Other Related Sectors*, Smart Nanomaterials Technology, https://doi.org/10.1007/978-981-96-6499-3_9

1 Introduction

Agriculture is a cornerstone of human society, providing food, raw materials, and livelihoods to billions worldwide. However, with the rapid increase in global population and the intensification of environmental challenges, the agricultural sector faces significant pressure to enhance productivity, optimize resource utilization, and ensure environmental sustainability. The adoption of cutting-edge technologies, such as nanosensors, represents a promising avenue for addressing these challenges by enabling precise, real-time monitoring and management of various agricultural parameters [1]. Among the different types of nanomaterials, gold-based nanomaterials and their composites have emerged as particularly effective candidates for developing high-performance nanosensors, due to their unique properties and versatility in diverse sensing applications.

Gold nanomaterials, known for their exceptional optical, electrochemical, and catalytic properties, are increasingly being employed in the agricultural sector [2]. Their tunable surface plasmon resonance, high biocompatibility, and excellent stability make them ideal for detecting a wide range of agricultural analytes, including soil nutrients, water pollutants, pathogens, and pesticides [3, 4]. Gold-based nanocomposites further enhance the functionality of these materials by combining them with polymers, carbon-based materials, and other metal oxides, thereby improving their sensitivity, selectivity, and stability under challenging environmental conditions [5, 6]

The applications of gold nanomaterials in agriculture are vast [7, 8]. From soil and water quality monitoring to real-time detection of plant health and pathogen exposure, gold-based nanosensors have the potential to revolutionize the way agricultural processes are managed. For example, gold nanosensors can detect heavy metal contaminants in soil at trace levels, ensuring that crops grow in safe, non-toxic conditions [9]. Additionally, they offer solutions for rapid, in-field detection of nutrient deficiencies and pest infestations, allowing farmers to make informed decisions that promote crop health and yield [10–12]. Despite the promise of gold nanomaterials in agriculture, several challenges remain. These include the high cost of production, potential environmental impacts, and issues related to scalability and regulatory approval [13, 14] Overcoming these obstacles will be essential for the widespread adoption of gold-based nanosensors and their integration into smart farming systems that leverage big data and Internet of Things (IoT) technologies.

This chapter provides an in-depth exploration of gold nanomaterials and composites-based nanosensors in agricultural applications. It examines their unique properties, synthesis methods, sensing mechanisms, and specific applications across different aspects of agriculture. The chapter also addresses the advantages, limitations, and future prospects of these nanosensors, emphasizing their transformative potential in modernizing agriculture to address the needs of a rising population and changing climate.

2 Properties and Synthesis of Gold Nanomaterials

Gold nanomaterials have gained substantial attention in agricultural nanosensor applications due to their exceptional physical, chemical, and optical properties. These characteristics not only contribute to their high sensitivity and specificity in detection but also facilitate their incorporation into a variety of sensing platforms, ranging from electrochemical to optical sensors [2].

2.1 Key Physical and Chemical Properties of Gold Nanomaterial Relevant to Sensor Applications

Optical Properties: Gold nanomaterials exhibit a unique phenomenon known as localized surface plasmon resonance (LSPR), where collective oscillation of conduction electrons occurs when these materials are exposed to light. This property leads to strong absorption and scattering in the visible spectrum, which is highly sensitive to changes in the surrounding environment, making gold nanomaterials effective for detecting molecular interactions on their surfaces [15, 16].

Electrochemical Properties: Gold nanoparticles possess excellent conductivity and electron-transfer capabilities, allowing for efficient electrochemical transduction. These properties enable the development of highly sensitive sensors for detecting analytes like pesticides, heavy metals, and soil nutrients, crucial for maintaining agricultural productivity [17, 18].

Stability and Biocompatibility: Gold nanomaterials are chemically inert and exhibit high stability against oxidation [19]. Their biocompatibility makes them suitable for environmental applications, including in situ field testing, where harsh environmental factors may affect other materials. The inert nature of gold minimizes interference with biological systems, making these nanomaterials favorable for agricultural applications where safety and environmental compatibility are paramount [2].

Surface Functionalization: Gold nanomaterials can be functionalized with various biomolecules, polymers, and ligands, enhancing their selectivity and allowing for targeted sensing of specific analytes. This flexibility in surface modification allows for the development of tailored sensing platforms suited to the diverse needs of agricultural monitoring [20, 21].

2.2 Common Synthesis Methods of Gold Nanoparticles and Nanomaterials

The synthesis method significantly affects the shape, size, and functional properties of gold nanomaterials, which in turn influence their performance as nanosensors. Here are some common approaches used to synthesize gold nanomaterials for agricultural applications:

Chemical Reduction

In this widely used synthesis method, gold ions (Au^{3+}) are reduced to gold atoms using chemical reductants like sodium borohydride or citrate, leading to the formation of gold nanoparticles. Adjusting the concentration of the reducing agents and stabilizers allows control over particle size and morphology [22]. The Turkevich method, which uses citrate as a reducing and capping agent, is particularly popular for synthesizing spherical gold nanoparticles. It is cost-effective and straightforward, making it a viable option for producing gold nanosensors on a larger scale [23, 24].

Biological Synthesis

Biological synthesis, or "green synthesis," uses plant extracts, bacteria, fungi, or other biological materials to reduce gold ions into nanoparticles (Fig. 1). This approach is environmentally friendly and often results in biocompatible gold nanomaterials, which are desirable in agriculture due to lower toxicity [25–29]. For example, various plant extracts have been successfully employed as reducing and stabilizing agents, yielding nanoparticles with unique shapes and surface properties [30, 31]. The method is regarded as sustainable and cost-effective, though scalability remains a challenge. Green synthesis provides greater control over particle size and shape, producing gold nanomaterials that are highly stable and uniform. These qualities are beneficial for developing reliable and reproducible nanosensors for agricultural monitoring [19].

2.3 Structural Characteristics and Their Implications for Sensor Efficiency

Gold nanomaterials can be synthesized in various shapes, including nanoparticles, nanorods, nanoshells, and nanostars. Each shape possesses unique properties, making it ideal for particular sensing applications:

Nanoparticles (Spherical): Gold nanoparticles, particularly spherical ones, are commonly used in plasmonic and electrochemical sensors. Their high surface area-to-volume ratio enhances their sensitivity and provides ample space for surface functionalization, which is useful for selective detection of agricultural pollutants [32].

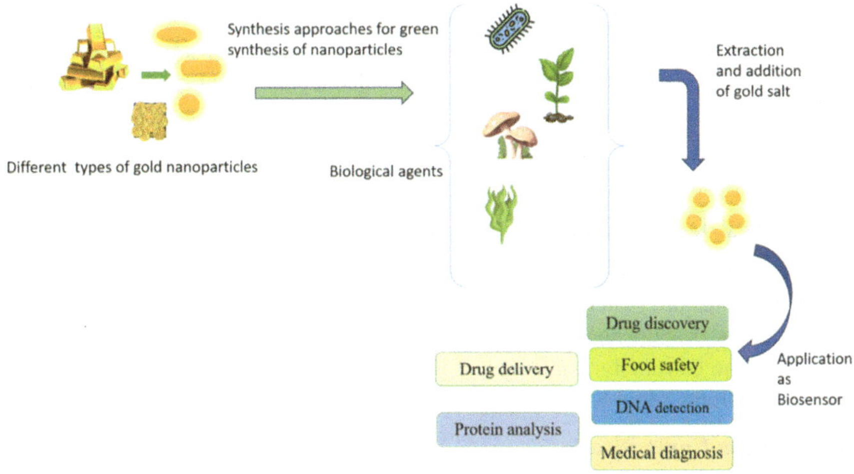

Fig. 1 Green synthesis and various applications of gold nanoparticles

Nanorods: Gold nanorods exhibit anisotropic optical properties, resulting in two plasmon bands corresponding to transverse and longitudinal oscillations [33]. This feature makes them highly sensitive to the refractive index of the surrounding medium, which is beneficial for detecting minute changes in analyte concentration. Nanorods are often used in colorimetric sensors for rapid, visual detection of contaminants [34].

Nanoshells and Nanostars: Gold nanoshells consist of a dielectric core surrounded by a gold shell, which provides a tunable plasmon resonance spectrum. These structures are suitable for applications requiring near-infrared (NIR) light, as their absorption can be tuned to this range [35] Nanoshells and nanostars are often employed in optical sensing where deeper penetration and minimal scattering are advantageous [36].

Ahmad et al. [37] optimized for various sensor applications in agriculture, facilitating accurate, real-time monitoring of soil quality, water contaminants, and crop health indicators.

3 Gold-Based Composites in Nanosensors

Gold-based composite materials have become increasingly important in the development of nanosensors for agricultural applications [38]. By integrating gold with other materials, such as carbon, polymers, and metal oxides, these composites enhance the sensing capabilities of gold-based sensors, yielding improved sensitivity, selectivity, and stability [37]. This section discusses the definition and roles of composite materials, the types of composites employed in gold-based nanosensors, and the fabrication techniques that enable effective material integration for advanced sensing applications.

3.1 Types of Gold Nano-Composites:

Nanocomposites are materials made from two or more combined phases, where at least one phase is at the nanoscale. These materials stand out from earlier composites because of the extremely small size of their structural components and their high surface-to-volume ratio, leading to unique behaviors [39].

Gold-Carbon Composites:

Gold-carbon composites are widely used in sensor applications due to the conductive and adsorptive properties of carbon-based materials, such as activated carbon, grapheme and carbon nanotubes (CNTs). These composites are highly effective in electrochemical sensing, as the conductive carbon network facilitates electron transfer, enhancing signal strength.

Graphene-based gold composites, are known for their large surface area, mechanical stability, and flexibility, which allow for sensitive detection of pesticides, pathogens, and heavy metals in soil and water samples. The gold nanoparticles enhance the conductivity and serve as binding sites, while the graphene matrix offers mechanical support and flexibility [40]. To detect glyphosate, a herbicide harmful to the environment and human health, a graphene-based electrochemical sensor using a composite on screen-printed gold electrodes was developed, achieving high sensitivity with a detection limit around 0.08 ppb in river water samples [41].

Gold-Polymer Composites:

Polymers, when combined with gold, create versatile composite materials with tunable surface properties that can be tailored for specific applications. Polymers like polyaniline (PANI), polypyrrole (PPy), and polydopamine (PDA) are often employed in these composites due to their biocompatibility and ease of functionalization.

Gold-polymer composites are especially valuable for selective sensing because the polymer layer can be modified to recognize specific molecules. For example, polydopamine-coated gold nanoparticles have been used to detect specific analytes by binding target molecules in agricultural samples through hydrogen bonding and electrostatic interactions [42]. These composites are applied in detecting plant pathogens and environmental pollutants, allowing precise sensing through the polymer's selective interactions with analytes.

Gold-Metal Oxide Composites:

Metal oxides, such as zinc oxide (ZnO), titanium dioxide (TiO_2), and iron oxide (Fe_2O_3), are frequently used with gold to enhance the catalytic and optical properties of the sensor. Gold-metal oxide composites benefit from the high surface area, chemical stability, and photocatalytic activity of metal oxides, making them effective for detecting environmental pollutants and monitoring nutrient levels [5]. For instance, gold-TiO_2 composites are widely used in photocatalytic sensors, where the TiO_2 component absorbs light, generating electron–hole pairs that enhance sensor performance. These composites are beneficial for detecting nitrates and phosphates in soil, crucial indicators of nutrient levels in agriculture [43].

3.2 Fabrication Techniques and Material Integration for Composite-Based Sensors

The fabrication of gold-based composites involves a combination of physical and chemical techniques that enable the precise assembly of materials with specific properties. Key fabrication techniques include:

In Situ Growth and Deposition:

In situ growth involves synthesizing gold nanoparticles directly onto a substrate, such as carbon or polymer, to ensure strong interaction between the materials. This method is beneficial for producing uniform composites with controlled particle size and distribution [44]. For example, in the synthesis of gold-graphene composites, gold nanoparticles are grown on the graphene surface in a controlled environment, ensuring a stable interface that enhances electron transfer in electrochemical sensors [45].

Sol–Gel Method:

The sol–gel process is a versatile approach for fabricating gold-metal oxide composites, where a gold precursor is mixed with a metal oxide precursor solution and subsequently dried and annealed to form a solid network. This method provides uniform particle dispersion and excellent adhesion between gold and metal oxides, enhancing sensor durability and performance [46]. Gold-TiO_2 composites prepared via the sol–gel method have been widely applied in sensors due to their stable photocatalytic properties, which facilitate the detection of analytes in complex agricultural environments [47].

Electrochemical Deposition:

In electrochemical deposition, gold nanoparticles are deposited onto a conductive substrate, such as a carbon or polymer-based matrix, by applying an electric current. This technique allows precise control over the thickness and uniformity of the gold layer and can create highly conductive composites [48]. This method is particularly useful for preparing gold-polymer composites, as it allows gold to be deposited on polymer films or matrices with minimal loss of structural integrity, leading to stable and reproducible sensors for monitoring soil and water contaminants [49].

Physical Vapor Deposition (PVD):

PVD methods, such as sputtering and evaporation, can be used to coat substrates with gold to form composite materials. This method is beneficial for producing thin gold layers with high purity, though it may be more costly than other techniques [50]. Gold-carbon composites formed through PVD have shown promising results in sensors due to their enhanced sensitivity, particularly in gas and humidity sensing, which can be essential for understanding environmental conditions in agricultural settings.

By carefully selecting the appropriate fabrication technique and composite materials, gold-based composites can be customized for specific agricultural sensing applications, providing enhanced performance for field-ready, robust sensing devices.

4 Role of Composite Materials in Enhancing Sensing Performance

The properties of nanocomposites are greatly affected by the extent of interaction or mixing between the different phases. In nanosensors, the combination of gold with other materials can yield a synergistic effect, enhancing sensitivity, stability, and functionality. The unique properties of gold, such as its high conductivity and biocompatibility, are often complemented by the other material's characteristics [51]. For instance, carbon-based materials like graphene provide high surface area and mechanical strength, while polymers offer flexibility and can be tailored for specific binding interactions [6, 52, 53]. These enhancements make gold-based composites especially useful for applications requiring rapid and reliable detection of agricultural analytes. The role of composite materials in nanosensors extends to Improved Signal Amplification; Enhanced Selectivity; Stability and Durability, making them more suitable for use in outdoor agricultural environments where they may be exposed to varying temperatures and pH levels [54].

5 Applications of Gold Nanomaterials and Composites-Based Nanosensors in Agriculture

Gold nanomaterials and composites are making significant contributions to agricultural monitoring and quality control (Fig. 2). With their high sensitivity, stability, and versatile functionalization capabilities, gold-based nanosensors offers transformative potential for monitoring and managing crop health, soil and water quality, and food safety. This section outlines key applications in soil and water quality monitoring, crop health assessment, pesticide residue detection, and food safety assurance, highlighting how gold-based nanosensors address crucial agricultural needs.

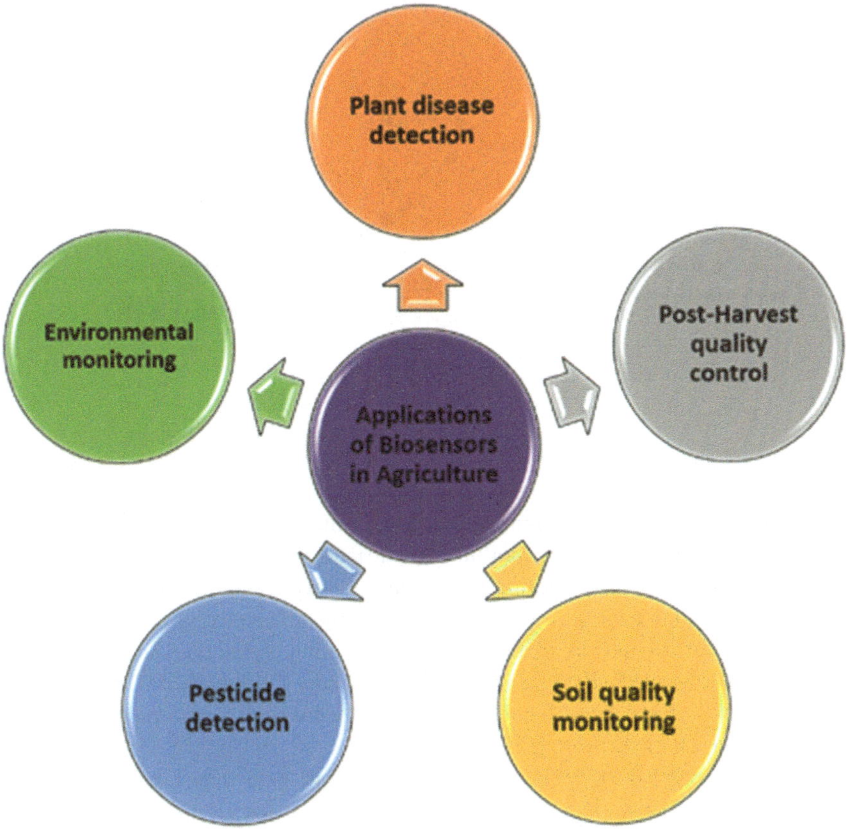

Fig. 2 Applications of gold nanomaterials and composites-based biosensors in agriculture

Gold nanomaterial/composite based biosensors detect pollutants, heavy metal, stress markers or nutrient by utilizing functionalized gold nanoparticles (GNPs) that interact selectively with metal ions. These interactions cause either aggregation or redox reactions, leading to changes in the localized surface plasmon resonance (LSPR) of the GNPs. This results in a detectable color alteration, which directly correlates with the concentration of heavy metals present (Fig. 3) [21]. This colorimetric response enables rapid and cost-effective detection of pollutants in environmental or biological samples.

5.1 In Soil and Water Quality Monitoring

Monitoring soil and water quality is essential for sustainable agriculture, ensuring optimal conditions for crop growth and mitigating environmental impacts. Gold

Fig. 3 Mechanism of interaction gold nanomaterial with analytes

nanomaterials have emerged as pivotal tools in the development of biosensors for soil and water quality management due to their exceptional chemical and physical properties [21]. Gold nanomaterials are commonly employed in fabricating nanosensors designed to detect pollutants, nutrients, and biochemical activities in agricultural environments, offering a powerful means to monitor and manage soil and water health. For example, a graphene-based sensor was ceated, which is capable of detecting contaminants as small as 0.6 nm, including organophosphate pesticides at levels well below U.S. EPA standards. Project leader Dr. Jonathan Claussen highlighted its cost-effectiveness for identifying pathogens and contaminants in water, soil, and livestock monitoring [55].

Gong et al. [56] developed an AChE-based system to detect paraoxon using nitrogen-doped carbon dots (CDs) and gold nanoparticles (AuNPs). The CDs, featuring surface –NH groups, interact with AuNPs to form a shell-like structure, causing fluorescence quenching due to Förster Resonance Energy Transfer (FRET). In paraoxon-free conditions, AChE hydrolyzes acetylthiocholine (ACTh) into thiocholine, whose –SH groups replace the CD-Au interactions, releasing CDs and restoring fluorescence. However, in the presence of paraoxon, reduced thiocholine production limits fluorescence recovery, correlating paraoxon concentration with fluorescence intensity. The system's detection limit is 1 ng/L, with nanoparticle aggregation playing a critical role, especially at higher paraoxon concentrations.

5.2 In Detection of Soil Nutrients

The rapid and accurate detection of soil nutrients and pollutants is essential to ensure sustainable agricultural practices. Gold nanoparticles (Au NPs)-based nanobiosensors have been developed to detect urea concentration and urease activity, which are critical indicators of soil health. These sensors utilize the unique optical and electronic properties of Au NPs to achieve high sensitivity and precision. For instance, nanobiosensors fabricated with Au NPs can detect urease inhibition, enabling real-time monitoring of nutrient bioavailability and the efficiency of fertilizer application [57].

Fluorescent-based nanosensors incorporating strigolactone receptors (DAD2 from *Petunia hybrida* and HTLT from *Striga hermonthica*) embedded with green fluorescent proteins have been developed for detecting agrochemicals involved in strigolactone signaling pathways. These nanosensors provide valuable insights into plant–microbe interactions and nutrient dynamics in the rhizosphere, ensuring better soil management and crop productivity [58]. The integration of Au NPs into nanobiosensors extends beyond detection to include multifunctional roles such as enhancing nutrient delivery and reducing environmental toxicity. Gold nanoparticle-based sensors help in reducing fertilizer leaching and eutrophication, thereby minimizing environmental impacts and supporting sustainable agriculture [59].

Gold nanomaterials-based biosensors represent a transformative approach in agricultural management, offering precision, speed, and cost-effectiveness. Their role in monitoring soil and water quality is pivotal for achieving ecological sustainability, enhancing crop productivity, and protecting the environment.

5.3 In Detection of Pesticides Residues

Gold nanomaterials (AuNMs) have emerged as highly effective tools for pesticide detection due to their exceptional optical, electronic, and catalytic properties. Their high surface area-to-volume ratio and ease of functionalization make them particularly suitable for enhancing the selectivity and sensitivity of biosensors, which are crucial for monitoring pesticide residues in agricultural systems. Gold nanoparticles (AuNPs) play a critical role in pesticide detection through various sensing mechanisms, including colorimetric, electrochemical, and surface-enhanced Raman scattering (SERS) techniques. These mechanisms exploit the unique properties of AuNPs, such as surface plasmon resonance (SPR) and high conductivity, to detect pesticides with remarkable precision [60].

In colorimetric assays, the aggregation of AuNPs upon interaction with pesticide molecules induces a visible color change, allowing rapid, on-site detection. For example, citrate-capped AuNPs have been used to detect organophosphate pesticides like chlorpyrifos, fenamiphos, and azinphos-methyl at concentrations as low as 120–400 ng/mL. This method offers simplicity and cost-effectiveness, making it ideal for field applications [61]. Electrochemical biosensors based on AuNPs also exhibit high sensitivity and selectivity. They function by immobilizing enzymes like acetylcholinesterase (AChE) on AuNP-modified electrodes, which interact with organophosphates, leading to measurable changes in electrical signals. These sensors have successfully detected pesticides such as carbofuran and methyl parathion, with detection limits in the nanomolar range. This electrochemical approach is advantageous for its rapid response and potential for miniaturization [62].

The ability to detect harmful agrochemicals is vital for maintaining soil and water quality. A molecularly imprinted polymer (MIP) electrochemical sensor arrayed with reduced graphene oxide and gold nanoparticles (rGO@Au) was reported for the fast estimation of carbofuran (CBF) pesticides. This sensor demonstrated a high adsorption capacity and excellent selectivity, enabling the accurate detection of pesticides at low concentrations. Such advancements facilitate effective intervention to mitigate pesticide contamination in agricultural soils and water systems [63].

The integration of AuNPs with other nanomaterials, such as graphene nanosheets and chitosan, further enhances their performance. For instance, a chitosan-gold nanoparticle composite biosensor demonstrated effective detection of organophosphate pesticides in soil samples, highlighting the potential for large-scale agricultural monitoring [64]. Additionally, a ratiometric fluorescent nanobiosensor using bovine serum albumin-stabilized gold nanoclusters (BSA-AuNCs) detected glyphosates with a detection limit of 3.4 ng/mL, showcasing its application in food safety [65].

Advanced techniques such as surface-enhanced Raman scattering (SERS) have further expanded the capabilities of gold-based nanosensors. SERS-based sensors provide ultra-sensitive detection of pesticides, including carbaryl and chlorpyrifos, by amplifying the Raman signal of pesticide molecules adsorbed on the AuNP surface. For example, an electrochemical biosensor utilizing Au–S bonds was developed to detect chlorpyrifos in apple and pak choi samples, achieving a detection limit of 36 ng L^{-1}. This method can detect trace amounts of pesticides, often in the parts-per-trillion range, which is critical for ensuring food safety and environmental protection [66].

The continuous application of pesticides in agriculture poses significant environmental and health risks. Gold nanomaterial-based biosensors offer a promising solution for real-time, in-field monitoring of pesticide residues, reducing the reliance on conventional methods like gas chromatography-mass spectrometry (GC–MS) and high-performance liquid chromatography (HPLC), which are time-consuming and costly [67]. These biosensors enable early detection and intervention, minimizing the impact of pesticide contamination on ecosystems and human health. In conclusion, gold nanomaterials have revolutionized pesticide detection by providing highly sensitive, selective, and cost-effective tools for monitoring agricultural environments. Their integration with other nanomaterials and advanced sensing techniques holds great promise for enhancing food safety and promoting sustainable agricultural practices.

5.4 In Detection of Heavy Metals:

Heavy metals, such as lead, arsenic, cadmium, and mercury, can accumulate in soil and water, posing risks to crop health and food safety. Gold-based electrochemical nanosensors can detect these heavy metals with high sensitivity. Due to the strong electron affinity of gold, these sensors show rapid response and low detection limits, providing real-time data on contamination levels (Zhang et al., 2020). Portable gold-based sensors offer on-site analysis, enabling farmers to mitigate contamination before it affects crops.

Heavy metals such as cadmium (Cd), lead (Pb), mercury (Hg) and arsenic (As) pose significant risks to environmental and human health by accumulating in soil and water, threatening food safety and ecosystems. Conventional methods for detecting these contaminants, such as atomic absorption spectroscopy (AAS) and high-performance liquid chromatography (HPLC), are effective but often time-consuming, costly, and require sophisticated instrumentation. Gold-based nanosensors provide a sensitive, cost-effective, and portable alternative, enabling rapid, on-site detection of heavy metals in agricultural settings [3].

Gold nanoparticles (AuNPs) are widely used in heavy metal detection due to their strong electron affinity, high surface-to-volume ratio, and remarkable surface plasmon resonance (SPR) properties. These attributes facilitate interactions with metal ions, leading to measurable optical or electrochemical changes. For instance, gold-based electrochemical nanosensors can detect heavy metals like Pb and Cd with high sensitivity and minimal detection limits. The strong binding affinity of AuNPs for metal ions enables rapid response times and accurate real-time monitoring of contamination levels [68]. A recent study demonstrated a novel method using gold nanoparticles modified on ion exchange membranes, combined with laser-induced breakdown spectroscopy (NMIEM-LIBS), to detect lead and cadmium in soil samples. This approach achieved detection limits of 2.62 mg/kg for Pb and 0.24 mg/kg for Cd, providing a fast and efficient strategy for environmental monitoring [69].

Colorimetric sensing systems based on gold nanoparticles have also shown significant promise for detecting heavy metals. These systems exploit the visible color changes that occur when AuNPs interact with specific metal ions. For example, green-synthesized AuNPs prepared using *Annona muricata* fruit pulp extract exhibited selective detection of Cd^{2+} ions in water and juice samples, with detection limits as low as 1.13×10^{-10} M [70]. This method offers a sustainable, cost-effective alternative to chemical synthesis, enhancing both sensitivity and environmental compatibility. Similarly, biogenic AuNPs synthesized using *Cistanche deserticola* extract and stabilized on poly(styrene-co-maleic anhydride) (PSMA) demonstrated excellent sensitivity and selectivity for Pb^{2+} ions, achieving a detection limit of 0.03 μM. These nanoparticles showed enhanced stability and catalytic efficiency due to the PSMA matrix, making them suitable for practical applications in soil and water analysis [71].

In addition to colorimetric and electrochemical methods, surface-enhanced Raman scattering (SERS) techniques utilizing gold nanomaterials have been developed for ultra-sensitive detection of heavy metals. SERS-based sensors amplify the Raman signal of metal ions adsorbed on AuNP surfaces, allowing detection at trace levels. For instance, lignin-functionalized gold nanoparticles exhibited a highly selective colorimetric response for Pb^{2+} ions, with a detection limit of 1.8 μM. The visible change in color from red to purple indicated the presence of lead, demonstrating the potential for rapid, on-site analysis in environmental samples [72].

5.5 In Detection of Diseases and Plant Protection

Gold nanomaterials (AuNMs) have become essential tools in the field of plant protection against pathogens due to their remarkable physicochemical properties and high biocompatibility. These nanomaterials play a crucial role in developing advanced biosensors for the rapid detection of bacterial, fungal, and viral pathogens affecting crops. Prompt and precise identification of plant pathogens is critical for minimizing agricultural losses and ensuring food security. Conventional methods such as enzyme-linked immunosorbent assays (ELISA) and polymerase chain reaction (PCR) are reliable but time-consuming, costly, and require sophisticated laboratory equipment. In contrast, gold-based nanosensors offer portable, cost-effective, and highly sensitive alternatives, facilitating real-time pathogen detection in the field [12].

One of the most promising applications of AuNMs in pathogen detection involves the use of gold nanoparticles (AuNPs) conjugated with nucleic acid probes. For instance, single-stranded DNA (ssDNA) probes specific to *Acidovorax avenae*, the pathogen responsible for bacterial fruit blotch, were labeled with AuNPs to create a strip-based DNA sensor. This sensor demonstrated high sensitivity and could qualitatively detect as low as 4 nM of target DNA, with a detection limit of 0.48 nM [73]. This rapid, in-situ detection method provides farmers with a valuable tool for early intervention, reducing the spread of infections and mitigating crop losses. A similar approach was employed to detect *Ralstonia solanacearum*, a soil bacterium causing wilt disease in potatoes. A colorimetric assay based on AuNP probes allowed direct detection of unamplified DNA. In the presence of target DNA, the AuNPs remained dispersed, maintaining a red color, whereas aggregation in negative samples led to a color change to purple. This assay demonstrated high sensitivity and specificity, enabling direct detection from soil samples without the need for DNA amplification, thus simplifying the diagnostic process [74], 75. In addition to bacterial detection, AuNP-based biosensors have shown efficacy in identifying fungal pathogens. For example, a sandwich ELISA strip immunoassay was developed to detect *Pantoea stewartii* subsp. *stewartii*, a pathogen causing Stewart's wilt in maize. Monoclonal detection antibodies labeled with colloidal AuNPs provided a detection limit of 1×10^5 cfu/mL without cross-reactivity with other pathogens, highlighting the sensor's specificity and reliability. This immunoassay's portability and rapid results make it particularly useful for on-site diagnostics, facilitating prompt disease management decisions [76] (Table 1)

Gold nanomaterials are also integrated into advanced techniques such as surface-enhanced Raman scattering (SERS) for ultra-sensitive detection of pathogens. SERS-based biosensors amplify the Raman signal of pathogen-specific biomolecules adsorbed on the AuNP surface, enabling the detection of trace amounts of pathogens. For example, lignin-functionalized AuNPs provided a selective colorimetric response for detecting bacterial DNA, demonstrating their potential for rapid and precise pathogen identification [83].

Table 1 Gold- nanomaterial based biosensors employed for the detection of plant pathogens in different crops

Species	Plant disease/ pathogen	Types of gold nano-biosensors used	Sensing mechanism	Key references
Tomato	Tomato yellow leaf curl virus (TYLCV)	Aunps with colorimetric nano-biosensing	Localized surface plasmon resonance	[77]
Tomato soybean rice tobacco	*Pseudomonas syringe* pathovars	Thiol-linked DNA–gold nanoparticle probes	Colorimetric detection	[78]
Orchid leaves	*Odontoglossum Ringspot* virus	Anodic aluminum oxide (AAO) with aunps	Electrochemical impedance spectroscopy (EIS)	[79, 80]
Wheat	Karnal bunt disease	Gold electrode-based SPR immunosensor	SPR immunosensing	[81]
Lime	Witches' broom disease (*Candidatus Phytoplasma aurantifolia*)	Quantum dot (QD)-based nano-biosensor	Fluorescence resonance energy transfer (FRET)	[82]

5.6 *In Real Time Plant Nutritional Analysis*

Gold nanomaterials (AuNMs) have emerged as essential tools for plant nutrient detection, offering enhanced sensitivity and accuracy in monitoring macronutrients and micronutrients critical for crop growth. The application of gold-based nanosensors in agriculture addresses the need for real-time, precise nutrient assessment, which is key to improving fertilization practices and supporting sustainable crop production. AuNMs enable the development of portable, user-friendly biosensors that can provide rapid, on-site analysis of nutrient levels in soil, water, and plant tissues.

Electrochemical biosensors incorporating gold nanoparticles (AuNPs) are particularly effective for detecting essential nutrients like phosphorus (P), nitrogen (N) and potassium (K) [10]. These sensors function by converting biochemical interactions into measurable electrical signals, facilitated by the excellent conductivity and biocompatibility of AuNPs. For instance, an electrochemical sensor based on gold nanostructures has been developed to detect nitrate levels in plant sap, offering high sensitivity and specificity [84]. This sensor utilizes the nitrate reductase enzyme immobilized on a gold-modified electrode, allowing farmers to assess nitrate concentration with minimal sample preparation. Such tools are invaluable for optimizing nitrogen fertilization, reducing environmental pollution, and preventing nutrient leaching into water bodies. Gold-based optical biosensors also play a significant role in nutrient detection. These sensors leverage the surface plasmon resonance (SPR) properties of AuNPs to detect nutrient ions with high precision. For example, SPR-based nanosensors can monitor phosphorus levels in soil and water, providing real-time feedback on nutrient availability [85]. The interaction between phosphorus ions and functionalized AuNPs induces a measurable shift in the SPR signal, enabling quantitative analysis. This approach not only enhances detection sensitivity but also offers a non-destructive method for assessing nutrient status, facilitating informed decision-making in precision agriculture.

The integration of gold nanomaterials with other nanostructures, such as graphene and carbon nanotubes, further improves sensor performance. These composite sensors exhibit enhanced stability, conductivity, and selectivity, making them suitable for detecting multiple nutrients simultaneously. For instance, a gold-graphene composite sensor has been reported to detect potassium ions in soil samples with high accuracy, demonstrating the potential for multi-nutrient monitoring [86]. Such advancements enable comprehensive nutrient profiling, allowing farmers to tailor fertilization strategies to the specific needs of their crops.

5.7 *In Abiotic Stress Management*

Gold nanomaterials (AuNMs) have shown immense potential in managing abiotic stress in plants by offering advanced detection and mitigation strategies. Abiotic stresses such as drought, salinity, temperature extremes, and heavy metal toxicity significantly impact plant growth, metabolism, and productivity. These stresses induce various physiological and biochemical responses, including the generation of reactive oxygen species (ROS), hormonal imbalances, and alterations in nutrient uptake. Gold nanomaterial-based sensors and delivery systems provide real-time monitoring and targeted intervention, helping plants cope with these challenges more effectively [87, 88].

One of the primary mechanisms through which gold nanomaterials contribute to abiotic stress management is by detecting stress-related biomarkers. For instance, ROS and hydrogen peroxide (H_2O_2) play a central role in plant stress signaling pathways. Gold nanoparticle (AuNP)-based sensors have been developed to detect H_2O_2 levels in plants, providing insights into their stress responses. These sensors utilize the catalytic properties of AuNPs to facilitate electron transfer, enabling the detection of ROS with high sensitivity and specificity. [89] reported the development of HeAptDNA-SWCNT sensors capable of detecting H_2O_2 signaling molecules, which are crucial for monitoring plant health under environmental and pathogen-induced stresses. Similarly, gold nanoparticles (AuNPs) modified with cholesterol oxidase (ChOx) utilize 11-mercaptoundecanoic acid (MUDA) as a linker molecule. MUDA facilitates the attachment of ChOx to the AuNPs, enabling a reaction with free cholesterol that generates hydrogen peroxide (H_2O_2) [90]. The produced H_2O_2 serves as an indicator, allowing for the accurate determination of cholesterol levels.

In addition to ROS detection, gold-based electrochemical sensors have been employed to monitor phytohormones that regulate plant responses to abiotic stress. For example, an Au nano-electrode coupled with copper nanoparticles was used to detect salicylic acid levels in oilseeds infected with fungal pathogens. Salicylic acid is a key hormone involved in plant defense mechanisms and stress tolerance [91]. By providing real-time data on phytohormone levels, these sensors enable early intervention and targeted stress management strategies. Moreover, gold nanomaterials also play a crucial role in managing drought stress, one of the most significant abiotic challenges in agriculture. Drought stress affects plant water relations, leading to reduced growth and yield. A study demonstrated that gold nanoparticle-based biosensors could monitor miRNA concentrations in plant tissues, which serve as specific indicators of drought stress [92]. These sensors provide a reliable method for assessing drought-induced changes at the molecular level, facilitating the development of drought-resistant crop varieties and optimizing irrigation practices. Salinity stress is another major abiotic factor that hampers agricultural productivity, particularly in arid and semi-arid regions. Gold nanomaterial-based sensors have been developed to detect salinity-induced changes in plant physiology. For instance, spinach-assisted gold nanoparticles (S-AuNPs) based biosensor significantly contributes to salt stress endurance in spinach plants by modulating various physiological attributes [93]. These sensors help identify early signs of salinity stress, enabling farmers to take corrective measures before irreversible damage occurs (Hu et al., 2020).

Moreover, gold nanomaterials have been used in combination with other nanostructures to enhance plant stress tolerance. Composite nanomaterials, such as gold-graphene hybrids, exhibit improved stability and conductivity, making them effective for detecting and mitigating multiple stress factors simultaneously. These nanocomposites can deliver stress-relief agents directly to plant cells, enhancing their resilience to adverse environmental conditions [45].

6 Challenges in the Development of Nano Sensors in Agriculture

Gold nanomaterials (AuNMs) hold great promise for advancing agricultural diagnostics, especially in pathogen detection and abiotic stress management. However, several challenges hinder their widespread implementation in agriculture. One significant obstacle is the cost associated with gold, which, despite being used in nanoscale quantities, can still affect the scalability and economic feasibility of large-scale agricultural applications. Efforts are ongoing to explore alternative nanocomposites that incorporate smaller amounts of gold with more cost-effective materials to address this issue [21]. For nanosensors to be widely adopted, their economic benefits must outweigh the initial investment, making affordability a key consideration for farmers and agricultural stakeholders.

Integration with existing agricultural systems also presents a challenge. Ensuring consistent synthesis of nanosensors with uniform size, shape, and surface functionality remains difficult. Variations in these parameters can lead to fluctuations in sensor performance, reducing their reliability in real-world applications. Developing precise characterization methods to assess nanosensor properties is essential, but these methods are often expensive and require specialized expertise [94]. Additionally, integrating nanosensors with data acquisition and analysis systems adds another layer of complexity. Effective miniaturization of sensors and their electronic components is crucial for field deployment, where devices must be lightweight, unobtrusive, and capable of real-time monitoring. Reliable and sustainable power sources, particularly for remote or wireless applications, are also essential for enhancing the practicality of nanosensors in agriculture [94].

Another challenge is the durability and resilience of nanosensors in harsh agricultural environments. These devices must withstand extreme temperatures, high humidity, and exposure to chemicals while maintaining their functionality. Factors such as oxidation, photodegradation, and biofouling can degrade nanosensors over time, necessitating strategies to enhance their stability and extend their operational lifespan [95]. Ensuring long-term performance is critical for their effectiveness in pathogen detection and stress management applications. Furthermore, nanosensors must exhibit high selectivity to accurately detect specific pathogens or stress markers while minimizing interference from other substances in the environment. Achieving this level of specificity is particularly important for timely detection of diseases or nutrient deficiencies, which can significantly impact crop health and productivity [94].

Environmental and regulatory concerns also pose significant challenges to the development and deployment of nanosensors in agriculture. The potential release of nanoparticles into the environment raises questions about their long-term impact on ecosystems and human health. Safe disposal methods and environmentally sustainable practices must be developed to mitigate these risks [96]. Regulatory frameworks governing the use of nanomaterials in agriculture are still evolving, and inconsistencies in these regulations can hinder the adoption of nanosensor technologies. Addressing these challenges requires a multidisciplinary approach, involving collaboration between researchers, policymakers, and industry stakeholders to ensure that nanosensors are both effective and safe for agricultural applications [97].

In summary, while gold nanomaterials offer transformative potential for pathogen detection and abiotic stress management in agriculture, several challenges must be overcome to realize their full benefits. These include high costs, integration complexities, durability issues, and environmental concerns. Addressing these obstacles through continued research and innovation will be essential for harnessing the power of nanotechnology to enhance agricultural sustainability and productivity.

7 Future Prospects of AuNM-Based Biosensors in Agriculture

The future of nanobiosensors in agriculture holds immense promise for enhancing sustainability and productivity, but several critical challenges need to be addressed for their successful implementation. One significant concern is the potential toxicity and ecotoxicity of nanomaterials. Due to their unique properties, including small size, high surface-to-volume ratio, and varied compositions, nanomaterials can interact unpredictably with living organisms and ecosystems. Therefore, it is essential to comprehensively assess their environmental impact, focusing on factors such as nanoparticle size, dose, accumulation behavior, immune response, and retention time [98]. A clear understanding of these aspects will help mitigate risks and ensure the safe application of nanobiosensors in agriculture.

Another challenge lies in the miniaturization and portability of nanobiosensors. Developing compact and portable sensors is crucial for real-time, on-site monitoring, particularly in remote agricultural regions. If integrated effectively with information technology, these sensors can significantly benefit the agricultural and food industries by reducing costs, improving productivity, and enabling early detection of disease outbreaks. Such integration would also enhance the efficient use of natural resources, including soil, water, and climatic conditions [99]. However, the high cost of fabricating nanobiosensors remains a barrier. To address this, researchers are exploring alternative, cost-effective biological components such as enzymes and cells, as well as novel matrices like chitosan for stabilizing and immobilizing nanomaterials. These innovations could improve the overall stability, reusability, and affordability of nanobiosensors, making them more accessible for widespread agricultural use.

Transitioning nanobiosensors from prototype stages to commercial products presents its own set of challenges. Field-scale trials are necessary to evaluate their performance under real-world conditions, which often involve complex and variable environmental factors. These trials are essential to validate the reliability, accuracy, and robustness of nanobiosensors in diverse agricultural settings. Furthermore, raising awareness among end-users, including farmers and agricultural stakeholders, is crucial for the successful adoption of nanobiosensors. Comprehensive training programs and educational initiatives can bridge the knowledge gap, ensuring that users understand the benefits and limitations of these advanced technologies.

Regulatory compliance is another critical aspect that must be addressed. Nanobiosensors used in food and agriculture must meet strict regulatory standards to ensure safety and efficacy. However, inconsistencies in regulations across different regions can complicate the approval process. Thakur et al. [38] highlighted that some food manufacturers are hesitant to disclose their use of nanobiosensors due to concerns about negative consumer perceptions. This lack of transparency can lead to mistrust and conflicts between consumers and manufacturers. Therefore, establishing clear regulatory frameworks and promoting transparency in the use of nanotechnology are essential for building consumer confidence and fostering industry-wide acceptance.

Looking forward, environmentally friendly production methods for nanomaterials will play a pivotal role in sustainable agriculture. Organic nanomaterials, such as nano-emulsions derived from natural compounds, are gaining attention for their eco-friendly attributes. These materials not only reduce environmental impact but also support the principles of a circular economy, promoting sustainable agricultural practices. By addressing these challenges through interdisciplinary research, technological innovation, and regulatory alignment, nanobiosensors can revolutionize agricultural practices, enhancing productivity, resource efficiency, and environmental stewardship.

8 Conclusion

Gold nanomaterials and composite-based nanosensors hold tremendous potential to revolutionize agriculture by enabling precise monitoring and targeted treatments. Gold nanoparticles (AuNPs) are particularly well-suited for developing sensitive and selective nanosensors due to their unique properties, such as high conductivity, biocompatibility, and durability. When integrated into composite nanosensors, AuNPs enhance detection capabilities, allowing real-time monitoring of critical agricultural parameters like soil nutrients, pH, moisture levels, and plant health indicators.

The application of gold nanoparticles in nanosensor design leads to highly sensitive, robust instruments suitable for field conditions. These advanced tools support a shift towards data-driven, resource-efficient farming practices. Farmers can optimize the use of resources such as water, fertilizers, and pesticides, reducing environmental impacts while improving crop yields and quality. This precision-driven approach addresses key challenges in sustainability and food security.

However, the widespread adoption of gold-based nanosensors in agriculture faces challenges, including high costs, scalability issues, and environmental concerns. Ongoing research and technological advancements are expected to address these barriers, making these innovations more accessible. As the field progresses, gold nanomaterials and composite-based nanosensors are poised to play a pivotal role in future sustainable and high-efficiency agriculture.

Ensuring the environmental safety of gold nanoparticles and understanding their impact on soil health, water systems, and ecosystems are essential areas for further investigation. Future interdisciplinary research that combines nanotechnology, materials science, biotechnology, and agricultural sciences will drive significant breakthroughs. These advancements could lead to the development of versatile, compact, and user-friendly sensors, potentially integrated with IoT systems for remote monitoring and data analysis. Such innovations would provide farmers, agronomists, and policymakers with real-time, comprehensive insights into crop and soil health, transforming decision-making processes across the agricultural supply chain.

In conclusion, gold nanomaterials and composite-based nanosensors are set to be foundational in sustainable agriculture. These technologies offer precise, efficient, and environmentally friendly solutions, helping to meet future global food demands while preserving natural resources for generations to come.

Competing Interests: The authors declare that they have no competing interest.

Declaration of generative AI and AI-assisted technologies in the writing process:

During the preparation of this work the book chapter author(s) have not taken any assistance from AI, and the book chapter author(s) reviewed and edited the content as needed and take(s) full responsibility for the content of the publication. All authors read and approved the final version of the manuscript.

References

1. Yadav A, Yadav K, Ahmad R, Abd-Elsalam K (2023a) Emerging frontiers in nanotechnology for precision agriculture: advancements, Hurdles and Prospects. Agrochemicals 2(2):220–256. https://doi.org/10.3390/agrochemicals2020016
2. Hassan H, Sharma P, Hasan MR, Singh S, Thakur D, Narang J (2022) Gold nanomaterials—the golden approach from synthesis to applications. Mater Sci Energy Technol 5:375–390. https://doi.org/10.1016/j.mset.2022.09.004
3. Mondal R, Dam P, Chakraborty J, Paret ML, Katı A, Altuntas S, Sarkar R, Ghorai S, Gangopadhyay D, Mandal AK, Husen A (2022) Potential of nanobiosensor in sustainable agriculture: the state-of-art. Heliyon 8(12):e12207. https://doi.org/10.1016/j.heliyon.2022.e12207
4. Zhou T, Huang J, Zhao W, Guo R, Cui S, Li Y, Zhang X, Liu Y, Zhang Q (2022) Multifunctional plasmon-tunable Au nanostars and their applications in highly efficient photothermal inactivation and ultra-sensitive SERS detection. Nanomaterials 12(23):4232. https://doi.org/10.3390/nano12234232
5. Bharti K, Sadhu KK (2022) Syntheses of metal oxide-gold nanocomposites for biological applications. Res Chem 4:100288. https://doi.org/10.1016/j.rechem.2022.100288
6. Ramanathan S, Lau WJ, Goh PS, Omar MF, Breadmore MC, Ismail AF, See HH (2024) Emerging interest of gold-carbon nanocomposites in water remediation: a new way forward. J Environ Chem Eng 12(3):112931. https://doi.org/10.1016/j.jece.2024.112931
7. Husen A (2024) Plant response to gold nanoparticles. Springer Nature Singapore Pte Ltd., Singapore https://doi.org/10.1007/978-981-97-7174-5
8. Siddiqi KS, Husen A (2016) Engineered gold nanoparticles and plant adaptation potential. Nanoscale Res Lett 11(400):1–10
9. Fu X, Gou Y, Li G, Zhao S, Ma S, Zhao C (2024) Detection of available heavy metals in soil using gold nanoparticles-modified ion exchange membrane with laser-induced breakdown spectroscopy. Microchem J 204:110922. https://doi.org/10.1016/j.microc.2024.110922
10. Bharti A, Jain U, Chauhan N (2024) From lab to field: nano-biosensors for real-time plant nutrient tracking. Plant Nano Biol 9:100079. https://doi.org/10.1016/j.plana.2024.100079
11. Kulabhusan PK, Tripathi A, Kant K (2022) Gold Nanoparticles and plant pathogens: an overview and prospective for biosensing in forestry. Sensors 22(3):1259. https://doi.org/10.3390/s22031259

12. Shaw DS, Honeychurch KC (2022) Nanosensor applications in plant science. Biosensors 12(9):675. https://doi.org/10.3390/bios12090675
13. Rana L, Kumar M, Rajput J, Kumar N, Sow S, Kumar S, Kumar A, Singh SN, Jha CK, Singh AK, Ranjan S, Sahoo R, Samanta D, Nath D, Panday R, Raigar BL (2024) Nexus between nanotechnology and agricultural production systems: challenges and future prospects. Discover Appl Sci 6(11):555. https://doi.org/10.1007/s42452-024-06265-7
14. Saritha GNG, Anju T, Kumar A (2022) Nanotechnology–big impact: how nanotechnology is changing the future of agriculture? J Agric Food Res 10:100457. https://doi.org/10.1016/j.jafr.2022.100457
15. Kravets VG, Kabashin AV, Barnes WL, Grigorenko AN (2018) Plasmonic surface lattice resonances: a review of properties and applications. Chem Rev 118(12):5912–5951. https://doi.org/10.1021/acs.chemrev.8b00243
16. Thipe VC, Karikachery AR, Çakılkaya P, Farooq U, Genedy HH, Kaeokhamloed N, Phan D-H, Rezwan R, Tezcan G, Roger E, Katti KV (2022) Green nanotechnology—an innovative pathway towards biocompatible and medically relevant gold nanoparticles. J Drug Deliv Sci Technol 70:103256. https://doi.org/10.1016/j.jddst.2022.103256
17. Daniel M-C, Astruc D (2004) Gold nanoparticles: assembly, supramolecular chemistry, quantum-size-related properties, and applications toward biology, catalysis, and nanotechnology. Chem Rev 104(1):293–346. https://doi.org/10.1021/cr030698+
18. Hua Z, Yu T, Liu D, Xianyu Y (2021) Recent advances in gold nanoparticles-based biosensors for food safety detection. Biosens Bioelectron 179:113076. https://doi.org/10.1016/j.bios.2021.113076
19. Hammami I, Alabdallah NM, Jomaa AA, Kamoun M (2021) Gold nanoparticles: synthesis properties and applications. J King Saud Univ Sci 33(7):101560. https://doi.org/10.1016/j.jksus.2021.101560
20. Amina SJ, Guo B (2020) A review on the synthesis and functionalization of gold nanoparticles as a drug delivery vehicle. Int J Nanomed 15:9823–9857. https://doi.org/10.2147/IJN.S279094
21. Kumalasari MR, Alfanaar R, Andreani AS (2024) Gold nanoparticles (AuNPs): a versatile material for biosensor application. Talanta Open 9:100327. https://doi.org/10.1016/j.talo.2024.100327
22. Daruich De Souza C, Ribeiro Nogueira B, Rostelato MECM (2019) Review of the methodologies used in the synthesis gold nanoparticles by chemical reduction. J Alloy Compd 798:714–740. https://doi.org/10.1016/j.jallcom.2019.05.153
23. Oliveira AEF, Pereira AC, Resende MAC, Ferreira LF (2023) Gold nanoparticles: a didactic step-by-step of the synthesis using the turkevich method, mechanisms, and characterizations. Analytica 4(2):250–263. https://doi.org/10.3390/analytica4020020
24. Turkevich J, Stevenson PC, Hillier J (1951) A study of the nucleation and growth processes in the synthesis of colloidal gold. Discuss Faraday Soc 11:55. https://doi.org/10.1039/df9511100055
25. Attaran N, Eshghi H, Rahimizadeh M, Mashreghi M, Bakavoli M (2016) Genetically modified luminescent bacteria Ralostonia solanacerum, Pseudomonas syringae, Pseudomonas savastanoi, and wild type bacterium Vibrio fischeri in biosynthesis of gold nanoparticles from gold chloride trihydrate. Artif Cells Nanomed Biotechnol 44(1):263–269. https://doi.org/10.3109/21691401.2014.942457
26. Husen A, Rahman QI, Iqbal M, Yassin MO (2019) Plant-mediated fabrication of gold nanoparticles and their applications. In: Husen A, Iqbal M (eds) N*anomaterials* and plant potential. Springer International Publishing AG, Cham, pp 71–110. https://doi.org/10.1007/978-3-030-05569-1_3
27. Husen A (2017) Gold nanoparticles from plant system: synthesis, characterization and their application. In: Ghorbanpourn M, Manika K, Varma A (eds) Nanoscience and plant–soil systems. Springer International Publishing AG, Cham, pp 455–479
28. Lee KX, Shameli K, Yew YP, Teow S-Y, Jahangirian H, Rafiee-Moghaddam R, Webster T (2020) Recent developments in the facile bio-synthesis of gold nanoparticles (AuNPs) and their biomedical applications. Int J Nanomed 15:275–300. https://doi.org/10.2147/IJN.S233789

29. Siddiqi KS, Husen A (2017) Recent advances in plant-mediated engineered gold nanoparticles and their application in biological system. J Trace Elem Med Biol 40:10–23
30. Santhosh PB, Genova J, Chamati H (2022) Green synthesis of gold nanoparticles: an eco-friendly approach. Chemistry 4(2):345–369. https://doi.org/10.3390/chemistry4020026
31. Zeiri Y, Elia P, Zach R, Hazan S, Kolusheva S, Porat Z (2014) Green synthesis of gold nanoparticles using plant extracts as reducing agents. Int J Nanomed 4007. https://doi.org/10.2147/IJN.S57343
32. Khan A, Rashid R, Murtaza G, Zahra A (2014) Gold nanoparticles: synthesis and applications in drug delivery. Trop J Pharm Res 13(7):1169. https://doi.org/10.4314/tjpr.v13i7.23
33. Bang NA, Thom PT, Nhat HN (2013) A comparative study of classical approaches to surface plasmon resonance of colloidal gold nanorods. Gold Bull 46(2):91–96. https://doi.org/10.1007/s13404-013-0082-0
34. Xiao N, Yu C (2010) Rapid-response and highly sensitive noncross-linking colorimetric nitrite sensor using 4-aminothiophenol modified gold nanorods. Anal Chem 82(9):3659–3663. https://doi.org/10.1021/ac902924p
35. Shi W, Sahoo Y, Swihart MT, Prasad PN (2005) Gold nanoshells on polystyrene cores for control of surface plasmon resonance. Langmuir 21(4):1610–1617. https://doi.org/10.1021/la047628y
36. Sharma M, Easha P, Tapasvi G, Reetika R (2020) Nanomaterials in biomedical diagnosis. In: Nanomaterials in diagnostic tools and devices. Elsevier, pp 57–83. https://doi.org/10.1016/B978-0-12-817923-9.00002-X
37. Ahmad R, Griffete N, Lamouri A, Felidj N, Chehimi MM, Mangeney C (2015) Nanocomposites of gold nanoparticles@molecularly imprinted polymers: chemistry, processing, and applications in sensors. Chem Mater 27(16):5464–5478. https://doi.org/10.1021/acs.chemmater.5b00138
38. Thakur M, Wang B, Verma ML (2022) Development and applications of nanobiosensors for sustainable agricultural and food industries: recent developments, challenges and perspectives. Environ Technol Innov 26:102371. https://doi.org/10.1016/j.eti.2022.102371
39. Sen M (ed) (2020) Nanocomposite materials. In: Nanotechnology and the environment. IntechOpen. https://doi.org/10.5772/intechopen.93047
40. Adhikari B-R, Govindhan M, Chen A (2015) Carbon nanomaterials based electrochemical sensors/biosensors for the sensitive detection of pharmaceutical and biological compounds. Sensors 15(9):22490–22508. https://doi.org/10.3390/s150922490
41. Thanh CT, Binh NH, Duoc PND, Thu VT, Van Trinh P, Anh NN, Van Tu N, Tuyen NV, Van Quynh N, Tu VC, Thao BTP, Thang PD, Abe H, Van Chuc N (2021) Electrochemical sensor based on reduced graphene oxide/double-walled carbon nanotubes/octahedral Fe_3O_4/chitosan composite for glyphosate detection. Bull Environ Contam Toxicol 106(6):1017–1023. https://doi.org/10.1007/s00128-021-03179-7
42. Mao X, Wang Y, Jiang L, Zhang H, Zhao Y, Liu P, Liu J, Hammock BD, Zhang C (2022) A polydopamine-coated gold nanoparticles quenching quantum dots-based dual-readout lateral flow immunoassay for sensitive detection of carbendazim in agriproducts. Biosensors 12(2):83. https://doi.org/10.3390/bios12020083
43. Tang KY, Chen JX, Legaspi EDR, Owh C, Lin M, Tee ISY, Kai D, Loh XJ, Li Z, Regulacio MD, Ye E (2021) Gold-decorated TiO_2 nanofibrous hybrid for improved solar-driven photocatalytic pollutant degradation. Chemosphere 265:129114. https://doi.org/10.1016/j.chemosphere.2020.129114
44. Vinnacombe-Willson GA, Conti Y, Stefancu A, Weiss PS, Cortés E, Scarabelli L (2023) Direct bottom-up in situ growth: a paradigm shift for studies in wet-chemical synthesis of gold nanoparticles. Chem Rev 123(13):8488–8529. https://doi.org/10.1021/acs.chemrev.2c00914
45. Khalil I, Julkapli N, Yehye W, Basirun W, Bhargava S (2016) Graphene-gold nanoparticles hybrid—synthesis, functionalization, and application in a electrochemical and surface-enhanced raman scattering biosensor. Materials 9(6):406. https://doi.org/10.3390/ma9060406

46. Bokov D, Turki Jalil A, Chupradit S, Suksatan W, Javed Ansari M, Shewael IH, Valiev GH, Kianfar E (2021) Nanomaterial by Sol-Gel method: synthesis and application. Adv Mater Sci Eng 2021(1):5102014. https://doi.org/10.1155/2021/5102014
47. Rodríguez-González V, Terashima C, Fujishima A (2019) Applications of photocatalytic titanium dioxide-based nanomaterials in sustainable agriculture. J Photochem Photobiol, C 40:49–67. https://doi.org/10.1016/j.jphotochemrev.2019.06.001
48. Bhattacharyya B (2015) Design and developments of microtools. In Electrochemical micromachining for nanofabrication, MEMS and nanotechnology. Elsevier, pp 101–122. https://doi.org/10.1016/B978-0-323-32737-4.00006-2
49. Zabihollahpoor A, Rahimnejad M, Najafpour G, Moghadamnia AA (2019) Gold nanoparticle prepared by electrochemical deposition for electrochemical determination of gabapentin as an antiepileptic drug. J Electroanal Chem 835:281–286. https://doi.org/10.1016/j.jelechem.2019.01.039
50. Lee SH, Jung HK, Kim TC, Kim CH, Shin CH, Yoon T-S, Hong A-R, Jang HS, Kim DH (2018) Facile method for the synthesis of gold nanoparticles using an ion coater. Appl Surf Sci 434:1001–1006. https://doi.org/10.1016/j.apsusc.2017.11.008
51. Khazaei M, Hosseini MS, Haghighi AM, Misaghi M (2023) Nanosensors and their applications in early diagnosis of cancer. Sens Bio-Sens Res 41:100569. https://doi.org/10.1016/j.sbsr.2023.100569
52. Kumar S, Kumar S, Sengar M, Kumari P (2021) Gold-carbonaceous materials based heterostructures for gas sensing applications. RSC Adv 11(23):13674–13699. https://doi.org/10.1039/D1RA00361E
53. Zhang R-C, Sun D, Zhang R, Lin W-F, Macias-Montero M, Patel J, Askari S, McDonald C, Mariotti D, Maguire P (2017) Gold nanoparticle-polymer nanocomposites synthesized by room temperature atmospheric pressure plasma and their potential for fuel cell electrocatalytic application. Sci Rep 7(1):46682. https://doi.org/10.1038/srep46682
54. Saha K, Agasti SS, Kim C, Li X, Rotello VM (2012) Gold nanoparticles in chemical and biological sensing. Chem Rev 112(5):2739–2779. https://doi.org/10.1021/cr2001178
55. Sahub C, Tuntulani T, Nhujak T, Tomapatanaget B (2018) Effective biosensor based on graphene quantum dots via enzymatic reaction for directly photoluminescence detection of organophosphate pesticide. Sens Actuat B Chem 258:88–97. https://doi.org/10.1016/j.snb.2017.11.072
56. Gong NC, Li YL, Jiang X, Zheng XF, Wang YY, Huan SY (2016) Fluorescence resonance energy transfer-based biosensor composed of nitrogen-doped carbon dots and gold nanoparticles for the highly sensitive detection of organophosphorus pesticides. Anal Sci 32(9):951–956. https://doi.org/10.2116/analsci.32.951
57. Deng H-H, Hong G-L, Lin F-L, Liu A-L, Xia X-H, Chen W (2016) Colorimetric detection of urea, urease, and urease inhibitor based on the peroxidase-like activity of gold nanoparticles. Anal Chim Acta 915:74–80. https://doi.org/10.1016/j.aca.2016.02.008
58. Chesterfield RJ, Whitfield JH, Pouvreau B, Cao D, Alexandrov K, Beveridge CA, Vickers CE (2020) Rational design of novel fluorescent enzyme biosensors for direct detection of strigolactones. ACS Synth Biol 9(8):2107–2118. https://doi.org/10.1021/acssynbio.0c00192
59. Yadav A, Yadav K, Abd-Elsalam K (2023b) Nanofertilizers: types, delivery and advantages in agricultural sustainability. Agrochemicals 2(2):296–336. https://doi.org/10.3390/agrochemicals2020019
60. Umapathi R, Sonwal S, Lee MJ, Mohana Rani G, Lee E-S, Jeon T-J, Kang S-M, Oh M-H, Huh YS (2021) Colorimetric based on-site sensing strategies for the rapid detection of pesticides in agricultural foods: new horizons, perspectives, and challenges. Coord Chem Rev 446:214061. https://doi.org/10.1016/j.ccr.2021.214061
61. Bao J, Hou C, Chen M, Li J, Huo D, Yang M, Luo X, Lei Y (2015a) Plant esterase–chitosan/gold nanoparticles-graphene nanosheet composite-based biosensor for the ultrasensitive detection of organophosphate pesticides. J Agric Food Chem 63(47):10319–10326. https://doi.org/10.1021/acs.jafc.5b03971

62. Yang Y, Asiri AM, Du D, Lin Y (2014) Acetylcholinesterase biosensor based on a gold nanoparticle–polypyrrole–reduced graphene oxide nanocomposite modified electrode for the amperometric detection of organophosphorus pesticides. Analyst 139(12):3055. https://doi.org/10.1039/c4an00068d
63. Tan X, Hu Q, Wu J, Li X, Li P, Yu H, Li X, Lei F (2015) Electrochemical sensor based on molecularly imprinted polymer reduced graphene oxide and gold nanoparticles modified electrode for detection of carbofuran. Sens Actuat B Chem 220:216–221. https://doi.org/10.1016/j.snb.2015.05.048
64. Bao Y, De Keersmaecker H, Corneillie S, Yu F, Mizuno H, Zhang G, Hofkens J, Mendrek B, Kowalczuk A, Smet M (2015b) Tunable ratiometric fluorescence sensing of intracellular pH by aggregation-induced emission-active hyperbranched polymer nanoparticles. Chem Mater 27(9):3450–3455. https://doi.org/10.1021/acs.chemmater.5b00858
65. Yan Z, Wang H, Wu S, Peng Z, Lai J, Qiu P (2023) Bovine serum albumin-stabilized gold nanoclusters as fluorescent probe for enzyme-free detection of glyphosate. Chem Pap 77(4):2183–2192. https://doi.org/10.1007/s11696-022-02621-2
66. Hou L, Zhang X, Kong M, Jiang G, Sun Y, Mo W, Lin T, Ye F, Zhao S (2020) A competitive immunoassay for electrochemical impedimetric determination of chlorpyrifos using a nanogold-modified glassy carbon electrode based on enzymatic biocatalytic precipitation. Microchim Acta 187(4):204. https://doi.org/10.1007/s00604-020-4175-1
67. Anh NH, Doan MQ, Dinh NX, Huy TQ, Tri DQ, Ngoc Loan LT, Van Hao B, Le A-T (2022) Gold nanoparticle-based optical nanosensors for food and health safety monitoring: recent advances and future perspectives. RSC Adv 12(18):10950–10988. https://doi.org/10.1039/D1RA08311B
68. Mahmoud R, Abdel-Hady EE, Mohamed HFM, Ibrahim M, Abd El-Fatah G, Zaher A, Gadelhak Y (2024) Nanomaterials for electrochemical sensing of heavy metals in wastewater streams. In: Ali GAM, Chong KF, Makhlouf ASH (eds) Handbook of nanosensors. Springer Nature Switzerland, pp 1219–1242. https://doi.org/10.1007/978-3-031-47180-3_48
69. Annadhasan M, Muthukumarasamyvel T, Sankar Babu VR, Rajendiran N (2014) Green synthesized silver and gold nanoparticles for colorimetric detection of Hg2+, Pb2+, and Mn2+ in aqueous medium. ACS Sustain Chem Eng 2(4):887–896. https://doi.org/10.1021/sc400500z
70. Singh K, Kukkar D, Singh R, Kukkar P, Kim K-H (2018) Exceptionally stable green-synthesized gold nanoparticles for highly sensitive and selective colorimetric detection of trace metal ions and volatile aromatic compounds. J Ind Eng Chem 68:33–41. https://doi.org/10.1016/j.jiec.2018.07.026
71. Nguyen THA, Le TTV, Huynh BA, Nguyen NV, Le VT, Doan V-D, Tran VA, Nguyen A-T, Cao XT, Vasseghian Y (2022) Novel biogenic gold nanoparticles stabilized on poly(styrene-co-maleic anhydride) as an effective material for reduction of nitrophenols and colorimetric detection of Pb(II). Environ Res 212:113281. https://doi.org/10.1016/j.envres.2022.113281
72. Yu Y, Naik SS, Oh Y, Theerthagiri J, Lee SJ, Choi MY (2021) Lignin-mediated green synthesis of functionalized gold nanoparticles via pulsed laser technique for selective colorimetric detection of lead ions in aqueous media. J Hazard Mater 420:126585. https://doi.org/10.1016/j.jhazmat.2021.126585
73. Zhao W, Lu J, Ma W, Xu C, Kuang H, Zhu S (2011) Rapid on-site detection of Acidovorax avenae subsp. Citrulli by gold-labeled DNA strip sensor. Biosens Bioelectron 26(10):4241–4244. https://doi.org/10.1016/j.bios.2011.04.004
74. Khaledian S, Nikkhah M, Shams-bakhsh M, Hoseinzadeh S (2017) A sensitive biosensor based on gold nanoparticles to detect Ralstonia solanacearum in soil. J Gen Plant Pathol 83(4):231–239. https://doi.org/10.1007/s10327-017-0721-z
75. Yang T, Luo Z, Tian Y, Qian C, Duan Y (2020) Design strategies of AuNPs-based nucleic acid colorimetric biosensors. TrAC Trends Anal Chem 124:115795. https://doi.org/10.1016/j.trac.2019.115795
76. Feng M, Kong D, Wang W, Liu L, Song S, Xu C (2015) Development of an immunochromatographic strip for rapid detection of Pantoea stewartii subsp. Stewartii. Sensors 15(2):4291–4301. https://doi.org/10.3390/s150204291

77. Razmi A, Golestanipour A, Nikkhah M, Bagheri A, Shamsbakhsh M, Malekzadeh-Shafaroudi S (2019) Localized surface plasmon resonance biosensing of tomato yellow leaf curl virus. J Virol Methods 267:1–7. https://doi.org/10.1016/j.jviromet.2019.02.004
78. Vaseghi A, Safaie N, Bakhshinejad B, Mohsenifar A, Sadeghizadeh M (2013) Detection of Pseudomonas syringae pathovars by thiol-linked DNA–gold nanoparticle probes. Sens Actuat B Chem 181:644–651. https://doi.org/10.1016/j.snb.2013.02.018
79. Jian Y-S, Lee C-H, Jan F-J, Wang G-J (2018) Detection of odontoglossum ringspot virus infected phalaenopsis using a nano-structured biosensor. J Electrochem Soc 165(9):H449–H454. https://doi.org/10.1149/2.0351809jes
80. Lin W-P, Wang W-J, Lee C-H, Jan F-J, Wang G-J (2022) A two-in-one immunoassay biosensor for the simultaneous detection of Odontoglossum ringspot virus and Cymbidium mosaic virus. Sens Actuat B Chem 350:130875. https://doi.org/10.1016/j.snb.2021.130875
81. Karaman C, Bölükbaşı ÖS, Yola BB, Karaman O, Atar N, Yola ML (2022) Electrochemical neuron-specific enolase (NSE) immunosensor based on CoFe2O4@Ag nanocomposite and AuNPs@MoS2/rGO. Anal Chim Acta 1200:339609. https://doi.org/10.1016/j.aca.2022.339609
82. Ariffin SAB, Adam T, Hashim U, Faridah Sfaridah S, Zamri I, Uda MNA (2013) Plant diseases detection using nanowire as biosensor transducer. Adv Mater Res 832:113–117. https://doi.org/10.4028/www.scientific.net/AMR.832.113
83. Peng H, Chen IA (2019) Rapid colorimetric detection of bacterial species through the capture of gold nanoparticles by chimeric phages. ACS Nano acsnano.8b06395. https://doi.org/10.1021/acsnano.8b06395
84. Ben Messaoud N, Dos Santos MB, Espiña B, Queirós RB (2024) Development of electrochemical sensors based on silver nanoparticles electrodeposited on gold screen-printed electrodes: application to nitrate trace analysis in water. J Sens Sens Syst 13(1):135–145. https://doi.org/10.5194/jsss-13-135-2024
85. Philip A, Kumar AR (2022) The performance enhancement of surface plasmon resonance optical sensors using nanomaterials: a review. Coord Chem Rev 458:214424. https://doi.org/10.1016/j.ccr.2022.214424
86. Gokhale AA, Lu J, Weerasiri RR, Yu J, Lee I (2015) Amperometric detection and quantification of nitrate ions using a highly sensitive nanostructured membrane electrocodeposited biosensor array. Electroanalysis 27(5):1127–1137. https://doi.org/10.1002/elan.201400547
87. Li Z, Yu T, Paul R, Fan J, Yang Y, Wei Q (2020) Agricultural nanodiagnostics for plant diseases: recent advances and challenges. Nanoscale Adv 2(8):3083–3094. https://doi.org/10.1039/C9NA00724E
88. Wahab A, Muhammad M, Ullah S, Abdi G, Shah GM, Zaman W, Ayaz A (2024) Agriculture and environmental management through nanotechnology: eco-friendly nanomaterial synthesis for soil-plant systems, food safety, and sustainability. Sci Total Environ 926:171862. https://doi.org/10.1016/j.scitotenv.2024.171862
89. Wu H, Nißler R, Morris V, Herrmann N, Hu P, Jeon S-J, Kruss S, Giraldo JP (2020) Monitoring plant health with near-infrared fluorescent H_2O_2 nanosensors. Nano Lett 20(4):2432–2442. https://doi.org/10.1021/acs.nanolett.9b05159
90. Nirala NR, Saxena PS, Srivastava A (2018) Colorimetric detection of cholesterol based on enzyme modified gold nanoparticles. Spectrochim Acta Part A Mol Biomol Spectrosc 190:506–512. https://doi.org/10.1016/j.saa.2017.09.058
91. Wang Z, Wei F, Liu S-Y, Xu Q, Huang J-Y, Dong X-Y, Yu J-H, Yang Q, Zhao Y-D, Chen H (2010) Electrocatalytic oxidation of phytohormone salicylic acid at copper nanoparticles-modified gold electrode and its detection in oilseed rape infected with fungal pathogen Sclerotinia sclerotiorum. Talanta 80(3):1277–1281. https://doi.org/10.1016/j.talanta.2009.09.023
92. Asefpour Vakilian K (2019) Gold nanoparticles-based biosensor can detect drought stress in tomato by ultrasensitive and specific determination of miRNAs. Plant Physiol Biochem 145:195–204. https://doi.org/10.1016/j.plaphy.2019.10.042

93. Amir M, Raheem A, Yadav P, Kumar V, Tewari RK, Jalil SU, Danish M, Ansari MI (2024) Phytofabricated gold nanoparticles as modulators of salt stress responses in spinach: implications for redox homeostasis, biochemical and physiological adaptation. Front Plant Sci 15:1408642. https://doi.org/10.3389/fpls.2024.1408642
94. Yang T, Duncan TV (2021) Challenges and potential solutions for nanosensors intended for use with foods. Nat Nanotechnol 16(3):251–265. https://doi.org/10.1038/s41565-021-00867-7
95. Zain M, Ma H, Nuruzzaman M, Chaudhary S, Nadeem M, Shakoor N, Azeem I, Duan A, Sun C, Ahamad T (2023) Nanotechnology based precision agriculture for alleviating biotic and abiotic stress in plants. Plant Stress 10:100239. https://doi.org/10.1016/j.stress.2023.100239
96. Iavicoli I, Leso V, Beezhold DH, Shvedova AA (2017) Nanotechnology in agriculture: opportunities, toxicological implications, and occupational risks. Toxicol Appl Pharmacol 329:96–111. https://doi.org/10.1016/j.taap.2017.05.025
97. Chakraborty U, Kaur G, Chaudhary GR (2021) Development of environmental nanosensors for detection monitoring and assessment. In: Kumar R, Kumar R, Kaur G (eds) New Frontiers of nanomaterials in environmental science. Springer Singapore, pp 91–143. https://doi.org/10.1007/978-981-15-9239-3_5
98. Prasad R, Bhattacharyya A, Nguyen QD (2017) Nanotechnology in sustainable agriculture: recent developments, challenges, and perspectives. Front Microbiol 8:1014. https://doi.org/10.3389/fmicb.2017.01014
99. Almansoori LH, Nazir A (2023) Future of nanomaterials in agricultural practices. In: The impact of nanoparticles on agriculture and soil. Elsevier, pp 385–401. https://doi.org/10.1016/B978-0-323-91703-2.00012-9

Silver Nanomaterial and Composite-Based Nanosensors for the Agriculture Sector

Sneh Lata and Nasreen Ghazi Ansari

Abstract India has been primarily an agricultural economy that employs more than 60% of the labour force with an effectual approximate output of 18%, which leads to excessive traditional cultivation. Further, various unsustainable farming practices lead to environmental deterioration, soil contamination, and declining water quality. Technology is making solid and rapid strides, such as nanosensors that have been developed to address these issues and limitations of contemporary agriculture. Nanosensors are utilized to identify dynamic changes based on behavioral of nano-sized molecules to notify of alterations in the environment. Silver nanomaterials and composite-based nanosensors have paved a paradigm shift in agricultural precision and sustainable management of crops. Due to their unique physiochemical properties, silver nanomaterials have a high affinity towards sensing that boosts sensor selectivity and performance. The created sensors provide on-the-spot monitoring of soil quality, nutrient levels, crop health, and environmental conditions, with the data instantly accessible for decision-making. This chapter provides insights on silver-based nanomaterials and composite nanosensors in terms of their synthesis, types, designs, sensing mechanisms, and applications in agriculture.

Keywords Silver-nanoparticle · Silver-nanocomposite · Sensing · Agriculture

Abbreviations

NPs	Nanoparticles
Ag NMs	Silver nanomaterials
SPR	Surface plasmon resonance
ROS	Reactive oxygen species

S. Lata · N. G. Ansari (✉)
Analytical Chemistry Laboratory, Analytical Sciences and Accredited Services Group, ASSIST Division, CSIR-Indian Institute of Toxicology Research (CSIR-IITR), Lucknow, Uttar Pradesh, India
e-mail: nasreen@iitr.res.in

Academy of Scientific and Innovative Research (AcSIR), Ghaziabad, India

A. Husen (ed.), *Nanobiosensors for Agricultural and Other Related Sectors*, Smart Nanomaterials Technology, https://doi.org/10.1007/978-981-96-6499-3_10

Ag NPs	Silver nanoparticle
GO	Graphene oxide
SERS	Surface-Enhanced Raman Scattering
AgP	Silver nanoplates
MEF	Metal enhanced fluorescence

1 Introduction

Recently, agriculture has confronted several obstacles that jeopardize sustainability and production. The early diagnosis becomes crucial as crop illnesses brought on by bacterial, viral, and fungal pathogens may result in poor yield and quality losses [1]. Additionally, environmental stresses such as drought, extreme temperatures, and salinity further strain agricultural systems, reducing crop resilience and contributing to fluctuating yields. Another crucial issue is nutrient management, as excessive or insufficient fertilizer usage may stunt agricultural development, degrade soil health, and exacerbate environmental issues like eutrophication and water pollution [2]. People's population causes all these demands for grain and food products, less arable land, and the necessity to minimise the environmental impacts of farming. Adopting nanotechnology along with other cutting-edge technologies usually implies the way forward for the technologists to relieve the aggravation. This approach is based on the fact that advanced equipment can convey the necessary data to farmers for optimum utilization of inputs, thereby improving the quality of the final product without compromising the environment [3]. Nanotechnology's popularity has expanded gamut of nanoparticles. These are very miniscule particles of nanoscale (1–100 nm) size that are invisible. The global use of nanoparticles (NPs) and related materials has been stretched to numerous fields because of its trailblazing scientific exertion [4].

Owing to their minuscule size, their physiochemical characteristics vary significantly from their macro or micro counterparts [5]. Silver nanoparticles, or AgNPs, are preferred among other metallic nanoparticles because of their unique combination of optical and electrical properties like huge surface area-to-volume ratio, incomparable electrical and thermal conductivity, chemical stability, catalytic activity, and surface plasmon resonance with localization, surface-enhanced Raman spectroscopy, and Förster/fluorescence resonance energy transfer. They have diverse applications, including textiles, biosensing, bioimaging, bioremediation, crop protection, food, healthcare, etc. [6–9]. These particles have shown very high conductivity stem from the free electrons within silver's atomic structure, making them ideal for sensing applications [10, 11].

SPR evidence of modifications in the intersectional refractive index, FRET gives parallel data on cross-molecular distances and interactions, and SERS provides information on molecular vibration and structure. The optical properties of silver nanoparticles are based on light absorption and scattering phenomena, which are applied in sensing applications [12, 13].

Nanosensors are nanoscale devices in which at least one sensing component should be in the size range of 1–100 nm. This fosters high selectivity and sensitivity by having a molecular-level interaction. Being nanoscale, these sensors utilize nanomaterials' chemical and physical properties for sensing applications. The nanomaterials for sensing applications include metal nanoparticles with transitions and other metals, carbon nanotubes incorporating nanocomposites, nanowires, and quantum dots.

A typical nanosensor device has (a) a sample in any physical state, (b) an analyte, which means the target substance, (c) a recognition element that identifies the analyte, (d) the transducer that converts the recognition event into signals, and (e) the signal processor that translates those signals into actionable data. Nanosensors have high sensitivity, detection speed, cost-effectiveness, and applicability in diverse environments. As an impetus to precision farming, nanosensors in agriculture facilitate real-time monitoring of crop conditions, pathogens, pesticides, moisture, and environmental leaching. This chapter covers various types of nanosensors and their applications in modern agriculture; hence, it elaborates on the future potential of the technology.

In agriculture, nanosensors effectively solve venerable challenges [14]. These nanosensors help perceive early detection of crop diseases and pest infestations and are also used to monitor soil and water quality or detect nutrient deficiencies [15]. The farmers can accumulate real-time data on some crucial parameters such as pH, moisture levels, and nutrient availability by implanting them into the soil and leachate water or plants, which allows the farmers to enable precise farming practices while minimising waste to enhance crop production and reduce the environmental burden of excessive chemical use, such as fertilisers and pesticides, which can pollute water sources and degrade soil quality [16]. The worldwide marketplace for nanosensors was valued at $ 0.65 billion in 2022, but now it is rising compounded at a rate of 9.18% from $ 0.87 billion to $1.77 billion by 2032 [17] (Fig. 1)

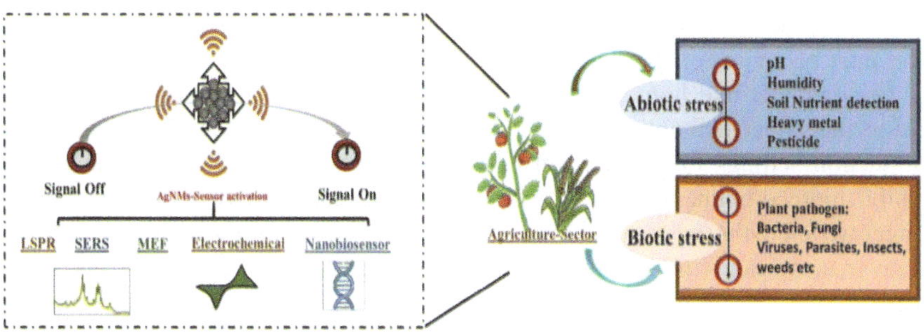

Fig. 1 Silver nanomaterial and composite-based nanosensor application

2 Why Silver-Based Nanomaterials or Composite Materials for Nanosensors?

The suitability of silver for nanoformulation is due to its electropositive reduction potential ($E^{\circ} = + 0.799$ V), which allows for easy reduction by various compounds. The inherent properties of AgNPs, such as a higher extinction coefficient, sharper extinction bands, a higher scattering/extinction ratio, and a more robust local electromagnetic field showing substantial energy bandgap, make them selective over other metallic particles for sensing applications. These phenomena are explained by classical electromagnetism theory. Another advantage of silver-based nanomaterials is that the synthesis procedure is easier, faster, and less expensive than other materials [18].

2.1 Silver Nanomaterials

Silver nanomaterials can be produced in different morphologies, such as monodisperse nanospheres, triangular prisms, nanoplates, nanocubes, nanowires, and nanorods (shown in Fig. 2), by monitoring the reaction conditions. The silver nanoparticle's surface chemistry, shape, and optical characteristics must be precisely regulated to achieve the intended functioning in the specified environment. Since specific wavelengths of light cause the electrons on the metal surface to vibrate collectively, silver and other noble metal nanoparticles show significant light-based interactions. Silver nanoparticles absorb and scatter far more powerfully than non-plasmonic nanoparticles of the same size due to an oscillation known as SPR. To alter the absorption and scattering properties of silver nanoparticles, one may change the particle's shape, size, and localized refractive index at the particle surface. Changing the surface chemistry of the particles, which includes the size, functional groups, and binding strength of the capping agents, can provide an additional level of control over their behavior.

The diameter and homogeneity of sphere-shaped silver nanoparticles are vital factors that affect their optical characteristics, and these may be achieved by adjusting production settings to produce particles of a regulated size. Larger nanospheres exhibit more prominent spectrum peaks, more scattering, and peak intensities at longer wavelengths, whereas smaller spheres mostly absorb light at around 400 nm. Many nanoparticles in aqueous fluids become electrostatically stabilised when charged species are added to their surface. The colloid's zeta potential may be measured to ascertain the type and density of charges. Taking some examples of compounds like citrate that are bonded to the surface, silver nanoparticles often have a negative zeta potential. Through surface-bound binding, novel capping agents may modify the nanoparticles' zeta potential and chemical activity by subjecting them to more closely coordinating ligands, often amine or thiol-containing molecules. Short-chain methoxy-terminated polyethene glycol (mPEG) molecules provide an almost

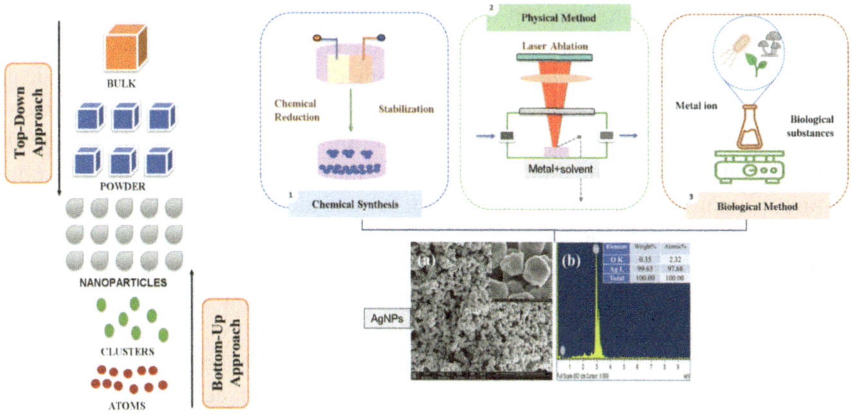

Fig. 2 Synthesis approaches of silver nanoparticle

neutral zeta potential when used in place of the citrate ions, but coating the particles with branched Poly Ethylene Imine (BPEI) creates an amine-dense surface with a markedly positive zeta potential [19].

The sensitivity and selectivity of silver nanomaterials in various applications can be achieved by functionalisation techniques that involve surface modification of nanomaterials. Specific ligands, polymers, or biomolecules work as surface modifiers by increasing interaction with target analytes for sensing applications. For example, a ligand-based thiol coating of silver nanoparticles increases the stability and specificity of the binding because the sulfur atoms are quickly bonded to silver, thus allowing the selective detection of metal ions or biomolecules [20]. Coatings based on polymers, including PEG or polydopamine, confer further secondary steric stabilisation and decrease nonspecific interactions, thereby increasing sensitivity in complex matrices. The conjugation of plant-derived biological molecules such as enzymes or DNA aptamers allows for selective recognition of specific biological and chemical targets by exploiting excellent biosensing applications by silver nanomaterials [21]. In addition, modification with functional groups such as amines, carboxyls, or alkynes provides further scope in conjugation chemistry that allows fine-tuning of selectivity at the nanomaterial level. This functionalisation strategy greatly improves the functionality of silver nanomaterials, allowing ultrasensitive detection limits and high selectivity from environmental, medical, and industrial application perspectives [22].

2.1.1 Synthesis

Silver nanoparticle synthesis can be classified based on standard practices and variations, including reactants and conditions. Numerous synthesis approaches have been recorded, including the top-to-bottom approach, conventional against non-conventional, and green methods [23]. The conventional method utilised organic reducers for synthesis, while the latest method consisted of electrochemical, microwave, microemulsion, photochemical, laser ablation, and evaporation–condensation techniques [24]. Top-down and bottom-up approaches include chemical, physical, and biological methods. This method has the advantage of macroscopic starting structures that are controllable from the outside. Some instances of the top-down approach include the nanoparticles produced by mechanically grinding bulk metals and adding colloidal protective agents. The bottom-up techniques involve additional self-assembly processes, namely metal reduction and electrochemical processes for downsizing material components up to the atomic level [25]. Physical techniques include laser ablation and evaporation–condensation with the advantage of uniformity of NP distribution and the lack of solvent contamination. At the same time, this method has the disadvantage of requiring large furnace tubes, high energy consumption, and lengthy time required to reach thermal stability. To regulate particle size and avoid aggregation during the synthesis, the chemical synthesis approach employed the reduction of silver salts by reducing agents in the presence of stabilising agents [26]. The most widely used method for creating silver nanoparticles is chemical reduction using organic and inorganic reducing chemicals. In general, different reducing agents such as sodium citrate, hydrazine, sodium ascorbate, sodium borohydride, elemental hydrogen, polyol approach, and tollens reagent are used for the reduction of silver ions into the formation of metallic silver in any medium. This inexpensive method can be employed for mass production in quicker reaction times. Its outstanding efficiency and precision provide control over size, shape, and morphology. Yet, these methods have significant disadvantages to the environment and human health due to chemical waste production. Green synthesis methods involve plant extracts, bacteria, fungi, or other biological entities to reduce silver salts compared to other methods. These methods reduce and stabilize AgNPs using green to provide an environmentally producing technique [18, 27]. However, they have disadvantages concerning time efficiency or forming irregular shapes particles[28]. The choice of production methods for efficiency scalability with environmental investigations is becoming more difficult [29].

2.2 *Silver Nanocomposites*

Nowadays, composites are gaining more attention in several industries, as it is noticeable that a single component does not always have all the qualities necessary for its existence. A composite is an engineered material combining two or more constituent elements with specific, unique physical–chemical and mechanical properties that make it ideal for sensing applications [30]. Composite material is an amalgamation of reinforcement and matrix. Composite materials such as ceramic matrix, metal matrix, polymer matrix, fibre-reinforced, particle-reinforced, and nanocomposite [31]. By selecting different combination of matrix and reinforcement, it can be customised to various applications.

2.2.1 Silver Nanoparticles onto Graphene Stabilizers

The exact process of nanosilver modification onto graphene is still unknown, but it is thought that (1) AgNPs nucleation is facilitated by reactive sites on the graphene surface due to oxygen capabilities; (2) Ag^+ ions reduced by ethylene glycol and a solution of sodium attach to the graphene surface; and (3) Ag-CCG nanocomposites are produced by further reduction of graphene to chemically converted graphene (CCG). Ag@rGO nanocomposites give a 7 × strong Raman signal and show less aggregation than native AgNPs. Also, graphene adsorbs bacteria as a non-specific substrate that increases the sterilising effect of these particles. Cobos and his colleagues synthesised graphene-AgNP composites (~2–5 nm) within an aqueous graphene oxide (GO) solution with ethylene glycol, PVP, and APTMS as stabilising agents, hydrazine and sodium borohydride as a reducing agents [32].

Moreover, the vacuum filtration method was employed on several filter membranes for in-situ synthesis of reduced graphene oxide silver nanocomposites [33]. Alternatively, a diverse approach without a solvent phase was applied to prepare Ag-GO nanofillers nanocomposite (Muqoyyanah et al. 2023). These nanocomposites demonstrated outstanding catalytic capability for detecting various contaminants, including cationic and ionic dyes and 4-nitrophenol [34].

2.2.2 Modification of Nanosilver onto Metals

A new technique was developed to fabricate antibacterial stainless steel with even scatterings of silver by using composites of Ag NPs and polyethyleneimine [35]. Skiba et al. synthesized Ag NPs on TiO_2 powder using an innovative single-step approach through atmospheric pressure discharge cold plasma treatment and thus avoided using any environmentally and biologically hazardous reducing agents. In related studies, TiAg coatings were deposited on titanium substrates by pulsed DC magnetron sputtering [36]. However, Ag-doped titanium dioxide nanotubes (TiO_2-NTs) by immersing TiO_2-NTs in $AgNO_3$ solution followed by UV irradiation which improves surface properties of titanium alloys, especially corrosion resistance. Pore content is decreased via DC sputtering plasma and anodic oxidation. In order to enhance the surface alloy for medical applications, additional biomedical surface engineering research entailed the preparation of an anodization of a Ti-13Nb-13Zr alloy, followed by the DC sputtering of a silver (Ag) coating and the development of a TiO_2 layer [37]. Plasma techniques and advanced coatings methods could potentially enhance biocompatibility and antimicrobial activities in titanium alloys in biomedical application [38].

2.2.3 Modification of Nanosilver onto Fibers

Cellulosic fibre extracted from scot pine wood using the green method was utilised for nanosilver surface modification [39]. Another example includes the production of Ag-impregnated antimicrobial poly(ethylene terephthalate) fibres by directly grafting silver nanoparticles onto microbes-coated PET fibre [40]. Ag@ACF fibre nanocomposite utilised a hydrothermal method to remove radioactive iodine; on the other hand, the same method was utilised in PU-HA-AgNPs production, which was used for wound healing applications [41, 42].

2.2.4 Modification of Nanosilver onto Ceramic Materials

The application of silver nanoparticles on ceramics is becoming a novel sintering additive, which gives YAG luminous ceramics the enhanced effect of optics [43]. Namely, in a recently conducted investigation, BN@C composites were synthesised by functionalising boron nitride nanosheets through surface polymerisation with phenolic aldehyde. Silver nanoparticle trace modification improved the impedance-matching characteristics of these composites. After being further annealed at temperatures over 700 °C, they were put into silicone rubber to enhance their ability to absorb microwaves and control heat [44]. Further developments in dental restorative materials include the identification of zirconia nanoparticle-modified 3D printing resins and halloysite nanotubes immobilised with silver nanoparticles, which exhibit improved mechanical and antibacterial qualities [45].

2.2.5 Modification of Nanosilver onto Polymers

Polymer-based nanocomposite can be customized according to desired physical, mechanical, and chemical properties. Layer-by-layer assembly is one of the polymer fabrication methods [46]. Insitu-synthesized polymer-Ag composite minimizes nanoparticle aggregation by dispersing silver ions on the matrix [47]. Nanocomposite films have the capability of hydrophilic properties that protect against corrosion by curbing biofilm, microbial attachment, and the deposition of electrolytes and proteins. Some of the recent developments in the field of gas-barrier and bacteria-resistant packaging materials are nanoboron nitride laminated poly(ethyl methacrylate)/poly(vinyl alcohol) composite films with silver nanoparticles [48]. Chloroxylenol-carboxyethyl chitosan-modified silver nanoparticle nanolayer films have also played a significant role in advancing surgical suture durability and antibacterial efficacy [49].

3 Design, Types and Sensing Mechanisms of Silver-Based Nanosensor

The design of a nanosensor requires careful consideration of the target analyte, sensor sensitivity, and operating environment. A nanoscale transducer is the core of a nanosensor design; it translates biological, chemical, or physical information into measurable outputs. The nanomaterials that make up this transducer can have a wide range of compositions, including metal nanoparticles, carbon nanotubes, and semiconductor quantum dots, all of which are chosen for specific desired properties such as a very high surface area, highly tunable optical or electronic characteristics, and chemical reactivity [50]. To give selectivity, such materials are often functionalized with molecular recognition components, such as antibodies, enzymes, or aptamers, which selectively bind to the analyte of interest. Signal amplification is required to detect very low analyte concentrations [51]. Nano-structured designs will improve analyte-sensor surface interactions, as will hybrid constructions that blend multiple materials to achieve maximum responsiveness. Microfluidic integration and electrical components with data collecting and processing capabilities are the last steps toward real-time and precise monitoring in various contexts in agricultural fields [52].

Mie's classical electromagnetic theory remains the core around which models are developed for describing nanoparticles' optical properties. According to this method, Localized Surface Plasmon Resonance (LSPR), the resonance condition, happens when the metal's dielectric function is equivalent to that of the surrounding medium [53]. Therefore, LSPR must depend very much on the nanoparticle and its environment. Such dependence endows the LSPR with the possibility of tuning by changing the composition, dimensions, and form of the nanoparticles while making the nanoparticles sensitive to changes in the local dielectric environment, thus allowing for LSPR shifts that can be detected [54]

Two main types of plasmonic sensors have emerged over the past ten years:

(1) LSPR frequency shift-based sensors, which are either aggregation sensors depending on the coupling of plasmons because of nanoparticles standing close to each other or they are refractive index sensors that are sensitive to the local refractive index perturbation sensors that detect the local refractive index changes as well to the systems,

(2) Surface-Enhanced Raman Spectroscopy, Metal-Enhanced Fluorescence, and Fluorescence Resonance Energy Transfer (FRET) are the three techniques that sharpen the electromagnetic fields at the nanoparticles' interfaces.

3.1 Chemical Sensing via LSPR Shift in Silver Nanoparticles

In some cases, aggregation of nanoparticles is caused by the analyte itself, so analyte-induced detection systems must be used for this application. For instance, a fast, selective electrochemical biosensor based on a screen-printed carbon electrode was constructed for real-time, field-based detection of E. coli. The colloidal stability of aptamer-conjugated silver nanoparticles in PBS buffer was assessed, which revealed the aggregation of AgNPs after conjugation. UV–visible spectrophotometry established that AgNP-aptamer complex binding to E. coli (DH5α) occurred in an AgNP-dependent manner and the absorbance increased linearly with increasing concentrations of E. coli. AgNPs were drop-casted onto SPCEs to immobilise E. coli-specific aptamers, and the electrode surface properties were verified by cyclic voltammetry. Differential pulse voltammetry revealed a linear decrease in current at increased E. coli concentrations after incubation for 15 min. The biosensor had a high sensitivity (LOD of 150 CFU/mL), good reproducibility (RSD of 6.91%), selectivity for E. coli, stability over four weeks, and recoveries of 95.27–107% in tap water testing [55].

However, this approach applies only to analytes that cause nanoparticle aggregation and other interferences from substances in the sample. Incorporating molecular recognition events addresses these issues and brings functional groups into analysing analytes for aggregation-based sensor selectivity. The nanoparticle surface is functionalised with molecular recognition ligands. Aggregation is then triggered explicitly by the target analyte, while the degree of aggregation and the plasmon band's associated broadening and redshift directly measure the target concentration. [56].

3.1.1 Based on Plasmon Shifts with Local Refractive Index

The silver nanoparticles' localised surface plasmon resonance sensitivity is quite high to changes in local refractive indices introduced by analyte binding close to the nanoparticle surface. Hence, refractive index sensors are possible through measurable LSPR shifts within the UV–Vis extinction spectrum, which depend on the analyte's concentration. Silver nanoparticles provide sharper resonances for LSPR, and as a result, the accuracy of the measurements is higher when compared to other noble metals [57]. Target molecules are distinguished by their unique properties to the nanostructures of different shapes. In regards to shapes used, triangles, rhombuses, and spheres are the possibilities considered for the sensor designs [58]

3.1.2 Based on Inelastic Light Scattering: SERS

Raman scattering is inefficient with cross sections smaller than 10^{-29} cm^2 sr^{-1}, so it is much less sensitive than infrared absorption or fluorescence. Its sensitivity for sensitive samples of biologicals or pollutants is therefore restricted. However, there is a much more sensitive technique called Surface Enhanced Raman Scattering, wherein the Raman signals are dramatically amplified when molecules are close to metal nanoparticles, especially silver. The two guiding approaches to SERS are the electromagnetic mechanism, which is a metal surface plasmon resonance mechanism that is the dominant factor, and the chemical mechanism that is connected to charge transfer interactions [59]. In the electromagnetic mechanism, the electric fields around metal nanoparticles become stronger and denser, thus enhancing the Raman scattering [60]. The method, for example, through the use of the optical properties of the nanoparticles and through the control of composition, size, shape, and spacing between the inter-particles is one of the tunings of the mechanism. Silver nanoparticle scattering is the SERS enhancement that is the best method for single-molecule detection [41].

The aggregation of silver nanoparticles often creates regions known as "hot spots" with localised plasmon coupling that greatly amplifies Raman signals [61]. More recent developments include stable aggregates attached to substrates that have greatly improved both reproducibility and durability for quantitative applications. Modified SERS substrates in the form of nanoparticle-functionalized molecules to enhance analyte adsorption and resolve the problem of low surface affinity. Tailored sensors allow for detecting molecules such as polycyclic aromatic hydrocarbons and glucose in vivo. Fiber-optic SERS sensors as well as SERS tags used in biological assays offer high spatial resolution with a possibility of multiplexed detection[62]. With nanoparticles containing high Raman cross-section molecules for SERS tags, applications include immunoassays and intracellular sensing. Recently, it has even been reported to integrate SERS with magnetic separation into a multifunctional platform for drug screening and diagnostic applications [63]. Herbicide named as Prometryn and atrazine in the soil can be detected by using AgNPs sensor based on SERS mechanism [64]. Another example of detection of thiram and methyl parathion (MP) includes Ag-nanoplates@graphene hybrids (Ag-NP@GH) as SERS substrate [65].

3.1.3 Based on Fluorescence/Foster Resonance Energy Transfer (FRET)

Fluorescence resonance energy transfer, or FRET, is a distance-dependent energy transfer process. This system will confirm the nanoscale distance between two interacting and co-localizing macromolecules. The energy transfer from the donor to the acceptor occurs because the donor's emission and the acceptor's absorption are in resonance due to the coupling of the emitter and absorber by a dipole–dipole interaction, which is, in turn, directly proportional to the efficiency of the FRET relationship of the inverse function of the sixth power of the distance between the donor and the acceptor. The distance should be less than 10 nm for a single fluorophore pair [66]. For instance, a CTAB-capped AgNP used the FRET mechanism to detect Hg (II) and pyrene dimer in the environment system. This probe is highly selective and sensitive for Hg (II) over other competing metal ions because of amalgam formation which leads to color change of pyrene solution from yellowish brown to colorless [67]. Moreover, the detection of E. coli (BL21) is done using the same mechanism, using carbon quantum dots incorporated in silver nanoparticles-based FRET sensors [68].

3.1.4 Chemical Sensing Based on Metal-Enhanced Fluorescence: MEF

MEF overcomes some of the fundamental problems associated with organic fluorophores, namely the relatively low signal-to-background ratio, poor photostability, and even photoblinking commonly found in single-molecule fluorescence detection. The phenomenon of MEF arises when the fluorescence emission of nearby fluorophores is enhanced by metal nanoparticles, primarily silver. Two effects are involved in the mechanism for the enhancement: enhanced local fields around the metal nanoparticle plus greater rates of radiative decay due to LSPR, which increases the quantum yield but diminishes fluorophore shelflife, thus enhancing photostability. It has been observed that silver nanospheres can increase the intensity of a fluorophore up to 1000 times in case the fluorophore is between 4 and 10 nm from the nanoparticle surface [69]. Research shows that silver nanoparticles improve the brightness and photostability of fluorophores and thus allow for faster and more sensitive bioassays. MEF has also been applied to create solid substrate-based assays, glucose sensors, and pH sensors. Additionally, MEF-enhanced molecular imaging agents have immense potential for fluorescent cell imaging, for instance, with improved brightness and reduced photobleaching [70, 71] (Fig. 3).

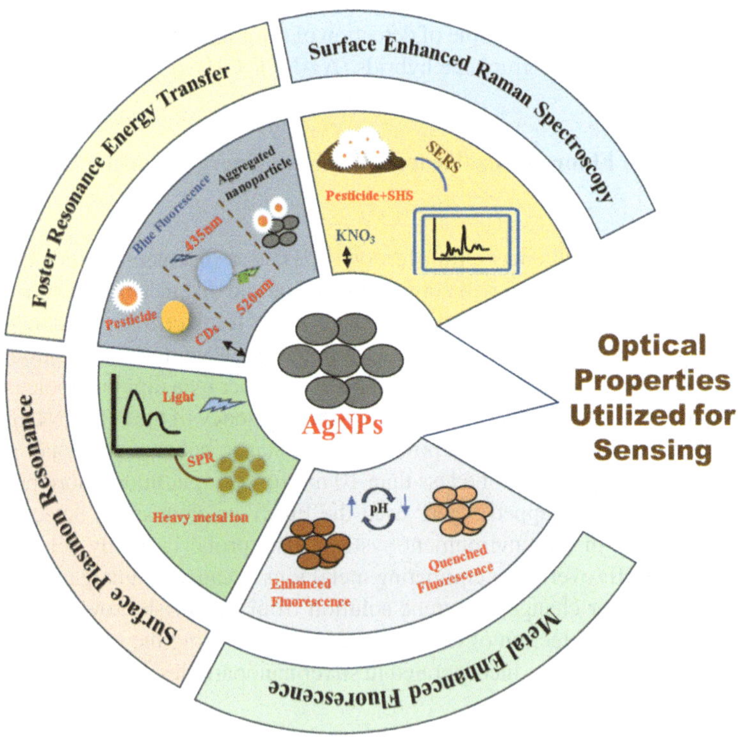

Fig. 3 Optical properties utilised for sensing application

3.2 *Electrochemical Nanosensors*

This exploits electrochemical principles to detect and transduce chemical signals as detectable electrical output. It forms two parts form it. The first part consists of a system for recognizing chemicals; the second part is a physicochemical transducer that interprets chemical responses into detectable signals. Some popular electrochemical techniques are potentiometry, conductometry, amperometry, and voltammetry. They promise to approach very high sensitivity and power efficiency with miniaturisation [72].

3.2.1 Based on Electrical Conductivity

Conductive nanomaterials show a significant change in electrical conductivity through the interaction or adsorption of analytes. This property makes the nanomaterials applicable for indirect measure of the responses to chemical and biological species. The aerogels of the ultralight and stretchable polyimide hybrid protected by the particles (of the scorpion pectin kind) were characterised as extremely sensitive, precise sensors that showed stable conductivity over a wide temperature range [73]. An electrocatalytic mechanism used for the detection of nitrate from an environmental matrix utilized screen-printed carbon electrodes (SPCEs), which are a hybrid of silver nanoparticles (Ag NPs) and copper (II)-terephthalate metal–organic frameworks (Cu-BBDCs). AgNPs/Cu-BDC MOF composite was manufactured through in-situ deposition of Ag NPs on surfaces of solvothermally synthesized cuboid shaped Cu-BDC MOF. This conjuction boost the electrical conductivity of the hybrid material make it suitable for nitrate sensing. The sensor showed outstanding sensitivity of 1159.8 $\mu A\ mM^{-1}\ cm^{-2}$ and detection limit of 0.24 μM wile operating in amperometric conditions (−0.80 V *vs.* Ag/AgCl) under a linear range of 0.5–1000 μM [74]. Another single-walled carbon nanotube has been anchored using silver nanoparticles through a coupled polymer for wide variety of sensing applications [75].

3.2.2 Catalytic Properties

Metal nanoparticles have a catalytic activity that catalyzes electrochemical reactions with target analytes. Electrochemical responses such as current or potential changes can be used to detect analytes. A electrochemical sensor based on silver nanoparticle/carboxylated multi-walled carbon nanotube infused on glassy carbon electrode was designed to detect carbendazim pesticides in agricultural products. The modified glassy carbon electrode (modified GCE/BP/MWCNT-COOH) shows excellent catalytic activity for CBZ, and the interface between the BP and MWCNT-COOH employs an antifouling effect so that no drift occurs and pinpoint detection of CBZ is carried out. Furthermore, it has the advantage of a wide linear range from 9.0

to 1000 ng/ml with the lowest detection limit at 4.0 ng/ml, which is far better than previous research results [76].

3.2.3 Ion-Gating Mechanisms

Transistor-based sensors exploit ion-gating effects using semiconducting nanomaterials such as Au coating on silver nanowires (AgNWs) embedded in polydimethylsiloxane (PDMS) electrodes. Adopting sodium chloride (NaCl) into the gold (Au) complex makes the galvanic replacement reaction effectively suppress unwanted precipitation, favouring homogeneous Au deposition over the AgNW surface. This coating could strongly decrease the contact resistance, hence enhancing the electrical performance of the devices in p-type organic transistors [77].

3.3 Other Detection Mechanisms:

3.3.1 Mechanical Transduction

The mechanical transducers, for instance, the nanomaterial-enhanced arch and resonators that are dealing with the analyte binding, may go to this limit of mass or surface stress differences for the captured cells. This leads to change in the resonating frequency or deflection of the cantilevers, which can then be translated into measurable signals for the presence and concentration of the analyte [78]. There was also promise in breakthroughs in applying physical stimuli that sensitise resistant bacteria to antimicrobials or reduce the dosage of required antimicrobials. AgNPs incoprated piezoelectric poly (vinylidene fluoride-co-trifluoroethylene) (PVDF-TrFE) nanocomposite was developed which creates electroactive enviroment which makes them suitable for antibacterial application against Escherichia coli and Staphylococcus epidermidis [79].

3.4 Thermal and Calorimetric Responses

Many nanomaterials exhibit a thermal or heat flow change, which is part of the binding event or reaction with the analyte. The detections of analyte things and contents are very sensitive when the altering of temperature or heat flow is monitored. The right mechanism of sensing and particular nanomaterial is selected as per the type of the analyte the construct targets and the sensitivities and specificity levels that are needed, as well as the specifications that are required for the application. A multicomponent nanomaterial with the corresponding sensing technique is essential to realize a versatile yet micro-cellular sensing system that is designed for the needed types of analysis [80].

3.5 Nanobiosensors

These are coupling the elements of biological recognition, for instance, antibodies and enzymes, with detectors that measure the modifications in bioactive molecules, such as pH or metabolic transformation, to name a few. These nanosensors are more often not only used for screening diseases and pests as well as abiotic and biotic stresses both in-field and laboratory. Plant wearables and nanomaterials are introduced to the plants and subsequently observed to monitor them to see whether they are in good condition, such as plant growth and even the quality of soils characteristic of a home garden or commercial greenhouse [81]

4 Application of Silver Nanomaterials in Agricultural Nanosensors

The food processing industry is of utmost importance in supporting socio-economic development within the agricultural sector, especially in economies with high purchasing power parity. In low and middle-income countries (LMICs), increasing agricultural productivity is directly related to meeting the primary needs of deprived populations [82]. Therefore, such countries are inclined to find new methods of promoting the agricultural sector to ensure quality food products.

However, toxic pollutants from diverse industrial application besides domestic garbage and sewage, sully the calibre of agricultural products [83]. The pesticide residues and heavy metals in yielded foodstuffs during the post-harvest period would expose people and animals to grave health risks. In the food processing industry, chemical preservatives are added to enhance shelf life and retain food colour [84]. Many such additives are safe, but there are exceptional cases where they may also adversely affect human health and the environment [85]. The identification and measurement of lethal contaminants in processed food products is essential before its market release. Most available monitoring techniques are expensive and time-consuming, presenting a high demand for devices that enable fast, transportable, and reliable monitoring of contaminated food samples. These innovative sensor devices, particularly nanosensors, hold the potential to significantly improve agricultural productivity, providing fast, transportable, and reliable monitoring of contaminated food samples [86].

Detecting soil quality using nanosensors is feasible for pH, nutrients, temperature, and moisture parameters. A variety of nanomaterials such as metal-based nanoparticles are applied to detect microbial diversity, soil moisture, and heavy metals that help track the soil's health. For measuring water quality, nanosensors such as nanowires and quantum dots provide ultralow detection limits for pollutants, such as nitrates and heavy metals, pesticide etc. In general, nanosensors can quickly detect traceable amounts of these contaminants in the environment. Plant stress detection can be done with the help of nanosensors that sense different signalling molecules like

H_2O_2, Ca^{2+}, and nitric oxide that change during the state of stress. Wearable sensors and optical nanosensors help monitor the responses of plants in real time. Plant wearables are fragile and lightweight nanosensors based on Microelectromechanical Systems (MEMS) technology that stick flexible sensors to the leaves and hence monitor the health status of the plant and the environment in real-time. On the other hand, quantum dots, which have a common size of approximately less than 10 nm, are semiconductor nanocrystallites made up of elements that belong to groups II to VI in the periodic table. Their peculiar photophysical properties, such as narrow emission spectra, broad absorption spectra, and emission tuned to any wavelength on demand, make them relatively efficient optical nanosensors for plant imaging and disease detection [87].

The advanced biosensors measure critical agricultural parameters with precision, improving crop yield and resource efficiency. Nitrogen sensors monitor real-time nitrogen species in soil and water, thus optimal fertiliser use without environmental pollution. Phosphorus sensors can monitor bioavailable phosphorus to ensure efficient use and minimal environmental impact [81]. Macronutrient and micronutrient sensors like potassium, calcium, boron, and zinc prevent overfertilisation [88]. The sensors of plant hormones detect the change in phytohormones under stress and offer the directed intervention: the possibility of capturing intercellular signalling without chlorophyll interference and fast and accurate SERS signals in nanosensors decorated with added AgNS@PDDA nanoshells [89]. Heavy metal sensors detect contaminants like lead and cadmium, which raises an alert concerning plant and human health risks with eco-friendly colourimetric indicators [90]. The current progress of silver nanoparticle (AgNP) synthesis and surface functionalisation have been developed as very sensitive, selective, and adaptive platforms for sensing a range of materials. Potdar et al. fabricated the optical fibre ammonium sensor using capped silver nanomaterials, showing excellent selectivity and sensitivity enhancement in detecting an ammonium ion [91]. Rossi et al. synthesised AgNPs coated with 11-MUA and proved their presence by UV–Vis spectroscopy. Particularly, the shift of its SPR absorption band between 390 and 420 nm verifies the coating process. Moreover, the confirmation was provided by Dynamic Light Scattering (DLS), SEM, EDX, and IR spectra, which reveal the presence of the 11-MUA layer. As synthesised AgNPs@11MUA was highly selective for detecting Ni^{2+} at the level of micromoles, and it still works in the presence of other cations eg. manganese (II) cobalt (II), cadmium(II) copper(II) zinc(II), Iron(II), mercury(II), lead(II) and chromium (III) [92]. Tian et al. also synthesised a green production of AgNPs on nitrogen-doped graphene, enhancing the application of such a system for H_2O_2 detection [93]. Liyanage et al. prepared electroconductive nanocomposite films with a single layer of thin polymer layers (silica/poly-4-vinylpyridine/polyvinylpyrrolidone) interfaced with AgNPs, enabling sensitive ammonia detection through conductivity changes arising due to hydrogen bonding and ammonia adsorption on the silver surface [94]. Yu et al. disclosed a colorimetric glyphosate sensor based on silver nanoparticle-functionalized copper porphyrin enzymes, in which glyphosate prevented catalytic oxidation, resulting in a distinctive colour change [95]. Electrochemical nitrate sensors designed by Ben et al. rely on observing water quality using silver nanoparticle-modified gold screen-printed electrodes. In comparison, the sensing has a detection limit of 7.7 mg/l, much below permissible drinking water levels [96]. Additionally, Hussain et al. synthesised RMF-capped AgNPs for dual-mode La(III) detection using a colourimetric and paper-based method [97]. Zhang et al. investigated a dual-modality fluorescence and colourimetric sensor for cobalt ions using the employment of rhodamine 6G-functionalized AgNPs [98]. A non-covalent sensor synthesised for phosphorus herbicides based on AgNPs combined with π-conjugated pyrimidine nitrogen and sulfur moieties. Such studies prove the versatility and scope of AgNP-based sensors for environmental, agricultural, and chemical detection[97]. A SERS nanosensor with sharp-edge and tip configurations composed of Ag decahedral nanoparticles for ultra-sensitive hydrogen peroxide detection at a minimum concentration of 10 nM, based on a novel application of H_2O_2-dependent abiotic stress responses in plants. Since the intracellular content of H_2O_2 is considered one of the indicators of abiotic stress responses in plants, the SERS-based sensor provides a non-destructive alternative to the traditional methods of detection of H_2O_2, usually plagued by the interference of chlorophyll-induced autofluorescence [99]. This plasmonic sensor makes further use of the silver nanoparticles' SPR signal enhancement properties, and due to the adenosine-imprinted PHEMA-MAC-AgNPs referred to as the MIP-Ag sensor, selects adenosine, with higher selectivity than that for other biomolecules such as dopamine and tyrosine. This dual approach emphasises the effectiveness of the sensor in terms of specific biomarker detection for stress monitoring and diagnostics in plant systems [100]. Bhandari and his colleagues developed ultra-sensitive silver nanoparticles imprinted flexible paper-based colourimetric sensors for real-time detection of paraquat in different matrices [101]. Some of the biotic and abiotic stress-sensing applications are highlighted in Table 1.

5 Conclusion

The silver nanomaterials and composite-based nanosensors will pave the way forward in the agriculture sector for precise monitoring and real-time analysis in the forthcoming times. The traits of nanosensors, such as sensitivity, selectivity, and stability, enhance their efficiency in detecting different agrochemicals, pathogens, and factors critical for crop yield. The multifunctional properties of silver nanoparticles encapsulate high antibacterial efficacy and good optical properties, enabling many applications, including monitoring nutrient levels in soil, pesticide residue detection, and disease outbreak identification. The other functional components added to silver-based composites are carbon nanostructures or polymers that enhance the capability of detection and environmental stability and, therefore, offer new applications in challenging agricultural environments. Silver nanomaterial-based nanosensors can be designed to promote sustainable agriculture through efficient management of resources, a lower intake of chemical inputs, and the least environmental impact. However, the prospect of silver nanomaterial-based nanosensors looks quite

Table 1 Application of silver nanomaterial and composite-based nanosensors

System	Abiotic/biotic stress	Material	LOD	References
Surface-enhanced raman scattering (SERS)				
Apple fruit	Parathion-ethyl	Plasmonic Ag nanoaggregates	1 mg/L	[102]
Barley, wheat	*Fusarium Graminearum*	Ag nanoshelled (Agnss)	-	[103]
Complex matrix	Urea	Silver nanosol (Agnp)	4.92 nm	[104]
Aqueous solution	Chloride anions (Cl^-)	Epoxy-silver nanocomposite	14 mM	[105]
Rice samples	Isocarbophos	Femof@OCTB-PEG400-AG(I)	0.010 nm	[106]
Water, apple, and carrot samples	Parathion methyl	Graphene quantum dots modified silver nanoparticles nanoprobe	0.017 mg/L	[107]
Electrochemical/optical genosensor				
GM tomato samples	Camv35s gene segment	Fe_3O_4-Au@Ag	1.26×10^{-17} M	[108]
Citrus	Salmonella	DNA-HRP Functionalized Agnp Probe	-	[109]
Soil and water	Glyphosate	Agnps-ergo nanocomposites	0.015 mg m l^{-1}	[110]
Complex matrix	Fe^{3+}	Surface immobilised silver nanoparticles (Agnps)	Optical:25 mM Electrochemical: 16 mM	[111]
Water, and grape extract,	Malathion	Ag-GO and Acetylthiocholine (Atch)	0.01 pM,	[112]
Other mechanism				
Aqueous solution	Mercury (II)	Au–Ag Bnps	0.526 + 0.001 mg L^{-1}	[113]
Aqueous solution	Mercury ion	Ag–S-gCN QDs	0.13 μM	[114]

promising; it also includes several challenges related to scale-up, biocompatibility, and regulatory compliance. It would be essential for various disciplines to interact and collaborate for better research to address the constraints and issues related to the effective use of silver nanomaterial-based nanosensors in field devices. Such research and development advancements will enhance agricultural processes, increase crop yields, minimise waste, and promote sustainable farming techniques.

6 Competing Interests:

The authors declare that they have no competing interest.

Declaration of generative AI and AI-assisted technologies in the writing process:

During the preparation of this work the book chapter author(s) have not taken any assistance from AI, and the book chapter author(s) reviewed and edited the content as needed and take(s) full responsibility for the content of the publication. All authors read and approved the final version of the manuscript.

References

1. Islam T, Danishuddin, Tamanna NT, Matin MN, Barai HR, Haque MA (2024) Resistance mechanisms of plant pathogenic fungi to fungicide, environmental impacts of fungicides, and sustainable solutions. Plants 13(19):2737
2. Karthik P, Saravanaraj A, Vijayalakshmi V, Ragavan K, Vaidyanathan VK (2024) Nanobiopesticides: significance, preparation technologies, safety aspects, and commercialization for sustainable agriculture. In: Nanohybrid fungicides. Elsevier, pp 367–393
3. Usman M, Farooq M, Wakeel A, Nawaz A, Cheema SA, ur Rehman H, Sanaullah M (2020) Nanotechnology in agriculture: current status, challenges and future opportunities. Sci Total Environ 721:137778
4. Pardeshi S, Mohite P, Rajput T, Puri A (2024) The nanotech potential of curcumin in pharmaceuticals: an overview. Curr Drug Discov Technol 21(2):e260723219113
5. Zahra M, Abrahamse H, George BP (2024) Green nanotech paradigm for enhancing sesquiterpene lactone therapeutics in cancer. Biomed Pharmacother 173:116426
6. Husen A (2025a) Plant response to silver nanoparticles. Springer Nature Singapore Pte Ltd., Singapore
7. Husen A, Siddiqi KS (2023) Advances in smart nanomaterials and their applications. Elsevier Inc., USA https://doi.org/10.1016/C2021-0-02202-1
8. Jin-Chul K, Madhusudhan A, Husen A (2021) Smart Nanomaterials in Biomedical Applications. Springer Nature Switzerland AG, Cham, Switzerland. https://doi.org/10.1007/978-3-030-84262-8
9. Siddiqi KS, Husen A (2022) Plant response to silver nanoparticles: a critical review. Crit Rev Biotechnol 42(7):773–990. https://doi.org/10.1080/07388551.2021.1975091
10. Almatroudi A (2024) Unlocking the potential of silver nanoparticles: from synthesis to versatile bio-applications. Pharmaceutics 16(9)
11. Diallo F, Seck I, Ndoye SF, Niang T, Dieng SM, Thiam F, Seck M (2024) Green synthesis and anti-inflammatory activity of silver nanoparticles based on leaves extract of Aphania senegalensis. Biochem Res Int 2024:3468868
12. Farooqi MA, Bae S, Kim S, Bae S, Kausar F, Farooqi HMU, Kang CU (2024) Eco-friendly synthesis of bioactive silver nanoparticles from black roasted gram (Cicer arietinum) for biomedical applications. Sci Rep 14(1):22922
13. Lima BGA, Silva RR, Meira HM, Durval IJB, Macedo Bezerra Filho C, Silva TAL, Luna JM (2024) Synthesis and characterization of silver nanoparticles stabilized with biosurfactant and application as an antimicrobial agent. Microorganisms 12(9)
14. Trivedi R, Shende P (2022) Nanotech-based food: an initiative for alternative pharmaceuticals. Curr Pharm Biotechnol 23(14):1739–1749
15. Hao D, Zhang Y, Ding Y, Yan Q (2024) Preparation and properties of silver-carrying nano-titanium dioxide antimicrobial agents and silicone composite. Sci Rep 14(1):18870

16. Bilkay M, Karatas B, Satana Kara HE (2024) Development of a new nanosensor for the determination of food coloring Sunset Yellow in powder drinks using L-cysteine coated copper nanoclusters. Turk J Chem 48(2):218–228
17. Dhapte A (2024) Nanosensors market research report information, by type (chemical and bacterial, optical, synthetic, electrochemical, mechanical), by application (robotics, health-care, food & beverage, automotive, petroleum), by technology (wind turbines, fuel cells, micro turbines)–forecast till 2032
18. Al-Dhabi NA, Ghilan AM, Esmail GA, Arasu MV, Duraipandiyan V, Ponmurugan K (2019) Environmental friendly synthesis of silver nanomaterials from the promising Streptomyces parvus strain Al-Dhabi-91 recovered from the Saudi Arabian marine regions for antimicrobial and antioxidant properties. J Photochem Photobiol B 197:111529
19. McCright J, Skeen C, Yarmovsky J, Maisel K (2022) Nanoparticles with dense poly (ethylene glycol) coatings with near neutral charge are maximally transported across lymphatics and to the lymph nodes. Acta Biomater 145:146–158
20. Bhattacharjee K, Prasad BL (2023) Surface functionalization of inorganic nanoparticles with ligands: a necessary step for their utility. Chem Soc Rev 52(8):2573–2595
21. Urmi R, Banerjee P, Singh M, Singh R, Chhillar S, Sharma N, Qamar I (2024) Revolutionizing biomedicine: aptamer-based nanomaterials and nanodevices for therapeutic applications. Biotechnol Rep e00843
22. Naaz F, Farooq U, Khan MM, Ahmad T (2020) Multifunctional efficacy of environmentally benign silver nanospheres for organic transformation, photocatalysis, and water remediation. ACS Omega 5(40):26063–26076
23. Jeon YN, Ryu SJ, Lee HY, Kim JO, Baek JS (2024) Green synthesis of silver nanoparticle using black mulberry and characterization, phytochemical, and bioactivity. Antibiotics (Basel) 13(8)
24. Shaverdi M, Rafiee Z, Razmjoue D, Oryan A, Ghaedi M, Abidi H (2024) Antibacterial, antioxidant, and anti-giardia properties of the essential oil, hydroalcoholic extract, and green synthesis of the silver nanoparticles of Salvia mirzayanii plant. Sci Rep 14(1):22866
25. Abdelaziz AM, Elshaer MA, Abd-Elraheem MA, Ali O, Haggag MI, El-Sayyad GS, Attia MS (2023) Ziziphus spina-christi extract-stabilized novel silver nanoparticle synthesis for combating Fusarium oxysporum-causing pepper wilt disease: in vitro and in vivo studies. Arch Microbiol 205(2):69
26. Ahmad S, Xu Q, Tariq M, Song M, Liu C, Yan H (2024) Assessing the potential of aconitum laeve extract for biogenic silver and gold nanoparticle synthesis and their biological and catalytic applications. Molecules 29(11)
27. Corciova A, Mircea C, Fifere A, Turin-Moleavin IA, Rosca I, Macovei I, Burlec AF (2024) Biogenic synthesis of silver nanoparticles mediated by aronia melanocarpa and their biological evaluation. Life (Basel) 14(9)
28. Asong JA, Frimpong EK, Seepe HA, Katata-Seru L, Amoo SO, Aremu AO (2023) Green synthesis of characterized silver nanoparticle using cullen tomentosum and assessment of its antibacterial activity. Antibiotics (Basel) 12(2)
29. Peng Y, Hu Y, Liu Y, Lin H (2024) Green synthesis of dracocephalum kotschyi-coated silver nanoparticles: antimicrobial, antioxidant, and anticancer potentials. Med Sci Monit 30:e944823
30. Cui D, Lei Y (2024) Exploring the potential of micro-nano composite structures for COVID-19 vaccines and beyond. Biomater Transl 5(1):86–88
31. Rashid A, Haque M, Islam S, Uddin Labib K (2024) Nanotechnology-enhanced fiber-reinforced polymer composites: recent advancements on processing techniques and applications, vol 10. Heliyon
32. Cobos M, De-La-Pinta I, Quindós G, Fernández MJ, Fernández MD (2020) Graphene oxide–silver nanoparticle nanohybrids: synthesis, characterization, and antimicrobial properties. Nanomaterials 10(2):376
33. Çelik Y, Akçil B, Aksoy E (2023) SERS performance of silver nanoparticle/reduced graphene oxide-coated filter membranes. Fullerenes Nanotubes Carbon Nanostruct 31(9):833–844

34. Das TK, Ghosh SK, Das NC (2023) Green synthesis of a reduced graphene oxide/silver nanoparticles-based catalyst for degradation of a wide range of organic pollutants. Nano-Struct Nano-Objects 34:100960
35. Chih Y-K, You J-L, Lin W-H, Chang Y-H, Tseng C-C, Ger M-D (2023) A novel method for the fabrication of antibacterial stainless steel with uniform silver dispersions by silver nanoparticle/polyethyleneimine composites. Materials 16(10):3719
36. Skiba M, Vorobyova V, Pasenko O (2022) Surface modification of titanium dioxide with silver nanoparticles for application in photocatalysis. Appl Nanosci 12(4):1175–1182
37. Wu C, Xu S, Wang W (2021) Synthesis and applications of silver nanocomposites: a review. J Phys Conf Ser
38. Hussein EY, Al-Murshdy JMS, Radhi NS, Al-Khafaji Z (2024) Surface modification of titanium alloy by titania/silver multilayers coating for biomedical application. J Adv Res Micro Nano Eng 20(1):66–78
39. Hasan KF, Champramary S, Al Hasan KN, Indic B, Ahmed T, Pervez MN, Hofmann T (2023) Eco-friendly production of cellulosic fibers from Scots pine wood and sustainable nanosilver modification: a path toward sustainability. Res Eng 19:101244
40. Gün GZ, Demiral A, Bozkaya O, Yiğitoğlu M (2021) In situ synthesis of silver nanoparticles on modified poly (ethylene terephthalate) fibers by grafting for obtaining versatile antimicrobial materials. Polym Bull 78:7241–7260
41. Chen X, Wang L, Ding C, Xie H, Zou H, Deng J, Ding Y (2024) Highly efficient removal of radioactive iodine anions by nano silver modified activated carbon fiber. Appl Surf Sci 643:158644
42. Rather AH, Khan RS, Wani TU, Rafiq M, Qureashi A, Rather S-U, Sheikh FA (2024) Hydrothermal modification of polyurethane fibers using hyaluronic acid and silver nanoparticles for wound healing. J Polym Environ 1–20
43. Kravtsov AA, Medyanik EV, Tarala VA, Lapin VA, Malyavin FF, Chikulina IS, Vakalov DS, Chapura OM, Tarala LV, Suprunchuk VE, Blinov AV (2024) Silver nanoparticles as a new sintering additive for the fabrication of YAG: Ce optical luminescent ceramics. J Am Ceram Soc 107(2):1020–1032
44. Luo C, Fei D, Zhou J, Tao J, Yao Z, Liu J, Tao X (2024) Silver particle-modified BN@ C composites for thermal management and microwave absorption. Ceram Int 50(13):22252–22265
45. Darbandi KR, Amin BK (2024) Innovation and evaluations of 3D Printing resins modified with Zirconia nanoparticles and Silver nanoparticle-immobilized Halloysite nanotubes for Dental Restoration. Coatings 14(3):310
46. Cheng J, Han J, Xu C, Zhang H, Dong X, Zhuang X, Qi M (2024) Nano silver composite hydroxyethyl methacrylate/vinylpyrrolidone hydrogel: tissue adhesives with antibacterial properties. Comp Part C Open Access 14:100498
47. Blosi M, Brigliadori A, Ortelli S, Zanoni I, Gardini D, Vineis C, Costa AL (2024) Re-designing nano-silver technology exploiting one-pot hydroxyethyl cellulose-driven green synthesis. Front Chem 12:1432546
48. Behera J, Patra S, Nazrul S, Sharma SK, Kumar D, Verma MK, Swain SK (2024) Nano boron nitride laminated poly (ethyl methacrylate)/poly (vinyl alcohol) composite films imprinted with silver nanoparticles as gas barrier and bacteria resistant packaging materials. J Appl Polym Sci 141(16):e55246
49. Chittratan P, Detsri E, Chalitangkoon J, Mathaweesansurn A, Monvisade P (2024) Antimicrobial nanolayer films of chloroxylenol–carboxyethylchitosan–modified silver nanoparticles for enhanced surgical suture performance. Colloids Surf A 693:133957
50. Ku C-A, Chung C-K (2023) Advances in humidity nanosensors and their application. Sensors 23(4):2328
51. Wang W, Yang R, Liu M (2024) Design and fabrication of micro/nano sensors and actuators, vol 15. MDPI, p 674
52. Riaz A, Din SU, Rakha A, Shad NA, Munawar A (2024) Nanotechnology-based sensors' design and fabrication. In: Nanotechnology-based sensors for detection of environmental pollution. Elsevier, pp 175–198

53. Alzoubi F, Ahmad AA, Aljarrah IA, Migdadi A, Al-Bataineh QM (2023) Localize surface plasmon resonance of silver nanoparticles using Mie theory. J Mater Sci Mater Electron 34(32):2128
54. Shaeel Ahmed A-T, Zaheer K (2020) Biogenic synthesis of silver nanoparticles, sensing and photo catalytic activities for bromothymol blue. J Photochem Photobiol 3–4:100010
55. Dabhade AH, Verma RP, Paramasivan B, Kumawat A, Saha B (2023) Development of silver nanoparticles and aptamer conjugated biosensor for rapid detection of E. coli in a water sample. 3 Biotech 13(7):244
56. Sardar S, Anas M, Maity S, Pal S, Parvej H, Begum S, Halder UC (2019) Silver nanoparticle modulates the aggregation of beta-lactoglobulin and induces to form rod-like aggregates. Int J Biol Macromol 125:596–604
57. Dutta R, Kundu T (2024) Plasmonic coupling dependent sensitivity in localized surface plasmon resonance based optical sensors. Opt Fiber Technol 88:103970
58. Alrahili M (2024) Single silver nanoparticles: local refractive index response to localized surface plasmon resonance and molar attenuation coefficient. Colloid J 86(1):138–152
59. Yu W, Lin X, Duan N, Wang Z, Wu S (2023) A fluorescence and surface-enhanced Raman scattering dual-mode aptasensor for sensitive detection of deoxynivalenol based on gold nanoclusters and silver nanoparticles modified metal-polydopamine framework. Anal Chim Acta 1244:340846
60. Jin C, Chen J, Du Z, Liu C, Liu F, Hu J, Han M (2022) Two orders of magnitude extra SERS enhancement on silver nanoparticle-based substrate induced by laser irradiation in nitrogen ambient. Spectrochim Acta A Mol Biomol Spectrosc 265:120372
61. Hang Y, Wang A, Wu N (2024) Plasmonic silver and gold nanoparticles: shape-and structure-modulated plasmonic functionality for point-of-caring sensing, bio-imaging and medical therapy. Chem Soc Rev
62. Tsao CW, Zheng YS, Sun YS, Cheng YC (2021) Surface-enhanced Raman scattering (SERS) spectroscopy on localized silver nanoparticle-decorated porous silicon substrate. Analyst 146(24):7645–7652
63. Hassan MM, Zareef M, Jiao T, Liu S, Xu Y, Viswadevarayalu A, Chen Q (2021) Signal optimized rough silver nanoparticle for rapid SERS sensing of pesticide residues in tea. Food Chem 338:127796
64. Rubira RJ, Camacho SA, Constantino CJ, Sanchez-Cortes S (2022) Increasing the sensitivity of surface-enhanced Raman scattering detection for s-triazine pesticides by taking advantage of interactions with soil humic substances. J Raman Spectrosc 53(1):40–48
65. Xiujuan W, Chuhong Z, Xiaoye H, Qiaoling X, Huaping Z, Guowen M, Yong L (2019) Highly sensitive surface-enhanced Raman scattering detection of organic pesticides based on Ag-nanoplate decorated graphene-sheets. Appl Surf Sci 486:405–410
66. Choi J-H, Ha T, Shin M, Lee S-N, Choi J-W (2021) Nanomaterial-based fluorescence resonance energy transfer (FRET) and metal-enhanced fluorescence (MEF) to detect nucleic acid in cancer diagnosis. Biomedicines 9(8):928
67. Walekar LS, Hu P, Molamahmood HV, Long M (2018) FRET based integrated pyrene-AgNPs system for detection of Hg (II) and pyrene dimer: applications to environmental analysis. Spectrochim Acta Part A Mol Biomol Spectrosc 198:168–176
68. Bai X, Hou X, Ga L, Ai J (2024) Aptamer-carbon quantum dots and silver nanoparticles construct a FRET sensor for sensitive detection of E. coli. Biomed Anal 1(2):162–173
69. Augustine S, Chinnamani MV, Mun CW, Shin JY, Trung TQ, Hong SJ, Lee NE (2024) Metal-enhanced fluorescence biosensor integrated in capillary flow-driven microfluidic cartridge for highly sensitive immunoassays. Biosens Bioelectron 248:115987
70. Jin Y, Petrescu FIT, Wang Y, Li X, Li Y, Shi G (2023) Spiropyran-based soft substrate with SPR, anti-reflection and anti-NRET for enhanced visualization/fluorescence dual response to metal ions. Materials (Basel) 16(10)
71. Thai DV, Pham VB, Sai CD, Nguyen THG, Tran TD, Tran TH, Bui HV (2024) Synthesis of SiO(2)@Ag nanocomposite for investigating metal-enhanced fluorescence and surface-enhanced Raman spectroscopy. J Fluoresc

72. Majeed S, Naqvi STR, ul Haq MN, Ashiq MN (2022) Electroanalytical techniques in biosciences: conductometry, coulometry, voltammetry, and electrochemical sensors. In: Analytical techniques in biosciences. Elsevier, pp 157–178
73. Xu H, Chen W, Wang C, Jia T, Wang D, Li G, Fan X (2023) Ultralight and flexible silver nanoparticle-wrapped "scorpion pectine-like" polyimide hybrid aerogels as sensitive pressor sensors with wide temperature range and consistent conductivity response. Chem Eng J 453:139647
74. Amali RKA, Lim HN, Ibrahim I, Zainal Z, Ahmad SAA (2022) Silver nanoparticles-loaded copper (II)-terephthalate framework nanocomposite as a screen-printed carbon electrode modifier for amperometric nitrate detection. J Electroanal Chem 918:116440
75. Ding J, Li Z, Kodra O, Couillard M, Ouyang J, Lapointe F, Malenfant PR (2023) Silver nanoparticles anchored on single-walled carbon nanotubes via a conjugated polymer for enhanced sensing applications. ACS Omega 8(15):14219–14232
76. Liao X, Luo X, Li Y, Zhou Y, Liang Q, Feng K, Xiong J (2023) An antifouling electrochemical sensor based on multiwalled carbon nanotubes functionalized black phosphorus for highly sensitive detection of carbendazim and corresponding response mechanisms analyses. Microchem J 190:108671
77. Park H, Kim S, Lee J, Kim K, Na H, Kim Y, Sim K (2024) Facile strategy for uniform gold coating on silver nanowires embedded PDMS for soft electronics. npj Flex Electron 8(1):63
78. Munawar A, Ong Y, Schirhagl R, Tahir MA, Khan WS, Bajwa SZ (2019) Nanosensors for diagnosis with optical, electric and mechanical transducers. RSC Adv 9(12):6793–6803
79. Moreira J, Fernandes MM, Carvalho EO, Nicolau A, Lazic V, Nedeljković JM, Lanceros-Mendez S (2022) Exploring electroactive microenvironments in polymer-based nanocomposites to sensitize bacterial cells to low-dose embedded silver nanoparticles. Acta Biomater 139:237–248
80. Yaoxin F, Tiegen L, Haonan W, Ziyihui W, Lili H, Junfeng J, Tianhua X (2024) Applications of nanomaterial technology in biosensing. J Sci Adv Mater Dev 9(2):100694
81. Sachan R, Sahni T, Gangwar K, Tiwari A, Sachan K (2022) Role of nanobiosensors for advancement of agriculture. Rec Innov Appr Agric Sci 193
82. Ridhi R, Saini G, Tripathi S (2024) Nanotechnology as a sustainable solution for proliferating agriculture sector. Mater Sci Eng B 304:117383
83. Agostini MG, Roesler I, Bonetto C, Ronco AE, Bilenca D (2020) Pesticides in the real world: the consequences of GMO-based intensive agriculture on native amphibians. Biol Conserv 241:108355
84. Yadav H, Kumar R, Sankhla M (2020) Residues of pesticides and heavy metals in crops resulting in toxic effects on living organism. In J Seybold Rep 15:1527–1541
85. Alengebawy A, Abdelkhalek ST, Qureshi SR, Wang M-Q (2021) Heavy metals and pesticides toxicity in agricultural soil and plants: ecological risks and human health implications. Toxics 9(3):42
86. Tarannum N, Gautam A, Chauhan T, Kumar D (2024) Nanomaterial based sensors for detection of food contaminants: a prospect. Sensing Technology 2(1):2373196
87. Yang Z, Cao L (2024) Electrochemical sensor based on silver nanoparticles/carboxylated multi-walled carbon nanotubes modified glassy carbon electrode for detection of carbamate pesticide in food. J Food Meas Charact 1–12
88. Dattatreya S, Khan AN, Jena K, Chatterjee G (2023) Conventional to modern methods of soil NPK sensing: a review. IEEE Sens J
89. Son WK, Choi YS, Han YW, Shin DW, Min K, Shin J, Lee MJ, Son H, Jeong DH, Kwak SY (2023) In vivo surface-enhanced Raman scattering nanosensor for the real-time monitoring of multiple stress signalling molecules in plants. Nat Nanotechnol 18(2):205–216
90. Ahmed F, Kabir H, Xiong H (2020) Dual colorimetric sensor for Hg^{2+}/Pb^{2+} and an efficient catalyst based on silver nanoparticles mediating by the root extract of *Bistorta amplexicaulis*. Front Chem 8:591958
91. Potdar RP, Khollam YB, Shaikh SF, More PS, Rana AUHS (2022) Polyvinylpyrrolidone-capped silver nanoparticles for highly sensitive and selective optical fiber-based ammonium sensor. Nanomaterials 12(19):3373

92. Rossi A, Zannotti M, Cuccioloni M, Minicucci M, Petetta L, Angeletti M, Giovannetti R (2021) Silver nanoparticle-based sensor for the selective detection of nickel ions. Nanomaterials 11(7):1733
93. Tian Y, Wang F, Liu Y, Pang F, Zhang X (2014) Green synthesis of silver nanoparticles on nitrogen-doped graphene for hydrogen peroxide detection. Electrochim Acta 146:646–653
94. Liyanage MV, Latour RA, Chumanov G (2024) Electroconductive silver nanoparticle composite films for sensing ammonia in the presence of water vapor. Sens Actuators B Chem 401:134990
95. Yu L-H, Jiang R, Yang C-L, Pang Y-H, Shen X-F (2024) Colorimetric sensing based on silver nanoparticle-functionalized copper porphyrin nanozyme for glyphosate. Food Biosci 61:104577
96. Ben MN, dos Santos MB, Espiña B, Queirós RB (2024) Development of electrochemical sensors based on silver nanoparticles electrodeposited on gold screen-printed electrodes: application to nitrate trace analysis in water. J Sens Sens Syst 13(1):135–145
97. Hussain K, Umar AR, Rasheed S, Hassan M, Laiche MH, Muhammad H, Shah MR (2024) Smartphone-Integrated resorcinarene macrocycle capped silver nanoparticles (RMF-AgNPs) probe for enhanced La (III) detection in diverse environments. J Indus Eng Chem
98. Zhang Y, Wang H, Lu M, Li G, Bai M, Yang W, Li G (2024) A dual-modality sensing probe of fluorescent and colorimetric for detection of cobalt ion based on silver nanoparticles functionalized rhodamine 6G derivatives. Chemosphere 362:142790
99. Tian Y, Cong L, Sun H, Wei W, Wu F, Li F, Niu W (2024) An in vivo surface-enhanced raman scattering nanosensor for detecting hydrogen peroxide in plants under abiotic stresses. ACS Appl Nano Mater 7(17):20308–20316
100. Göktürk I, Bakhshpour M, Çimen D, Yılmaz F, Bereli N, Denizli A (2022) SPR signal enhancement with silver nanoparticle-assisted plasmonic sensor for selective adenosine detection. IEEE Sens J 22(15):14862–14869
101. Bhandari S, Parihar VS, Kellomäki M, Mahato M (2024) Highly selective and flexible silver nanoparticles-based paper sensor for on-site colorimetric detection of paraquat pesticide. RSC Adv 14(39):28844–28853
102. Li H, Merkl P, Sommertune J, Thersleff T, Sotiriou GA (2022) SERS Hotspot engineering by aerosol self-assembly of plasmonic Ag nanoaggregates with tunable interparticle distance. Adv Sci 9(22):2201133
103. Son WK, Choi YS, Han YW, Shin DW, Min K, Shin J, Kwak S-Y (2023) In vivo surface-enhanced Raman scattering nanosensor for the real-time monitoring of multiple stress signalling molecules in plants. Nat Nanotechnol 18(2):205–216
104. Aihui L, Haolin W, Dongmei Y, Zhiliang J (2019) A simple and sensitive SERS quantitative analysis method for urea using the dimethylglyoxime product as molecular probes in nanosilver sol substrate. Food Chem 271:39–46
105. Franco A, Velásquez-Ordoñez C, Ojeda-Martínez M, Ojeda-Martínez M, Barrera-Calva E, Rentería-Tapia V (2022) Paper based colorimetric sensor for detection of chloride anions in water using an epoxy-silver nanocomposite. J Nanopart Res 24(3):54
106. Li C, Yu F, Yang J, Bai H, Ma X, Jiang Z (2023) SERS-and absorbance-based catalytic assay for determination of isocarbophos using aptamer-modified FeMOF nanozyme and in situ generated silver nanoparticles. Microchim Acta 190(1):4
107. Li Y, Chen S, Lin D, Chen Z, Qiu P (2020) A dual-mode nanoprobe for the determination of parathion methyl based on graphene quantum dots modified silver nanoparticles. Anal Bioanal Chem 412:5583–5591
108. Ye Y, Mao S, He S, Xu X, Cao X, Wei Z, Gunasekaran S (2020) Ultrasensitive electrochemical genosensor for detection of CaMV35S gene with Fe3O4-Au@ Ag nanoprobe. Talanta 206:120205
109. He H, Sun D-W, Wu Z, Pu H, Wei Q (2022) On-off-on fluorescent nanosensing: materials, detection strategies and recent food applications. Trends Food Sci Technol 119:243–256
110. Butmee P, Tumcharern G, Songsiriritthigul C, Durand MJ, Thouand G, Kerr M, Samphao A (2021) Enzymatic electrochemical biosensor for glyphosate detection based on acid phosphatase inhibition. Anal Bioanal Chem 413:5859–5869

111. Pedre I, González GA (2022) Silver nanoparticles-based sensing platform for Fe3+ ions in complex matrices. Miner Eng 185:107712
112. Deosarkar T, Chandrasekaran N, Mukherjee A (2021) An ultra-sensitive and selective AChE based colorimetric detection of malathion using silver nanoparticle-graphene oxide (Ag-GO) nanocomposite. Anal Chim Acta 1142:73–83
113. Ali S, Chen X, Ahmad S, Almehmadi M, Alsaiari AA, Allahyani M, Li L (2023) Morphology-controllable bimetallic gold nanostructures for mercury detection: recent developments, challenges and prospects. Arab J Chem 16(9):104997
114. Samarjit P, Ugrabadi S, Shubhalaxmi C, Garudadhwaj H (2022) Silver nanoparticles embedded sulfur doped graphitic carbon nitride quantum dots: a fluorescent nanosensor for detection of mercury ions in aqueous media. Colloids Surf A 648:129377

Zinc Oxide Nanosensors in Precision Agriculture: A Green Approach to Improve Global Food Security

Swati Sachdev and Mohammad Israil Ansari

Abstract Use of agro-chemicals, escalating pollution, and abrupt changes in climatic conditions are major threats to agricultural system, increasing soil degradation and reducing food production. To manage these challenges sustainably, nanosensors play a supportive role. Nanosensors fabricated from different nanomaterials due to extraordinary properties and small size show excellent performance in detection of factors affecting plant cultivation at micro levels. Addressing these issues at grassroots level help to optimize the agricultural activities and boost precision agriculture. Array of materials are employed for nanosensor development, among them ZnO nanosensors have been regarded as excellent choice. ZnO nanomaterials exhibit unique features such as wide band gap and high excitation energy, along with other characteristics that offer application in electrochemical and optical sensing. Electrochemical ZnO nanosensor are commonly researched nanosensor for application in agricultural fields. These sensors aid in detection of range of contaminants as well as environmental and biological stress in real-time, facilitating effective management. Due to immense potential of ZnO nanosensor in facilitating precision agriculture, the present chapter aims to explore the potential of zinc oxide nanosensors in executing agricultural activities sustainably.

Keywords Electrochemical sensor · Nanobiosensor · Nutrient level · Pesticides and heavy metal pollution · Nanomaterials

---

S. Sachdev
Department of Liberal Education, Era University, Sarfarazganj, Lucknow, Uttar Pradesh, India
e-mail: swati.sachdev@erauniversity.in

M. I. Ansari (✉)
Department of Botany, University of Lucknow, Lucknow, Uttar Pradesh, India
e-mail: ansari_mi@hotmail.com; ansari_mi@lkouniv.ac.in

A. Husen (ed.), *Nanobiosensors for Agricultural and Other Related Sectors*, Smart Nanomaterials Technology, https://doi.org/10.1007/978-981-96-6499-3_11

1 Introduction

Increasing susceptibility of agricultural productivity to biotic and abiotic stresses stemming from escalated pollution, climate change, and land degradation has upthrust the apprehension of food and nutrition availability. Agricultural activities in conjugation with industrial and urban ventures have formally injected toxic xenobiotic organic compounds and heavy metals in to the environmental matrixes that exhibit attributes to quench life-supporting characteristics of agro-ecosystem. Chemical pesticides and fertilizers are the epitome of sources of agro-pollution, exposing soil and its flora and fauna to risk of toxicity and degradation [1, 2], inflaming abiotic and biotic challenges, in particular, salinity, drought, and resistance development, and fueling climate change due to emission of greenhouse gases [3–5]. Confiscation of activities generating toxic compounds and halting fluctuations in macro- and micro-agro-climatic conditions, altogether are not worth considering. However, minimizing entry as well as accumulation of toxic compounds and ameliorating effect of physical factors on agro-environment is practically achievable approach. Focusing on monitoring and quantification of factors like temperature, salinity, soil moisture, nutrient level, pollutants, pests, crop growth parameters, and others in agricultural ecosystems in real-time can be foreseen as an aid in mitigating accumulation of contaminants and to arrest the damaging impacts, refining global food security and economic development.

Sensor is an analytical tool that aid in detection, identification, and quantification of analyte or changes in the environment and relay information in the form of signal, which can be electrical, thermal, mechanical or optical, to electronic device, generating measurable output [6, 7]. The performance of the sensors are rapid, rugged, and scalable, with some exhibiting high sensitivity providing observations for trace quantity of contaminants or minor change in the physical attributes [8, 9]. Real-time monitoring and quantification of physical, chemical, and/or biological factors influencing the agricultural activities can be undertaken using sensor devices. Use of sensors aid in direct detection of any change or contamination without addition of any reagent or chemical entity, reducing risk of pollution in the environment [10]. Eventually, this approach helps to optimize agricultural activities and conditions, which often referred as precision farming [8, 11], facilitating sustainable and efficient food production.

In aegis of extensive research and development, emerging technologies are replacing conventional techniques, designing ultrasensitive and highly efficient sensing devices like nanosensors that uses nanomaterials to detect the analyte at atomic scale with precision [8, 9] Nanosensors exhibit unique characteristics like small size, high surface/volume ratio, high surface reactivity, and good thermal stability that enable them to display specificity in target detection and monitoring [9, 12]. Such nanosensors have several applications in the field of agriculture including monitoring of edaphic physicochemical attributes, agro-climatic conditions, biotic constraints, and toxic compounds like heavy metals, pesticides, and other xenobiotics [8, 12, 13]. To fabricate nanosensors various nanomaterials like carbon nanosheets, carbon nanotubes, metal and metal oxide nanoparticles, nanocomposites, quantum dots, and many others are used [9, 14]. Among array of nanomaterials, zinc oxide based nanosensors that owe versatile characteristics like quantum size effect, and high specific surface area, and regarded safe as well as approved by Food and Drug Administration (FDA) for use in activities related to food, could have great opportunity in agriculture sector [15, 16]. The present chapter aims to explore the potential of zinc oxide nanosensors in agricultural ecosystems. This chapter will present the information on role and application of ZnO based nanosensors for precision farming.

2 Nanosensors

Nanosensors are novel tool designed at nano-level i.e., have at least one dimension ~100 nm, to measure the physical (temperature, soil moisture), chemical (heavy metals, pesticide residue), and biological (antibodies, microbial toxins, plant-emitted compounds) entities occurring within living body or in the environment [17, 18]. It can detect low level of analyte and transduce signal to produce result that can be easily analyzed [6]. However, to detect the target compound in trace amount, sensor require high sensitivity as well as selectivity that minimize the noise interference and tolerance to environmental factors like temperature, corrosion, and humidity [19]. The physical and chemical properties of material used for developing nanomaterials exhibit unique characteristics such as large surface area that facilitate better outcome. Moreover, formulating a nanomaterial improve the attributes of the bulk material, providing an edge in detection [6, 10]. Owing to nano-size, these sensors can interact with minute particles or biomolecules like protein with ease, which otherwise are difficult to measure or detect at macro- or large-scale [6]. Versatility in action allow application of nanosensors in vast areas including defence, agriculture, industry, environmental monitoring, aerospace, etc.

Nanosensor are designed by combing nanomaterial with transducer and detector (Fig. 1). Nanomaterial attached to sensor device recognize the analyte and send information to transducer, which further send signals, which are amplified and generate outcome. Nanosensors are classified on basis of recognizing principle, transducing mechanism used, and/or material used for sensing [10, 17]. On the basis of analyte recognition, nanosensor could be affinity-based or catalytic-based nanosensors that can recognize molecules tion like DNA, enzymes, etc. [17]. Transducing mechanisms based nanosensors are electrical, electrochemical, optical, piezoelectical, thermal, calorimetric and magnetic nanosensors [17, 18]. On basis of material used for sensing/recognizing analyte, nanosensors could be bionanosensor or chemical nanosensor. On the basis of transducer, biosensor are applied for detection of different analytes. The electrochemical transducers which works on assessing conductance, resistivity, and capacitance generated on surface of sensor result in electrical signals and eventually correspond the concentration of the analyte [20]. Such types of biosensors find application in pathogen detection due to no interference by luminescent compound

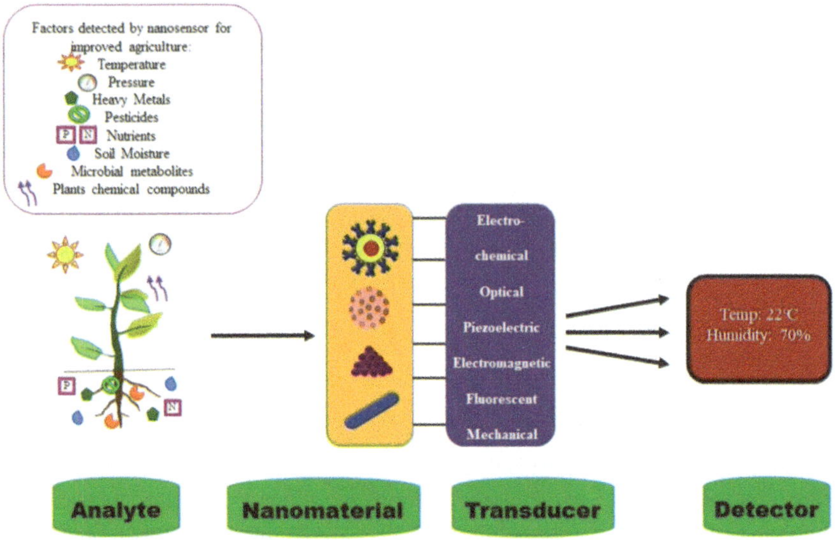

Fig. 1 Flowchart depicting structure of a nanosensor detecting analytes occurring in agricultural environment

and absorbents [18]. Optical biosensor work by sensing light and depends on luminescence and fluorescence signals [18]. This type of nanosensors can also be used for detection of pathogens [21].

Various nanostructures like nanowires, nanofilms, nanoparticles, and nanotubes are used to design nanomaterial used in nanosensors. These nanostructures are fabricated using vast array of materials, in particular, carbon, metals, metal oxides, polymers, and biomaterials [17]. Carbon-based nanomaterials like carbon nanotubes (CNT), graphene, and porous carbon exhibit large surface functionality, good conduction property, high stability, and are economical [17]. For instance, carbon nanotube based nanosensor developed by [22] to detect phenolic compounds was reported to possess high stability and rapid response. Biomaterial based nanosensors are highly efficient showing specificity toward target molecules and quick response, making detection process more accurate and rapid by minimizing effect of interfering compounds/gases. They include combination of biomolecule with nanomaterials, involving catalytic action of biological counterpart [17]. For example, a portable biosensor was developed by Li et al. [23] to detect *Escherichia coli* O157:H7 that exhibit 10^2 cfu/ml detection limit. The biosensor had short detection time requiring less than one hour for total detection. Due to ability to detect foodborne pathogens efficiently, the biosensor show good applicability in food industry [23]. Metal nanoparticles such as Au, Ag, Ti, Zn, Pt, and many others, display extraordinary features, particularly rare metals that are resistant to oxidation and corrosion [17]. Furthermore, metal nanosensor show good selectivity and sensitivity toward analyte that is always recommended and hence extensively preferred [17]. Gold nanoparticles are often applied in chemical and biological nanosensors owing to marvelous attributes like finely tunable optical, chemical, and catalytic properties [17]. Similarly, platinum nanoparticles with good catalytic attributes are utilized in electrochemical analysis [17]. Biosensor developed by modifying carbon film electrode using platinum nanoparticles showed sensitivity toward hydrogen peroxide, which was much higher than the bulk electrode [24] demonstrating better detection potential through utilization of metal nanoparticles.

Metal oxides (TiO_2, SnO_2, ZnO) based thin films and nanoparticles are also utilized for developing nanosensors. These nanomaterial display ultrahigh surface area along with other unique attributes such as high adsorptive capacity and electrochemical activity that increases their efficiency and application [17]. For instance, SnO_2 nanosensor in combination with graphene oxide due to good electrochemical activity reported to detect ultra-low amount of heavy metals in water [17]. Most of the metal oxide based nanosensors have wide range of detection ability with high sensitivity to any chemical alteration [17]. These nanosensors also have low cost of production and high stability. Metal oxide nanosensor developed to detect gases are based on change in electrical conductivity that takes place due to transfer in charge between surface complexes and the analyte. The efficiency of metal oxide based nanosensor is however affected by multiple factors like their morphology, particle size as well as surface to volume ratio and surface functionality [17]. Electrical resistance based gas nanosensor developed using metal oxides like ZnO, NiO, TiO_2, and SnO_2 have extensive application in detection of environmental pollutants including explosives and volatile organic compounds (VOCs) [17]. Among various metal oxide nanomaterials, ZnO nanomaterials have studied and used in many application due to its unique properties, low cost, and low toxic nature. Its application in nanosensor is appreciable that aid in detection of array of pollutants from various environmental matrixes, without compromising life-supporting properties and threatening existing life. Different nanomaterial-based sensors designed and used for detecting change in agricultural environment have been enlisted in Table 1.

3 Zinc Oxide Nanomaterials and Nanosensors

ZnO is an n-type semiconductor material with wide band gap and high excitation energy that facilitate application in electrochemical and optical sensing, and can be used at high working temperature ranging between 200 to 450°C [16, 17, 36]. The extraordinary physicochemical properties of ZnO like excellent electron transfer rate, pyroelectricity, and high electrochemical coupling coefficient, room-temperature luminescence, refractive index, and thermal expansion [37] facilitate fabrication of materials having application in diverse fields such as medicine, textile, agriculture, environment, cosmetics, and paints [16, 36]. In particular, the peizoelectric phenomenon displayed by hexagonal wurtzite ZnO crystal facilitate construction of transducer and electromechanical sensors [37]. Similarly, the polar surfaces of ZnO exhibiting self-repair as well as renewal process, supports formation of array of stable nanostructures [37].

Table 1 Application of nanomaterial based sensor in agricultural sector

Nanosensor	Target	Key references
Organophosphate hydrolase-gold nanoparticle conjugate with decoy mixture	Paraoxon pesticide. The change in fluorescence intensity detected was correlated with the concentration of pesticide	[25]
Gold electrode modified with copper nanoparticle	Detected salicyclic acid (signaling molecule) in oilseed rape infected with fungal pathogen (*Scleratinia sclerotiorium*)	[26]
Controlled immobilization of acetylcholinesterase on gold nanoparticle/ Prussian blue nanocomposite	Organophosphorus pesticides in environment and food	[27]
Electrochemical sensor-deposition of gold nanoparticles on carbon nanotubes modified glassy carbon	Triazophos pesticide in vegetable samples	[28]
Silver nanoparticle monolayer film	Methyl parathion (detection limit was upto 10^{-7} M)	[29]
Upconversion nanoparticle coupled with immuno-magnetic separation	Tomato ringspot virus, bean pod mottle virus, and arabis mosaic virus	[30]
Nanohybrid CdSe quantum dot	Cadmium ions in water samples (detection limit 25 nM)	[31]
Hollow gold-silver nanoparticle	Dry Clenbuterol from food samples (sentivity was low as 2 ppb within 15 min)	[32]
Electrochemical sensor-gold nanoparticle functionalized- and chitosan functionalized multiwalled carbon nanotube	Stem rot disease causing pathogen *Ganoderma boninense* (detection limit 0.0414 mg/L)	[33]
Optical nanosensor	H_2O_2 signals induced by wound in plants	[34]
Electrochemical sensor-Composite of ferrocene carboxylic acid functionalized metal organic framework and thermally reduce graphene oxide	Heavy metals (Cd^{2+}, Pb^{2+} and Cu^{2+})	[35]

ZnO nanoparticle exhibit biodegradability, biocompatibility, antimicrobial activity, and low toxicity, therefore found immense position in field of medicines, textile, as well in agriculture as nanopesticides, nanosensors, etc., and sustainable energy generation through dye-sensitized solar cells [36, 37]. Anticancer agent, wound dressing, antimicrobial activity, drug carrier, and biosensor are some biomedical applications of ZnO nanoparticles [38]. Additionally, good photoluminescent property of ZnO nanomaterials, along with other characteristics like high quantum yield, aqueous stability, and precise imaging aid its application in biomedical imaging [37]. ZnO nanomaterials can also help in controlling pathogen virulence, facilitating treatment of pathogenic diseases. In particular, ZnO nanospike with ability to reduce biosurfactant and violaceum production, demonstrated inhibitory impact on quorum sensing and biofilm development by human pathogenic bacteria *Pseudomonas aeruginosa* and *Chromobacterium violaceum* [39]. ZnO nanomaterials are fabricated as zero-dimensional (nanoparticles, nanodots), one-dimensional (nanorings, nanorods, nanocapsules), two-dimensional (nano-plates, nano-films, nanosheets), and three-dimensional nanostructures (nanoflakes, nanoflowers) [37, 40]. All different forms of ZnO nanomaterials display different characteristics and allow multiple applications [16, 18]. The ZnO nanosensors crafted from zero dimension nanomaterial have high sensitivity as they aid in better analyte and nanosensor interaction by providing high surface area, resulting in effective immobilization whereas ZnO nanosensors with one dimension have elongated structure and exhibit better precision due to easy transfer of electrons [18]. ZnO nanomaterials act as an ultra violet radiation filter and hence used in personal care products to protect humans from harmful rays [36]. Owing to good absorbing capacity, ZnO nanoparticles can be used for removing heavy metals from the wastewater. Different structure of nanoparticles have been reported to remove heavy metals from aqueous conditions efficiently. For instance, [41] reported efficient removal of mercury (Hg^{2+}), lead (Pb^{2+}), bismuth (Bi^{3+}), and cadmium (Cd^{2+}) using ZnO nanosphere from polluted solution. ZnO nanoparticles synthesized by Al-Mur [42] also reported removal of cadmium (Cd^{2+}) and lead (Pb^{2+}) from the aqueous condition. Similarly, removal of arsenic (As^{3+}), and chromium (Cr^{6+}) from wastewater by using ZnO nanoflower, nanocubes and nanorods was reported by [43].

Nanomaterials due to much smaller size (less than or equal to 100 nm) than counter bulk material exhibit unique properties, such as chemical and thermal stability, oxidation resistance, excellent conductivity, high catalytic activity, high isolelectric point, and good protein and metal absorption capacity, which extend their application as a nanosensor, that used for sensing gases, biomolecules, temperature, and pressure [16, 17, 37]. For instance, the 3D ZnO nanotetrapod exhibit ability to sense the ammonia vapor [44]. ZnO nanosensor also possess high sensitivity, low-cost, and fast response as well as recovery that enhance its arena of application [16]. ZnO nanosensor can be used to detect pressure as piezoelectric and piezoresistive phenomenon displayed by ZnO both allow to sense fluctuation in pressure. ZnO show piezoresistivity effect due to change in resistivity which occur when energy band of ZnO gets altered on exposure to any external force or strain [16]. Due to this piezoresistivity effect, the external pressure stimuli are sensed and converted into electrical signal [16]. As the wurtzite crystal of ZnO has non-central symmetry it shows piezoelectric effect due to which when force or strain is applied, adjustment in electrical charge occur resulting in signal detection [16]. The nanosenors also used for detecting change in temperature and works on principle of constant variation in resistivity and thermocouple potential with change in temperature. As carrier concentration as well as energy gap in ZnO varies constantly with fluctuation in temperature, it aids in sensing temperature signal and convert them into electric signal [16]. ZnO nanomaterial can also immobilize biomaterials efficiently that can be used to recognize the analytes present in the environment [21]. ZnO nanosensors have application in environmental and allied fields, and can be used for detecting gases like nitrogen dioxide, carbon dioxide, phenols, methane, formaldehyde, etc., for sensing humidity and refractive index, as well as biological analytes like pathogens and plant metabolites [36]. This ability of nanosensor/nanobiosensor to detect gaseous molecules produced by plants as signal molecules under stress aid in stress diagnosis rapidly and efficiently [45]. Owing to such extraordinary features ZnO nanosensors can be used efficiently in agricultural sector, improving food production and attenuating environmental damage.

4 Application of ZnO Nanosensors in Agriculture

Plant cultivation in natural environment routinely influenced by one or more biotic and/or abiotic factors, which hamper their growth and productivity. To manage these challenges, the measures used such as use of agro-chemicals, sewage sludge, plastics mulches, excessive irrigation, etc., result in further contamination of agro-environment, escalating magnitude to stress on plants. Thus, managing stress acting on plants and its micro-environment, it is crucial to monitor the real-time conditions accurately and precisely. The ZnO nanosensors owing to extraordinary attributes have demonstrated their potential to be utilized in agro-environment to sense or detect various contaminants and fluctuation in micro-climate such as temperature, metal ions, pesticides, nutrient level, and soil moisture. Nanosensors can help in addressing the problem of pest and pathogen induced stress by detecting infestation and diseases at early stage and monitoring plant health [46]. The optical biosensor fabricated using ZnO thin film was used to detect the *Grapevine virus A-type* (GVA) by Tereshchenko et al. [21]. Biosensor was developed by immobilizing antibodies on the surface of ZnO which was able to detect GVA-antigens by altering the photoluminescence due to the formation of a complex. The altered photoluminescence band was appeared in the region 400–550 nm [21]. ZnO and other forms of zinc can be used in agriculture sector as nanosensor for detection of contaminants such as pesticides. The photoelectrochemical biosensor fabricated using nanocomposite of AChE (acetylcholinesterase)-CdSe@ZnS/graphene by Li et al. [47] was able to generate photocurrent, inversely related to the concentration of organophosphorus pesticide. Through the application of this biosensor, the detection of dichlorovos and paraoxon within the concentration range of $10^{-12}-10^{-6}$ was achieved with very low detection limit of $10^{-14}-10^{-12}$ [47]. Array of ZnO based nanosensors/biosensors have been developed to detect the change in soil characteristics as well as to measure the level of biotic and abiotic stress on agroecosystem. Application of some of these sensors have been mentioned in Table 2. As ZnO nanomaterial based sensors are promising devices for management of agro-ecosystem, therefore, future research arena should focus on developing new low-cost nanosensors with high sensitivity, selectivity, and ability to detect multiple analytes simultaneously.

Table 2 Application of Zinc oxide nanosensors in detection of biotic and abiotic factors affecting agricultural activities and plant growth

Type of nanosensor	Nanomaterial used	Target and action	Key reference
Electrochemical immunosensors	Nano-ZnO film deposited onto indium–tin–oxide (ITO) glass plate	Ochratoxin-a detection (food contaminating mycotoxin)	[48]
Electrochemical DNA biosensor	DNA immobilization on ionic liquid, ZnO nanoparticles and a chitosan nanocomposite membrane on a modified gold electrode	The biosensor aid in rapid identification of species of *Trichoderma harzianum* by detecting hydridization reaction of a target DNA	[49]
Electrochemical DNA biosensor	Multi-walled carbon nanotube based zinc nanocomposite	Bio-recognition interface developed was used to sense the DNA sequence to detect the satellite molecule betasatellite associated with *Chili leaf curl virus*	[50]
Electrochemical sensor (Cyclic voltammetric sensor)	Zinc oxide nanoparticles on platinum electrode	Detection of nitrate (The detection limit of sensor was 10 nM)	[51]
Electrochemical sensor	Zinc oxide nanorods	Detection of bacteria	[52]
Electrochemical sensor	Zinc oxide nanorod	Measure pH with sensitivity of 1.06 nF/pH in the range of pH 4–10	[53]
Electrochemical sensor	Zinc oxide-multi walled carbon nanotube	Potassium level in soil	[54]

5 Conclusion and Future Prospective

Use of nanotechnology for fabrication of sensors with multiple applicability in field of agriculture, could facilitate improved plant growth and production, and reduce negative environmental externalities. Among various types of nanomaterial, zinc oxide nanomaterials has been reported with excellent properties that could develop new-generation sensors. ZnO nanomaterials exhibit improved features contrary to parent bulk material owing to small size, large surface area, wide band gap, and high excitation energy that aid in generation of more accurate result with high sensitivity and selectivity. These nanosenor can be used alone or in conjugation with other material in agro-ecosystems to detect nutrient level, pathogen stress, and pollutant concentration. Thus, data on various factors hampering soil fertility and plant growth could be modulated accordingly and precisely, encouraging precision agriculture. Though use of nanosenors are highly efficient, yet their use is limited due to high-cost of production, varying stability under different environmental conditions, biocompatibility issue, strict regulations, and potential risk of toxicity. Thus, the future research and policy formulation should be directed to focus on devising new approaches that reduce the limitation and encourage use of nanosensors. Application of nanosensor from grassroot level (farmer's level in small marginal fields) will not only support effective use of agro-chemicals but also reduce the financial burden on farmers and assist them to cultivate crop as per the soil conditions, paving a way to sustainable agriculture.

Competing Interests The authors declare that they have no competing interest.

Declaration of generative AI and AI-assisted technologies in the writing process During the preparation of this work the book chapter author(s) have not taken any assistance from AI, and the book chapter author(s) reviewed and edited the content as needed and take(s) full responsibility for the content of the publication. All authors read and approved the final version of the manuscript.

References

1. Alengebawy A, Abdelkhalek ST, Qureshi SR, Wang MQ (2021) Heavy metals and pesticides toxicity in agricultural soil and plants: ecological risks and human health implications. Toxics 9(3):42
2. Rajak P, Roy S, Ganguly A, Mandi M, Dutta A, Das K, Nanda S, Ghanty S, Biswas G (2023) Agricultural pesticides–friends or foes to biosphere? J Hazard Mater Adv 10:100264. https://doi.org/10.1016/j.hazadv.2023.100264
3. Biswas S, Das R (2024) Organic farming to mitigate biotic stresses under climate change scenario. Bull Natl Res Cent 48(1):71
4. Omuto CT, Kome GK, Ramakhanna SJ, Muzira NM, Ruley JA, Jayeoba OJ, Raharimanana V, Owusu AA, Khamis NA, Mathafeng KK, Elmobarak AA, Vargas RR, Koetlisi AK, Dembele D, Diwara M, Mbaikoubou M, Maria RM, Boukary IA, Malatji A, Nyamai M (2024) Trend of soil salinization in Africa and implications for agro-chemical use in semi-arid croplands. Sci Total Environ 951:175503. https://doi.org/10.1016/j.scitotenv.2024.175503
5. Zain M, Ma H, Nuruzzaman M, Chaudhary S, Nadeem M, Shakoor N, Azeem I, Duan A, Sun C, Ahamad T (2023) Nanotechnology based precision agriculture for alleviating biotic and abiotic stress in plants. Plant Stress 10:100239. https://doi.org/10.1016/j.stress.2023.100239
6. Javaid M, Haleem A, Rab S, Singh RP, Suman R (2021) Sensors for daily life: a review. Sens Int 2:100121
7. Queiros RB, Noronha JP, Sales MGF, Aguilar GG (2010) Sensors for the detection and quantification of bacterial contamination in water for human use. Adv Eng Mater 12(5):B175–B178
8. Beegum S, Das S (2022) Nanosensors in agriculture. In: Agricultural nanobiotechnology. Woodhead Publishing, pp 465–478
9. Kaur R, Sharma SK, Tripathy SK (2019) Advantages and limitations of environmental nanosensors. In: Advances in nanosensors for biological and environmental analysis. Elsevier, pp 119–132
10. Riu J, Maroto A, Rius FX (2006) Nanosensors in environmental analysis. Talanta 69(2):288–301
11. Pongnumkul S, Chaovalit P, Surasvadi N (2015) Applications of smartphone-based sensors in agriculture: a systematic review of research. J Sens 2015(1):195308
12. Srivastava AK, Dev A, Karmakar S (2018) Nanosensors and nanobiosensors in food and agriculture. Environ Chem Lett 16:161–182
13. Naaz S, Sachdev S, Husain R, Pandey V, Ansari MI (2023) Nanomaterials and nanocomposites: significant uses in plant performance, production, and toxicity response. In: Nanomaterials and nanocomposites exposures to plants: response, interaction, phytotoxicity and defense mechanisms. Springer Nature Singapore, Singapore, pp 1–18

14. Sharma P, Pandey V, Sharma MMM, Patra A, Singh B, Mehta S, Husen A (2021) A review on biosensors and nanosensors application in agroecosystems. Nanoscale Res Lett 16:1–24
15. Al Jabri H, Saleem MH, Rizwan M, Hussain I, Usman K, Alsafran M (2022) Zinc oxide nanoparticles and their biosynthesis: overview. Life 12(4):594
16. Que M, Lin C, Sun J, Chen L, Sun X, Sun Y (2021) Progress in ZnO nanosensors. Sens 21(16):5502
17. Abdel-Karim R, Reda Y, Abdel-Fattah A (2020) Nanostructured materials-based nanosensors. J Electrochem Soc 167(3):037554
18. Carrillo-Lopez LM, Villanueva-Verduzco C, Villanueva-Sánchez E, Fajardo-Franco ML, Aguilar-Tlatelpa M, Ventura-Aguilar RI, Soto-Hernández RM (2024) Nanomaterials for plant disease diagnosis and treatment: a review. Plants 13(18):2634
19. Tao N (2019) Challenges and promises of metal oxide nanosensors. ACS sens 4(4):780–780
20. Srivastava KR, Awasthi S, Mishra PK, Srivastava PK (2020) Biosensors/molecular tools for detection of waterborne pathogens. Waterborne Pathog 237–277
21. Tereshchenko A, Fedorenko V, Smyntyna V, Konup I, Konup A, Eriksson M, Yakimova R, Ramanavicius A, Balme S, Bechelany M (2016) ZnO films formed by atomic layer deposition as an optical biosensor platform for the detection of Grapevine virus A-type proteins. Biosens Bioelectron 92:763–769. https://doi.org/10.1016/j.bios.2016.09.071
22. Govindhan M, Lafleur T, Adhikari BR, Chen A (2015) Electrochemical sensor based on carbon nanotubes for the simultaneous detection of phenolic pollutants. Electroanalysis 27(4):902–909
23. Li X, Zheng Z, Liu X, Zhao S, Liu S (2015) Nanostructured photoelectrochemical biosensor for highly sensitive detection of organophosphorous pesticides. Biosens Bioelectron 64:1–5
24. Lebègue E, Anderson CM, Dick JE, Webb LJ, Bard AJ (2015) Electrochemical detection of single phospholipid vesicle collisions at a Pt ultramicroelectrode. Langmuir 31(42):11734–11739
25. Simonian AL, Good TA, Wang SS, Wild JR (2005) Nanoparticle-based optical biosensors for the direct detection of organophosphate chemical warfare agents and pesticides. Anal Chim Acta 534(1):69–77
26. Wang Z, Wei F, Liu S, Xu Q, Huang J, Dong X, Yu J, Yang Q, Zhao Y, Chen H (2010) Electrocatalytic oxidation of phytohormone salicylic acid at copper nanoparticles-modified gold electrode and its detection in oilseed rape infected with fungal pathogen Sclerotinia sclerotiorum. Talanta 80(3):1277–1281. https://doi.org/10.1016/j.talanta.2009.09.023
27. Wu S, Lan X, Zhao W, Li Y, Zhang L, Wang H, Han M, Tao S (2011) Controlled immobilization of acetylcholinesterase on improved hydrophobic gold nanoparticle/Prussian blue modified surface for ultra-trace organophosphate pesticide detection. Biosens Bioelectron 27(1):82–87. https://doi.org/10.1016/j.bios.2011.06.020
28. Li H, Xie C, Li S, Xu K (2012) Electropolymerized molecular imprinting on gold nanoparticle-carbon nanotube modified electrode for electrochemical detection of triazophos. Colloids Surf, B 89:175–181
29. Zhang L (2013) Self-assembly Ag nanoparticle monolayer film as SERS Substrate for pesticide detection. Appl Surf Sci 270:292–294
30. Zhang M, Chen W, Chen X, Zhang Y, Lin X, Wu Z, Li M (2013) Multiplex immunoassays of plant viruses based on functionalized upconversion nanoparticles coupled with immunomagnetic separation. J Nanomater 2013(1):317437
31. Wang J, Jiang C, Wang X, Wang L, Chen A, Hu J, Luo Z (2016) Fabrication of an "ion-imprinting" dual-emission quantum dot nanohybrid for selective fluorescence turn-on and ratiometric detection of cadmium ions. Analyst 141(20):5886–5892
32. Wang J, Zhang L, Huang Y, Dandapat A, Dai L, Zhang G, Lu X, Zhang J, Lai W, Chen T (2017) Hollow AU-AG nanoparticles labeled immunochromatography strip for highly sensitive detection of clenbuterol. Sci Rep 7(1). https://doi.org/10.1038/srep41419
33. Isha A, Akanbi FS, Yusof NA, Osman R, Mui-Yun W, Abdullah SNA (2019) An NMR metabolomics approach and detection of ganoderma boninense-infected oil palm leaves using MWCNT-based electrochemical sensor. J Nanomater 2019(1):4729706

34. Lew TTS, Koman VB, Silmore KS, Seo JS, Gordiichuk P, Kwak S, Park M, Ang MC, Khong DT, Lee MA, Chan-Park MB, Chua N, Strano MS (2020) Real-time detection of wound-induced H2O2 signalling waves in plants with optical nanosensors. Nature Plants 6(4):404–415. https://doi.org/10.1038/s41477-020-0632-4
35. Wang X, Qi Y, Shen Y, Yuan Y, Zhang L, Zhang C, Sun Y (2020) A ratiometric electrochemical sensor for simultaneous detection of multiple heavy metal ions based on ferrocene-functionalized metal-organic framework. Sens Actuators, B Chem 310:127756
36. Wojnarowicz J, Chudoba T, Lojkowski W (2020) A review of microwave synthesis of zinc oxide nanomaterials: reactants, process parameters and morphologies. Nanomater 10(6):1086
37. Gulab H, Fatima N, Tariq U, Gohar O, Irshad M, Khan MZ, Saleem M, Ghaffar A, Hussain M, Jan AK, Humayun M, Motola M, Hanif MB (2024) Advancements in zinc oxide nanomaterials: synthesis, properties, and diverse applications. Nano-Struct & Nano-Objects 39:101271. https://doi.org/10.1016/j.nanoso.2024.101271
38. Mishra PK, Mishra H, Ekielski A, Talegaonkar S, Vaidya B (2017) Zinc oxide nanoparticles: a promising nanomaterial for biomedical applications. Drug Discov Today 22(12):1825–1834
39. Khan MF, Husain FM, Zia Q, Ahmad E, Jamal A, Alaidarous M, Banawas S, Alam MM, Alshehri BA, Jameel M, Alam P, Ahamed MI, Ansari AH, Ahmad I (2020) Anti-quorum sensing and anti-biofilm activity of Zinc oxide nanospikes. ACS Omega 5(50):32203–32215. https://doi.org/10.1021/acsomega.0c03634
40. Shaba EY, Jacob JO, Tijani JO, Suleiman MAT (2021) A critical review of synthesis parameters affecting the properties of zinc oxide nanoparticle and its application in wastewater treatment. Appl Water Sci 11(2):48
41. Angelin KB, Siva S, Kannan RS (2015) Zinc oxide nanoparticles impregnated polymer hybrids for efficient extraction of heavy metals from polluted aqueous solution. Asian J Sci Technol 6(12):2139–2150
42. Al-Mur BA (2023) Green zinc oxide (ZnO) nanoparticle synthesis using mangrove leaf extract from aveicenna marina: properties and metal ions (Cd2+ and Pb2+). Water 15(3):455
43. Yuvaraja G, Prasad C, Vijaya Y, Subbaiah MV (2018) Application of ZnO nanorods as an adsorbent material for the removal of As (III) from aqueous solution: kinetics, isotherms and thermodynamic studies. Int J Ind Chem 9:17–25
44. Paulowicz I, Postica V, Lupan O, Wolff N, Shree S, Cojocaru A, Deng M, Mishra YK, Tiginyanu I, Kienle L, Adelung R (2018) Zinc oxide nanotetrapods with four different arm morphologies for versatile nanosensors. Sens Actuators, B Chem 262:425–435. https://doi.org/10.1016/j.snb.2018.01.206
45. Kalia A, Abd-Elsalam KA, Kuca K (2020) Zinc-based nanomaterials for diagnosis and management of plant diseases: Ecological safety and future prospects. J Fungi 6(4):222
46. Sachdev S, Ahmad S (2021) Role of nanomaterials in regulating oxidative stress in plants. Nanobiotechnology: Mitig Abiotic Stress Plants 305–326
47. Li Z, Fu Y, Fang W, Li Y (2015) Electrochemical impedance immunosensor based on self-assembled monolayers for rapid detection of Escherichia coli O157: H7 with signal amplification using lectin. Sens 15(8):19212–19224
48. Ansari AA, Kaushik A, Solanki PR, Malhotra BD (2010) Nanostructured zinc oxide platform for mycotoxin detection. Bioelectrochemistry 77(2):75–81
49. Siddiquee S, Rovina K, Yusof NA, Rodrigues KF, Suryani S (2014) Nanoparticle-enhanced electrochemical biosensor with DNA immobilization and hybridization of Trichoderma harzianum gene. Sens Bio-Sens Res 2:16–22
50. Tahir MA, Hameed S, Munawar A, Amin I, Mansoor S, Khan WS, Bajwa SZ (2017) Investigating the potential of multiwalled carbon nanotubes based zinc nanocomposite as a recognition interface towards plant pathogen detection. J Virol Methods 249:130–136
51. Gumpu MB, Nesakumar N, Ramachandra BL, Rayappan JBB (2017) Zinc oxide nanoparticles-based electrochemical sensor for the detection of nitrate ions in water with a low detection limit—a chemometric approach. J Anal Chem 72:316–326
52. Al-Fandi MG, Alshraiedeh NAH, Oweis RJ, Hayajneh RH, Alhamdan IR, Alabed RA, Al-Rawi OF (2018) Direct electrochemical bacterial sensor using ZnO nanorods disposable electrode. Sens Rev 38(3):326–334

53. Kumar A, SK NK, Aniley AA, Fernandez RE, Bhansali S (2019) Hydrothermal growth of zinc oxide (ZnO) nanorods (NRs) on screen printed IDEs for pH measurement application. J Electrochem Soc 166(9):B3264
54. Kumar AA, Kumar SN, Fernandez RE (2020) Real time sensing of soil potassium levels using zinc oxide-multiwall carbon nanotube-based sensors. IEEE Trans Nanobiosci 20(1):50–56

Various Metal-Oxide Based Nanosensors for Agricultural Practices

Himshikha Malviya, Diksha Singh, and Jay Singh

Abstract The advent of nanotechnology has brought about various developments in the fields of drug, food sectors, agriculture, and industrial work. The nanomaterials have unique physicochemical, and mechanical properties leading to a significant advancement in monitoring and managing agricultural processes. Nanotechnological applications have resulted in the development of smaller-size structures called nanosensors. These nanosensors interact with the targeted substance on a nanoscale, producing detectable signals that can be analyzed through various methods. Metal oxide-based nanosensors are a promising tool because of their high selectivity, sensitivity, and stability. Different metal oxides such ZnO, TiO_2, SnO_2, and Fe_2O_3 are used in biosensing technologies for diagnosing different parameters like environmental conditions, moisture content, pesticides, plant pathogen, pH levels, and soil nutrients. There are several applications for these metal oxide nano biosensors in farming operations. Hence, utilizing these nanosensors permits environmentally friendly and smart farming practices. This chapter will give us an overview of metal oxides-based nanosensors, their types, and their applications in agriculture. Various advantages and limitations are discussed thoroughly in this chapter to provide useful insights into nano biosensors in modern agricultural systems.

Keywords Metal oxides · Nano biosensor · Agriculture applications

1 Introduction

Agriculture is the foundation of human civilization, and it is continuously developing to satisfy the increasing demands of an expanding global population. Agricultural productivity is being affected by various consequences like climate change, soil degradation, river eutrophication, development of weeds, and groundwater pollution, thus a sustainable farming practice is very much needed for healthy crop productivity

H. Malviya · D. Singh · J. Singh (✉)
Department of Chemistry, Institute of Sciences, Banaras Hindu University, Varanasi, Uttar Pradesh, India
e-mail: jaysingh.chem@bhu.ac.in

A. Husen (ed.), *Nanobiosensors for Agricultural and Other Related Sectors*, Smart Nanomaterials Technology, https://doi.org/10.1007/978-981-96-6499-3_12

[48]. Over the decades, the agricultural sector has substantially increased production about 200% in Latin America, the US, Asia, and 68% in Western Europe. This can be accomplished by incorporating technological advancements in crop types, fertilizers, insecticides, and automated monitoring of soil, atmospheric interactions and plants. Hence, we can say that the agriculture sector is increasingly relying on technological advancements. One of the most promising advances in this field is the utilization of nanotechnology. Nanotechnology potentially transforms diagnostics in health, medicine, food, the environment, and agriculture by translating theoretical concepts into practical applications. The practices of nano-integrated technology in agriculture are:

- Pesticides nanoformulations.
- Nanosensors in the protection of crops for disease identification and agrochemical remnants.
- Nanodevices for genetically engineering of crops.
- Post-harvest management.
- Poultry production and animal health measures [98].

Nanosensors have gained acceptance and are now used extensively in agricultural applications. The nanosensors provide unparalleled possibilities for monitoring and managing agricultural operations with great precision and efficiency. Nanosensors are highly advanced devices that operate at a nanoscale, often less than 100 nm. These chemical sensors are used to detect physical, chemical, and biological parameters on a nanoscale. These sensors work by interacting with the targeted substance at a nanoscale, generating detectable signals that can be analyzed.

Metal oxide-based nanosensors have emerged as a useful tool in agriculture. They are widely used in the auxiliary processes. The chapter deals with the design and fabrication of metal oxides, key metal oxides in nanosensors for agricultural practices, their application, challenges, and future directions.

2 Design and Fabrication of Nanostructured Metal Oxide-Based Sensor

There has been a remarkable advancement in the development and applications of nanotechnology. Nanostructured metal-oxide materials can be manufactured using specialized synthesis and processing processes, which may be more limited than traditional approaches. These metal oxide-based sensors are the frontier of current sensing technology due to their enhanced sensitivity, selectivity, and stability. The sensors have significant applications and to improve their performance these are meticulously fabricated and integrated with nanostructured material. The nanorods and nanowires have a 1D arrangement that enables effective charge transfer and signal transduction. Due to the remarkable mechanical, electrical, and thermal characteristics of nanomaterials, the performance of biosensors is credibly enhanced.

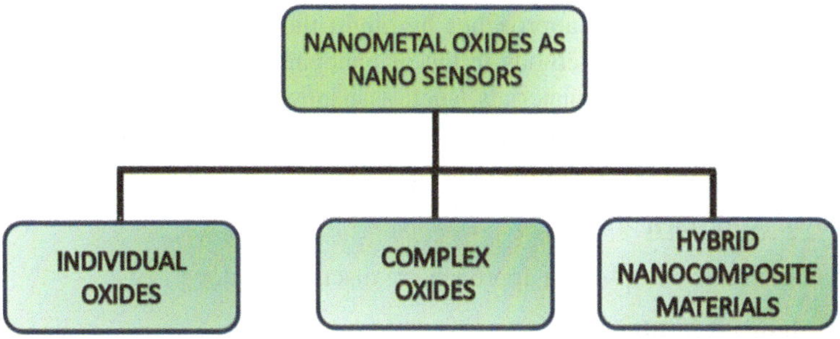

Fig. 1 Various classification of metal oxides

Researchers may advance biosensor technology and reinterpret its use in a variety of domains, from environmental monitoring to medical diagnostics and beyond, by utilizing the inherent benefits of these nanoscale structures [53].

The classification of nano metal-oxide can be done in three ways as shown in Fig. 1.

- Individual oxide.
- Complex oxide.
- Hybrid nanocomposite material.

2.1 Fabrication and Principles

There are several techniques for producing nanoparticles. The commonly used processes are physicochemical, and biological techniques. Based on the structural properties and quantity of the nanomaterial these techniques go through several phases [17]. Because of the large surface area-to-volume ratio and the size effects, nanostructured metal oxides like ZnO, SnO_2, TiO_2, and WO_3 exhibit unique electrooptical and physicochemical behaviors. These materials can be synthesized in various nanostructures, such as nanowires, nanotubes, nanorods, and nanoparticles, each having distinct advantages for sensing applications. The synthesis of the nanostructured materials can be achieved by two main strategies namely [58]:

1. Chemical method–Bottom-up.
2. Physical method–Top-down.

The top-down approach is basically utilised for the transformation of heavy material into nanoparticles. This method is easy to employ but ineffective in producing irregularly shaped or extremely tiny particles. The primary drawback of this strategy is the difficulty in acquiring accurate particle shapes and sizes. The bottom-up approach is the opposite of the top-down approach. Nanomaterials with precise shape size and structure are prepared through this approach. This method is considered to be the most beneficial due to its cost-effectiveness and simplicity [1, 75].

These wide varieties of approaches highlight the adaptability and intricacy of top-down and bottom-up methods in the of nanoparticles. The study and creation of metal oxide-based nanomaterials, such as nanowires (NWs), nanorods (NRs), carbon nanotubes (CNTs), quantum dots (QDs), and nanocomposite dendrimers, might greatly impact biosensor technology [53].

Top-Down Approach

Different methods of top-down approach are discussed below:

(a) **Mechanical milling**

Mechanical milling is a highly significant top-down technique employed for the production of diverse nanoparticles. This process is the most basic way to synthesize nanoparticles in powder form. The mechanical milling in industries offers several advantages including cheap cost, no precise management of the temperature or pressure in industrial manufacturing; no restrictions on a broad range of particle size distribution ranging from microns to nanometers and the liquid and solid phase of raw materials; and [97]. Mechanical milling aims to reduce particle size, mix or blend particles, modify their form, and create nanocomposite materials. It aims to reduce particle size, mix or blend particles, modify their shapes, and synthesize nanocomposite materials [69].

(b) **Laser ablation**

Laser ablation is the method of generating nanoparticles by irradiating a solid target material with a high-energy laser beam. It was first proposed by Einstein [72]. When the laser hits the target, it quickly warms it, causing the surface to evaporate or ablate. The evaporated material then cools and condenses, forming nanoparticles. This technology enables the manufacture of a diverse spectrum of nanoparticles with precise sizes and compositions. It is often utilized in research and industrial applications to produce metal, semiconductor, and other nanoparticles that may be employed in a variety of sectors including medicine, electronics, and material science [1]. In laser machining applications including high-precision drilling, laser beam milling, and laser cutting, the general process of laser ablation is stable. "Ablation" refers to a process that involves both the vaporisation process and melt ejection [72].

(c) **Lithography**

Lithography is a key component in micro and nanofabrication processes. Because highly functional miniaturized systems are required in various disciplines, including semiconductor development, chemical and biological analysis, and biomedical research, lithography methods have been developed and implemented accordingly [8]. Lithography is basically of two types: masked lithography and maskless lithography [66]. Masked lithography includes photolithography [88], soft lithography [101] and nanoimprint lithography [43]. While maskless lithography scanning probe lithography, electron beam lithography, and focused ion beam lithography are included.

(d) **Thermal decomposition**

Thermal decomposition is a process that uses heat and energy to break down chemical bonds in a substance. It is also known as thermolysis. Different metals that are produced by some chemical reactions are subsequently broken down at specific temperatures to form nanoparticles [55]. This procedure offers a one-step solution for creating nanostructures with regulated porosity, making it more efficient than previous methods. Yu and colleagues established a technique to produce nanoporous CdO from the microcrystals $CdCO_3$ [103, 104].

(e) **Sputtering**

The physical process of sputtering occurs when energetic ions, primarily noble gas ions, hit atoms in a solid state (the target) to cause them to be freed and move into the gas phase. In the process of sputtering solid surfaces are exposed to high-energy particles such as gas or plasma to produce nanomaterials. Sputtering is an effective method to create thin nanomaterial films [7]. Also, we can say that the target surface is bombarded by intense gaseous ions during the sputtering deposition process, which causes tiny atom clusters to physically eject depending on the incident gaseous ion energy [9].

Sputtering occurs in various techniques, including radio-frequency diode, direct current diode sputtering and magnetron. The sputtering method is more profitable as compared with the electron beam lithography [62]. In sputtering, alloy, ceramic, or a compound are used as a source material.

Bottom-Up Approach

Different methods of bottom-up approach are discussed below:

(a) **Green synthesis**

Green synthesis is more advantageous than the conventional synthesis method as it is cost-effective, produces less pollution, and enhances the environment and human health safety. The method employs an eco-friendly method of synthesis [102]. In green synthesis, the plant extract work as a reducing agent for reducing aqueous species in a solution. Green synthesis techniques use natural biological processes to produce nanomaterials. For example: $AgNO_3$ gets reduced to Ag^+ consuming plant extracts such as *Clitoria ternatea, Camellia sinensis, and Medicago sativa* [78]. For the synthesis of nanomaterials, the green method is considered the safest, cost-effective, and environmentally friendly approach. Various micro-organisms such as algae, bacteria, yeast, fungi, and certain plants are used as a substrate for the greener approach of synthesis of nanomaterials [32].

(b) **Sol–gel method**

The sol–gel technique is the most frequently used technique for the prepration of nanomaterials. The sol–gel technique is a wet chemical procedure that converts hydrolyzable precursors into oxide-based compounds with the help of condensation. Precursors often include ligands that are weaker than water, such as halides, nitrates, sulfates, alkoxides, or carboxylates. This process includes the transformation of a

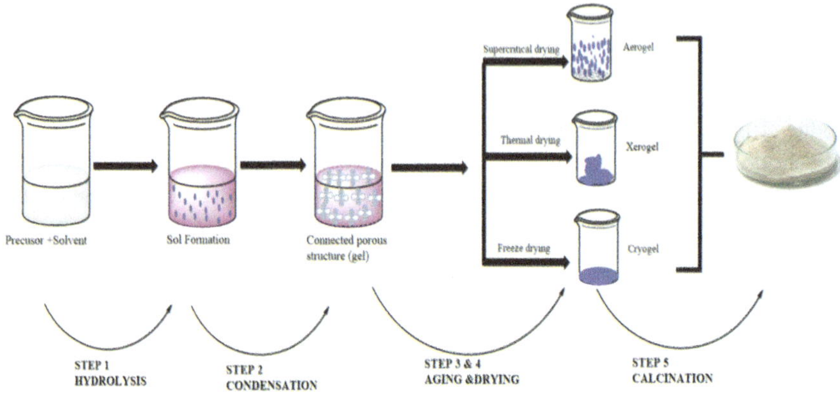

Fig. 2 Schematic representation of metal-oxide through sol–gel method

system from a liquid "sol" (usually colloidal) to a solid "gel" phase. It is particularly useful for generating materials of high purity and homogeneity. The technique may be utilized as a molding material in the ceramic manufacturing process, as well as an intermediate between thin layers of metal oxides in various applications. The sol–gel technique produces materials that are used in optical energy, electrical energy, surface engineering, biosensors, pharmaceuticals, separation technologies (such as chromatography), and a variety of other applications [12].

The sol–gel method includes the following steps, depicted in Fig. 2

- Hydrolysis
- Condensation
- Aging and drying
- Calcination [12].

(c) Chemical vapor deposition

CVD techniques are emerging technique for the preparation of 2D nanomaterials and thin films on solid substrates [67]. It contains the exposure of a substrate to volatile precursors that react and decompose on its surface, resulting in a thin film deposit.

The main steps involved in CVD techniques include reactant species i.e., precursor, the transportation of precursor in the reaction chamber, then comes the precursor adsorption, its decomposition and migration of atoms to the binding site. After the migration the process of nucleation, desorption and the removal of waste products takes place [11, 63].

CVD methods were used for the development of numerous 2D nanosheets like graphene [86] h-BN nanosheets [84], TMDs [35], metal carbides [25] etc. CVD offers superior control over other synthesis processes for producing 2D nanomaterials. The CVD technique allows us to produce high-quality 2D nanomaterials with few substrate errors and high crystal quality [67].

(d) **Hydrothermal**

The most common method for the preparation of nanomaterial is hydrothermal synthesis. This method is a solution reaction-based approach. Because water is employed as a solvent in this process, it is known as hydrothermal [1]. The hydrothermal method is used for the synthesis of nanomaterials from a temperature range of ambient to high temperatures. This method allows for fine control over the composition of nanoparticles and can produce high vapour pressure nanomaterials with minimal material loss [21]. Due to the homogeneous precipitation, low cost, and pure final product obtained through this method, it has a great contribution to modern science. Hydrothermal synthesis, crystal formation, as well as the preparation of useful ceramic powder, and the treatment of organic wastes, are other subcategories of the hydrothermal process [1].

(e) **Laser pyrolysis**

Laser pyrolysis is a vapor phase nanoparticle synthesis method that uses resonance energy transmission between laser photons and a gaseous reactant. It is an aerosol technique for nanomaterial synthesis [52]. This approach depends upon the excitation of a chemical that absorbs the energy of laser light and transfers it to a reaction medium, causing the temperature to rise fast. Decomposition occurs when reactants are heated to a high temperature. Nanoparticles are created once these precursor chemicals are dissociated, and they subsequently experience a quench (sudden cooling) action as they leave the flame. This abrupt reduction in temperature has the effect of preventing particle development. This results in the formation of nanosized particles [63].

According to [71], laser pyrolysis involves the functioning of laser photons, a sensitizer as well as a gaseous reactant species. The sensitizer acts as an energy transmission center which gets activated upon the absorption of CO_2 laser light and which is then transferred to the reactants via collision. He explained that $Fe(CO)_5$ breaks down into Fe and CO_2, triggering a chemical process, till a critical concentration of nuclei and homogeneous particle nucleation are achieved. The gas stream then retrains the nucleated particles, which are gathered at the outlet [71].

2.2 *Mechanism for Nanosensor in Agriculture*

Nanosensors in agriculture have emerged as important tools for enhancing precision farming, increasing crop yields, and providing sustainable agricultural practices. They detect certain environmental variables or substances at the nanoscale, providing real-time data that can help farmers determine soil nutrients, moisture levels, plant health, insect presence, and additional factors that can influence plant growth [28]. Here is a discussion of various mechanisms of nanosensors in agriculture.

(a) **Soil Nutrient Monitoring**:

Nanosensors detect and measure important elements in soil, including nitrogen, phosphorus, and potassium. These sensors are often functionalized with certain ligands or reactive materials that alter their electrical, optical, or chemical characteristics when in contact with the desired nutrients. This real-time monitoring enables precise fertilizer application, reducing waste and environmental effects. High temperatures, pressures, and humidity levels are just a few of the extreme environmental conditions that nanomaterials can tolerate. To increase their resilience and lifetime in the soil, nanosensors can be encapsulated or covered with protective. Nano-biosensors can measure soil organic matter (SOM) and microbial activity, indicating soil health and fertility [73]. SOM consists of plant and animal wastes, microbial biomass, and humus compounds at various phases of decomposition [40]. It regulates soil structure, water retention, nitrogen cycling, and carbon sequestration [81]. According to Lal et al., traditional wet chemistry techniques cannot accurately monitor SOM because of their great heterogeneity and dynamic nature [44]. Nano-biosensors detect SOMs by interactions between SOM components and receptors, such as enzymes or antibodies, resulting in electrical or optical signals [74]. Nanosensors detect and measure important elements in soil, including nitrogen, phosphorus, and potassium. These sensors are often functionalized with certain ligands or reactive materials that alter their electrical, optical, or chemical characteristics when in contact with the desired nutrients. This real-time monitoring enables precise fertilizer application, reducing waste and environmental effects.

(b) **Pest and Disease Detection**:

Agricultural production is limited by several factors, including weeds diseases, and insect pests [24]. Because of this, humans, animals, and plants are at risk from the presence of different heavy metal ions, such as Pb^{2+}, Li^{+}, Hg^{2+}, Cd^{2+}, Ag^{+}, and Cu^{2+}, which come from a variety of sources [83]. Nanosensors can detect volatile organic compounds (VOCs) released by plants in stressful conditions or attacked by pests and pathogens. For instance, carbon nanotubes or quantum dots can be engineered to bind with specific VOCs, causing a measurable change in the sensor's signal. Early diagnosis of ailments or insect infestations allows for prompt treatments, resulting in reduced crop losses [14]. By providing non-destructive, non-invasive, cost-effective, and user-friendly systems with enhanced detection limits, preciseness, on-site detection of plant pathogens, and sensitivity, nano biosensor-based technology adds a new level of complexity to plant disease diagnostic systems, as claimed by Kashyap et al. [39].

(c) **Water Quality and Irrigation Management**:

The excessive use of synthetic fertilizers can harm soil quality which leads to eutrophication, and damage groundwater, air, and water sources. Nanosensors can monitor water quality parameters like pH, salinity, and the presence of heavy metals or pollutants. These sensors often rely on nanomaterials that respond to specific chemical or physical properties of the water. The data from these sensors help to optimize irrigation schedules and ensure that the crop gets an adequate amount of water while minimizing water wastage [105].

(d) **Plant Health Monitoring**:

Nanosensors can measure plant physiological parameters such as chlorophyll content, moisture levels, and photosynthetic activity. These sensors can be embedded in plant tissues or used in the surrounding environment. The continuous monitoring of plant health allows for early detection of stress factors, enabling corrective actions to be taken before significant damage occurs. Besides this Johnson et al. have reviewed the selective nature of nanosensors for the targeted molecules using immobilized bioreceptor probes and a variety of specialized agricultural applications [36].

Environmental Monitoring:

Nanosensors detect and quantify environmental contaminants such as herbicides, pesticides, and greenhouse gases. They can be placed in soil, water, or air to measure pollution levels. This ensures that farming methods are environmentally friendly and helps to maintain regulatory compliance. Hence, nanosensors play a crucial role in environmental monitoring. Kumar et al., in his review article gave the overview of the research conducted on the application of DNA nanosensors for environment pollution detection as well as the difficulties encountered in regions that can be targeted for a safe environment [42].

2.3 Common Sensor Marker and Biomarker Utilized in Agriculture

In agriculture, sensor markers and biomarkers are critical for monitoring plant health, soil quality, and the environment. Some common examples of sensor markers and biomarker used in agriculture are discussed below.

(a) **Chlorophyll Content (SPAD)**

Chlorophyll concentration is an important measure of plant health and nitrogen levels. In the 1980s, Minolta Corporation, Ltd. Osaka, Japan, created a handheld absorbance-based dual-wavelength chlorophyll meter (SPAD models 501 and 502). A relative SPAD meter value can be derived through the SPAD-502 m that analyzes the transmission of red light radiation of wavelength 650 nm and infrared radiation of wavelength 940 nm through the leaf and derives a relative that corresponds to the amount of chlorophyll present in the sample leaf [91]. Netto et al. conducted chlorophyll concentration research to correlate the SPAD-502 results in *Coffea canephora* Pierre leaves with the quantity of the total nitrogen content, photosynthetic pigments extracted in DMSO, and the chlorophyll-a fluorescence characteristic [61].

(b) **NDVI (Normalized Difference Vegetation Index)**

NDVI is a frequently used indicator of vegetative health and biomass estimate. Remote sensing techniques, such as drones or satellite imaging, employ NDVI to monitor crop development, detect stress, and optimize resource utilization in precision agriculture. NDVI is the simplest and most widely used objective measure of plant density, and it is frequently used to assess sustainability exposure in urban areas for environmental health research [54]. One of the major vegetation indicators used to assess the photosynthetic size of a canopy is the NDVI, according to Kumar et al. The basis of it is the spectral reflectance in the electromagnetic spectrum's red and near-infrared regions (NIR) [41].

(c) **Soil Moisture Content**

Soil moisture is a crucial indicator of water availability and irrigation demands. Soil moisture sensors aid in precision irrigation by determining the volumetric water content of the soil, optimizing water consumption, and maintaining crop health. According to Adamchuk et al., on-the-go soil sensors have been manufactured by various researchers to measure the physicochemical and mechanical possessions of the soil [3]. In order to ascertain if the system operated effectively in dynamic field circumstances, a dielectric-based soil moisture sensor was integrated by Lui et al., into an instrumented shape and carried out ground experiments. They placed the moisture sensor right beneath the forefront of a soil mechanical resistance-mapping sensor in a covered nylon block that was isolated from the metal. Temperature, soil salinity, and maybe soil texture all impacted the moisture extent and the field experiments showed that this was a feasible method for quickly determining soil moisture content in-situ [51].

(d) **Leaf Area Index (LAI)**

The LAI is a dimensionless data that measures the entire one-sided area of photosynthetic tissue per unit ground surface [95]. LAI sensors are used to assess agricultural production potential and monitor how environmental factors affect crop growth. LAI is a measure of leaf canopy cover that reflects plant growth and well-being [37]. According to some reports, the LAI can be calculated by the remote sensing data using the statistical and physical approaches. The statistical approaches rely on empirical relations between the LAI and surface reflectance or vegetation indices. The LAI is determined using physical methods that rely on radiative light transmission mechanisms beneath the canopy [46].

(e) **Temperature and humidity sensors**

Plant growth and development are temperature-dependent, with each species having its own minimum, maximum, and optimal temperature range. Temperature and humidity are critical environmental variables that influence plant growth and disease development. These sensors detect microclimatic conditions within fields or greenhouses, which aids in disease prediction and management [29]. According to Ikram et al., soil temperature has an important role in agriculture, affecting germination, flowering, plant developments, microbiological and chemical activities. It is regulated by several factors, including thermal conductivity, water content, heat capacity, surface roughness, and cladding materials. Temperature probes are used in soil temperature sensors to transform temperature variations into electrical output, which is then converted into digital information by an electronic device [33]. Also, Lan et al. developed a viable technique for producing the large quantities of flexible humidity sensors. Capacitive plant humidity sensors were developed by the authors using laser direct writing technology. These sensor have high sensitivity, minimal hysteresis, and promising stability [45]. Moreover, Hackenberger et al. in his research article has explained the effects of soil moisture as well as soil temperature on biochemical indicators in earthworms and microbiological activity upon exposure to propiconazole and chlorantraniliprole [27].

(f) **Electrical Conductivity (EC)**

Soil salinity assessment is a result of arid-zone concerns about irrigated agricultural land and areas with shallow groundwater levels. It serves as the foundation for applying apparent soil EC to agriculture on a field-scale. A number of physicochemical characteristics influence apparent soil electrical conductivity, including bulk density, organic matter, soluble salts, clay concentration, temperature, water content, and mineralogy. Because of this, the EC is used to map the regional variation of a number of edaphic features, including the amount of clay, the salt and water content of the soil, the depth of flood-deposited sands, and the amount of organic matter. Furthermore, EC to agriculture has been utilized at the field scale to assess a range of artificial features, including irrigation patterns, percolation percentage, drainage and compaction patterns caused by agricultural mechanism [15]. Modern agriculture rely heavily on these markers and biomarkers, especially precision farming, where they help farmers make data-driven decisions to maximize crop productivity and sustainability.

3 Key Metal Oxides in Nanosensors for Agricultural Practices

Nanotechnology has an enormous effect on food and agriculture production change [92]. Agrochemicals practical to crops are often lost or may not reach their intended target due to causes such as hydrolysis, percolation, drifting, microbial degradation and also photolysis. Nanoparticles and nanocapsules effectively spread insecticides and fertilizers with great site specificity, limiting collateral harm. Nanotechnology is being used in farming to regulate and release pesticides, herbicides, and fertilizers more effectively [2]. Metal oxides have an important role in agricultural activities, especially in nanosensors, fertilizers, insecticides, and environmental monitoring devices. Their distinctive qualities, including large surface area, catalytic activity, and chemical durability, make them useful for enhancing agricultural efficiency and sustainability. There are several MONPs produced with possible future applications. Few of them are discussed below:

3.1 Zinc Oxide (ZnO)

ZnO is non-toxic and utilized for photocatalytic degradation of environmental contaminants. ZnO, both in bulk and thin films, is very sensitive to several hazardous gases. ZnO nanoparticles are highly significant since they are used in drug delivery systems, gas sensors, biosensors, cosmetics, and other applications. Because of their exceptional optical, physical, and antibacterial qualities, zinc oxide nanoparticles (ZnO NPs) have a lot of potential to improve agricultural practices [79]. ZnO nanoparticles potentially increase the yield and growth of food crops. An experiment was conducted on the peanut seeds where it was mix with several concentrations of the ZnO NPs. Zinc oxide nanoscale treatment at 1000 ppm concentration was used which ensures seedling viability, seed germination, and growth of plants. The ZnO nanoparticle was effective in root and stem growth in peanuts. When it comes to harmful pathogens, ZnO nanoparticles are an efficient antimicrobial agent. In simple terms, the primary mechanism of these metal oxide particles is antibacterial action due to the active oxygen species that have been discovered. The direct interaction between ZnO NPs and cell surfaces adjust the permeability of the cell membrane. These nanoparticles then enter bacterial cells and cause oxidative stress, which inhibits cell growth and ultimately leads to cell death in the plants [79]. The ZnO nanoparticles help in the CO_2 supply at the carboxylation site in chloroplasts by enhancing the carbonic anhydrase, which facilitates photosynthesis [18].

3.2 Copper Oxide (CuO)

Copper is an essential mineral existing in several proteins and metalloenzymes vital for plant health and nutrition. Cu NPs have exclusive belongings that make them more efficient than bulk copper particles in terms of activity and function. CuO NPs are critical micronutrients that promote the growth of plants. CuO NPs have antibacterial properties, making them advantageous for agriculture, healthcare, and industry. Gene expression, seed germination, seedling growth, and agricultural yield can all be impacted by CuO NPs. By boosting plant development, antioxidative enzyme levels, nutrient uptake, and yield potential in a variety of crop plants, its exposure can have a major effect on crop productivity. This covers internalisation, transport, absorption, and antioxidant action [56]. The CuO NPs are used in agrochemicals for the efficient delivery and controlled pesticide and fertilizer release [56].

3.3 Titanium Oxide (TiO_2)

TiO_2 nanoparticles help to enhance the seed stress resistance, water and oxygen intake by inducing active oxygen, such as hydroxide and superoxide anions, during photocatalytic processes in plants. Plants require TiO_2 to germinate quickly. Zheng et al. studied the impact of nano-TiO2 on the germination and development of naturally aged spinach seedlings by measuring their germination rate and vigour indicators as shown in Fig. 3. Nano-TiO_2 treated seeds from plants resulted in enhanced dry weight, photosynthetic rate, and chlorophyll production [106]. The morphology of TiO_2 plays an important role in photocatalytic activity. TiO_2 thin films can be used for photochemical disinfection of plant diseases, providing an alternative for their prevention [100]. The TiO_2 photocatalyst technology has potential for agricultural applications as it produces no harmful chemicals [60]. Scientists have examined the use of nano-TiO_2 photo semiconductors for the prevention of several agricultural diseases [94].

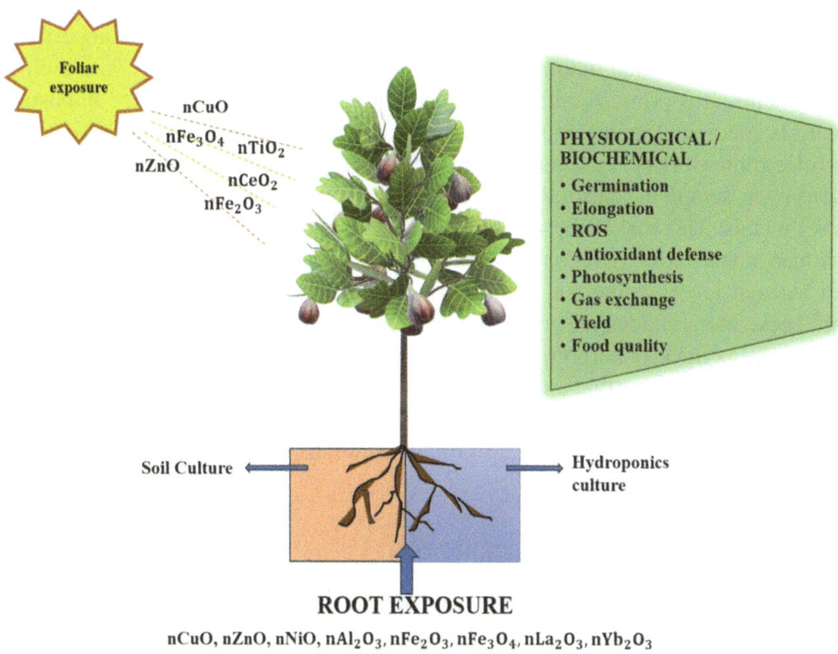

Fig. 3 The image depicts the foliar application of NMOs and their importance in several plant development processes

3.4 *Iron Oxide (FeO)*

FeO NPs can be produced as a Fe-fertilizer to boost agricultural growth, particularly in Fe-limiting soils that impede plant physiological development [77]. Furthermore, FeO NPs can enhance plant development through activities such as respiration, photosynthesis, enzyme activation, lignin production, nitrogen reduction, nitrogen fixation, and ribonucleic acid synthesis [13, 93]. The iron oxides have two oxidation states namely Fe_3O_4 and Fe_2O_3. Fe_3O_4 is used as a nano-fertilizer in agriculture to improve the quality of soil as iron contains the chlorophyll component which induces the development of plants and reduces the strength of chlorosis in plants [19]. Likewise, Fe_2O_3 can also be used for enhancing the growth of plants, oxidation–reduction reactions in plants, nodule formation, and chlorophyll synthesis in plants [85]. Other MONPs such as manganese oxide nanoparticles (MgO NPs) have the potential to be an efficient nanofertilizer for increasing agricultural output. It is the second most crucial micronutrient that plants require for optimal growth. MgO NPs, for example, considerably promote lettuce plant growth, and the impact of plant growth stimulation rises with concentration. It is also known to be less harmful than other MONPs, such as CuO and ZnO NPs [44, 49]. Similarly, CeO_2 nanoparticles are used because of their unique chemistry, high steadiness, and affinity with biological systems. It is primarily utilized in the manufacture of sensor cells, combustion catalysts, medicinal agents, drug delivery systems, anti-parasitic ointments, coatings, and electronic works [59]. CeO_2 nanoparticles can benefit plants under stressed situations by lessening the harmful impacts of water-soluble salts in the soil and also the heavy metals. This makes these elements more accessible to plants [6, 76]. Besides this manganese oxide (MnO_2) nanoparticles are used for sensing the soil moisture and nutrient levels in the soil because of its high surface area and redox properties. These metal oxides are frequently combined with other substances, including polymers, enzymes, or carbon-based nanostructures, to increase their efficacy and to develop hybrid systems that are suitable for certain sensing requirements. Nowadays smart agriculture is evolving as a result of ongoing research and development in this area, which promises sustainable, affordable solutions by enhancing the output and reducing environmental loss.

4 Metal Oxide Applications in Agricultural Practices

In India, agriculture is the backbone of the economy and the use of nanomaterials in the form of metal oxides for agriculture practices is a promising option for preventing and controlling plant diseases due to their high specific surface area and ability to interact with living cells [38, 57]. Due to their exceptional qualities, such as high sensitivity, chemical stability, and photocatalytic activity, a variety of metal oxides,

Fig. 4 Applications of metal oxides in agricultural practices

including CuO, FeO, ZnO, and TiO_2, are used in agricultural activities [64]. Various applications of metal oxides in agricultural practices are depicted in the Fig. 4.

4.1 Soil Quality Monitoring

The metal oxide nanoparticle-based sensors are advantageous for various applications in soil quality monitoring like nutrient sensing, pH sensing, moisture sensing, pesticide sensing, heavy metal sensing, etc. For optimizing crop yields, minimizing environmental pollution, and soil health maintenance, nutrient management is a very crucial step [96]. Potassium ions (K^+) are detected using metal oxide nanoparticle-based sensors, such as ZnO and SnO_2, which have a detection range of 0.1 to 100 mM [31]. Metal oxide nanoparticle-based sensors, with long-term stability of several months and a sensitivity of 0.01%, have been used to measure soil moisture. Examples of these sensors include SnO_2 and TiO_2 [5]. Quantum dots-based sensors can detect heavy metals like mercury (Hg), arsenic (As), and chromium (Cr) ions in soil with detection limits of 0.1, 0.5, and 1 nM, respectively in the soil [23].

The pH of the soil is measured using metal oxide-based sensors, such as IrO_x and RuO_2, which have a stability of several months and a sensitivity of 0.001 pH units [99]. Besides this soil moisture sensors have also been developed that measure the moisture content enabling efficient irrigation management and water conservation [20]. Similarly for greenhouse gas sensing in the soil, metal oxide nanoparticle-based sensors, such as SnO_2 and ZnO, can measure N_2O with a sensitivity of 0.01 ppm and long-term stability over several months [50].

4.2 Crop Health Assessment

Nanomaterials have been widely used for crop health assessment by monitoring plant parameters such as nutrient uptake, chlorophyll content, and stress level. The oxide-based NPs show beneficial roles in the biotic stress response of plants [90]. The majority of researchers used TiO_2 nanoparticles at the foliar level and it demonstrated beneficial effects on plants because of the photocatalytic properties of titanium NPs. For example: Solanum lycopersicum showed better growth results in fruit yields and chlorophyll concentration of plants when treated with TiO_2 NPs [22, 34, 70]. Furthermore, TiO_2 has also been reported to manage water stress in Triticum aestivum via increasing agronomic features [34]. Early flowering and increased leaf chlorophyll content were the results of treating the seeds with 1,000 ppm of ZnO NPs, which also increased seed germination and seedling vigour. Nevertheless, peanut development and production were adversely affected by greater ZnO NP concentrations (2,000 ppm) [68].

4.3 Pest and Disease Detection

Metal oxide nanoparticles exhibit strong antibacterial capabilities at low concentrations hence, preventing illnesses in plants. At different concentrations metal oxide nanoparticles inhibited antibacterial activity significantly. The development of nano pesticides involves the use of various metal oxides such as Zn, Cu, Fe, Ag, and Ti. CuO and Fe_2O_3. The oxidative nature of ZnO and TiO_2 NPs can function as insecticidal agents and its combination was evaluated on *Bactericera cockerelli* (a second-stage nymph) [26] Also, in sugar beet, ZnO nanoparticles suppress the growth of *Erysiphe betae* which is a fungal plant pathogen [4]. Several studies showed that the concentration of Fe_2O_3 based nanoparticles led to changes in the development of plants, antioxidant capacity, stress from oxidation, and defense mechanisms against foreign agents [80, 87, 107].

4.4 Water Quality Monitoring

Agriculture uses a lot of fresh water, but it also pollutes groundwater with harmful pesticides, fertilizers, and different agricultural chemicals. Besides this, the wastewater from numerous activities including home, industry, and other commercial places contains hazardous contaminants [30]. Major issues such as the quantity and quality of water, treatment, and reuse of the water, safety from chemical and biological entities monitoring, and sensors must all be considered while treating wastewater. To tackle this issue, creative, environmentally friendly, and affordable solutions are required to handle the vast volume of wastewater generated [82, 89].

A report indicates that the world's clean water shortage is worsening and affects over one billion people. It is expected that in the future, the average quantity of water accessible per person will drop by a factor of three, causing millions of people to die prematurely and needlessly. In the meanwhile, there is also a lack of clean water for appropriate farming methods [16].

In recent years, research and development in nanotechnology have enabled a significant interest in developing biosensors to detect environmental pollutants. Nanotechnology studies and research have enabled us to produce unique and cost-effective solutions for wastewater treatment and filtration. Visible light photocatalysis of transition metal oxides is a nanoscale technological development that has produced nanoparticles, nanoporous fibres, and nanoporous foams appropriate for microbial disinfection [47]. Nanoparticles have been widely used to extract heavy metals from water and wastewater due to their large surface area and high reactivity. MONPs with a broad surface area and unique bonding affinity for heavy metal uptake in aquatic settings include nanosized Fe_2O_3, MgO, Al_2O_3, TiO_2, MgO, and CeO_2.

4.5 Precision Farming

Various biological sources are employed to synthesize nanoparticles, as well as in agriculture for precision farming. Different types of aspects of precision agriculture are shown in the Fig. 5.

Precision farming uses computers, sensors, GPS, and remote sensing to measure local environmental conditions and identify crop growth efficiency or problems. Nanoparticles may be synthesized from several biological sources and are employed in agriculture for precision farming. Some of the nanoparticles are silver, zinc oxide, and titanium dioxide. Metal oxide-based solutions maximize agricultural yields while minimizing resource waste and environmental damage by facilitating real-time monitoring and focused interventions. More innovation is anticipated as a result of future developments in the synthesis, functionalization, and integration of these materials, opening the door to many environmentally friendly and fruitful farming methods.

5 Challenges and Future Directions

According to Bhalla et al., metal oxide nanoparticles (MONPs) help to boost agricultural production by increasing nutrient efficiency, plant protection, crop variety, disease detection, and monitoring. MONPs have the potential for agricultural growth, increasing food production and feeding the world's population in the next decades. MONPs can revolutionize agriculture by increasing crop productivity and feeding the population of the world in the next decades. Various metal-oxide nanosensors show difficulties and challenges in achieving high sensitivity and selectivity for specific

Fig. 5 Aspects of precision agriculture

agricultural analytes. Besides this the stability of sensors in the field in various environmental conditions; their scalability, cost, lifespan, and durability are different technical challenges that affect the application of metaloxide-based nanosensors in agriculture. There are several environmental challenges as well regarding the toxic effects on plants and soil microbes, its proper disposal system, and recycling aspects. The nanoparticles can affect the ecosystem and soil health as it leaches out into the soil and water system. The good news regarding biosensing technology is that most of the challenges are being rapidly overcome. The translational research has received significant funding globally, notably for healthcare applications. This brings industry and academics closer together to develop economically viable goods. On the other side, there has been a significant improvement in how scientists collaborate across boundaries [10]. The advanced functionalization of these metaloxide-based nanosensors has improvised the directions for the future development of these nanosensors in agriculture. Future research trends should prioritize multiplex and multi-functional nanosensors, wireless and networked systems, advanced data analytics and AI algorithm techniques, biodegradable and bioinspired nanosensors, and regulate proper protocols for soil health monitoring [65].

6 Conclusion and Summary

The best way to transform agricultural methods is to use metal oxide-based nanosensors, which allow for accurate, real-time monitoring of critical parameters including soil health, moisture content, nutrient content, and pathogen presence. These nanosensors can help with data-driven and sustainable farming by giving timely insights that help optimize inputs, eliminate resource waste, and reduce environmental impact. Various key findings of MONPs in agricultural practices are discussed below:

6.1 Key Findings

- **High sensitivity and selectivity**: Metal oxide-based nanosensors have shown promising results in precisely detecting agricultural analytes, making them useful for monitoring soil health, water quality, and crop diseases. The creation of selective sensors is critical for detecting certain chemicals or diseases, which might improve focused treatment.
- **Durability and Field Stability Challenges**: Despite their efficiency, these nanosensors confront stability issues in various field circumstances, such as temperature and humidity changes. Improving durability and sensitivity in field conditions is critical for wider agricultural use.
- **Environmental and Safety Considerations**: The usage of metal oxide nanoparticles may pose environmental hazards, including toxicity to plants and microorganisms and leaching into the soil. These issues require more study to address and minimize.
- **Scalability and Cost Issues**: Despite the potential of metal oxide nanosensors for precision farming, their large-scale deployment is hampered by their existing production costs and scalability issues. For a wide variety of agricultural enterprises to be able to use these sensors, innovations in economical production techniques are required.

Different implications of MONPs in agricultural practices are discussed below:

6.2 *Implications for Agricultural Practices*

- **Increased Precision and Sustainability**: Metal oxide-based nanosensors can decrease overuse of water, fertilizers, and pesticides by facilitating more accurate monitoring, which is consistent with resource efficiency and sustainable agricultural methods. This change can encourage a more renewable method of food production, improve soil health, and reduce agricultural runoff.
- **Helping Farmers Make Data-Driven Decisions**: Farmers are better able to make data-driven decisions by integrating these nanosensors with Internet of Things (IoT) networks and data analytics technologies. Increased agricultural yields, improved pest management, and more effective resource allocation are all possible outcomes of this, and they all support sustainability and productivity.
- **Innovation in Smart Agriculture in the Future**: As these sensors develop, they will probably be incorporated into smart farming systems and may even work in tandem with artificial intelligence (AI) to offer automated field reactions and predictive insights. This integration marks a significant step toward high-tech, self-sufficient agriculture capable of supplying future food demands responsibly.

In conclusion, metal oxide-based nanosensors represent an enormous advancement in agricultural technology, but further research and development are required to address technological, environmental, and economic problems. By improving accuracy, encouraging sustainability, and facilitating data-driven, intelligent farming methods, these nanosensors have the potential to completely transform agriculture if they are properly handled.

Competing Interests The authors declare that they have no competing interest.

Declaration of generative AI and AI-assisted technologies in the writing process During the preparation of this work the book chapter author(s) have not taken any assistance from AI, and the book chapter author(s) reviewed and edited the content as needed and take(s) full responsibility for the content of the publication. All authors read and approved the final version of the manuscript.

References

1. Abid N et al (2022) Synthesis of nanomaterials using various top-down and bottom-up approaches, influencing factors, advantages, and disadvantages: a review. Adv Colloid Interface Sci 300:102597. https://doi.org/10.1016/j.cis.2021.102597
2. Acharya A, Pal PK (2020) Agriculture nanotechnology: translating research outcome to field applications by influencing environmental sustainability. NanoImpact 19:100232. https://doi.org/10.1016/j.impact.2020.100232
3. Adamchuk V et al (2004) On-the-go soil sensors for precision agriculture. Comput Electron Agric 44(1):71–91. https://doi.org/10.1016/j.compag.2004.03.002
4. Amin HH et al (2023) Enhancing sugar beet plant health with zinc nanoparticles: a sustainable solution for disease management. J Soil, Plant Environ 2(1):1–20. https://doi.org/10.56946/jspae.v2i1.129

5. Atherton JJ, Rosamond MC, Zeze DA (2012) A leaf-mounted thermal sensor for the measurement of water content. Sens Actuators A: Phys 187:67–72. https://doi.org/10.1016/j.sna.2012.06.021
6. Ayub MA et al (2019) Cerium oxide nanoparticles: advances in synthesis, prospects and application in agro-ecosystem. pp 209–250. https://doi.org/10.1016/bs.coac.2019.10.003
7. Ayyub P et al (2001) Synthesis of nanocrystalline material by sputtering and laser ablation at low temperatures. Appl Phys Mater Sci & Process 73(1):67–73. https://doi.org/10.1007/s003390100833
8. Baek D et al (2021) Lithography technology for micro-and nanofabrication. pp 217–233. https://doi.org/10.1007/978-981-33-6158-4_9
9. Baig N, Kammakakam I, Falath W (2021) Nanomaterials: a review of synthesis methods, properties, recent progress, and challenges. Mater Adv 2(6):1821–1871. https://doi.org/10.1039/D0MA00807A
10. Bhalla N et al (2016) Introduction to biosensors. In: Estrela P (eds). Essays Biochem 60(1):1–8. https://doi.org/10.1042/EBC20150001
11. Binions R, P I (2011) Novel chemical vapour deposition routes to nanocomposite thin films. In: Advances in nanocomposites-synthesis characterization and industrial applications. https://doi.org/10.5772/14095
12. Bokov D et al (2021) Nanomaterial by sol-gel method: synthesis and application. In: Wang Z (ed) Advances in materials science and engineering, vol 2021, no 1. https://doi.org/10.1155/2021/5102014
13. Briat JF, Dubos C, Gaymard F (2015) Iron nutrition, biomass production, and plant product quality. Trends Plant Sci 20(1):33–40. https://doi.org/10.1016/j.tplants.2014.07.005
14. Cardoso RM (2022) Current progress in plant pathogen detection enabled by nanomaterials-based (bio)sensors. Sens Actuators Rep 4:100068. https://doi.org/10.1016/j.snr.2021.100068
15. Corwin DL, Lesch SM (2005) Apparent soil electrical conductivity measurements in agriculture. Comput Electron Agric 46(1–3):11–43. https://doi.org/10.1016/j.compag.2004.10.005
16. Cross KM et al (2009) Water decontamination using iron and iron oxide nanoparticles. In: Nanotechnology applications for clean water. Elsevier, pp 347–364. https://doi.org/10.1016/B978-0-8155-1578-4.50033-0
17. Dhand C et al (2015) Methods and strategies for the synthesis of diverse nanoparticles and their applications: a comprehensive overview. RSC Adv 5(127):105003–105037. https://doi.org/10.1039/C5RA19388E
18. DiMario RJ et al (2017) Plant carbonic anhydrases: structures, locations, evolution, and physiological roles. Molecular Plant 10(1):30–46. https://doi.org/10.1016/j.molp.2016.09.001
19. Dimkpa CO (2014) Can nanotechnology deliver the promised benefits without negatively impacting soil microbial life?. J Basic Microbiol 54(9):889–904. https://doi.org/10.1002/jobm.201400298
20. Fares A, Polyakov V (2006) Advances in crop water management using capacitive water sensors. pp 43–77. https://doi.org/10.1016/S0065-2113(06)90002-9
21. Gan YX et al (2020) Hydrothermal synthesis of nanomaterials. J Nanomater 1–3. https://doi.org/10.1155/2020/8917013
22. Gao F et al (2008) Was improvement of spinach growth by nano-TiO2 treatment related to the changes of Rubisco activase?. BioMetals 21(2):211–217. https://doi.org/10.1007/s10534-007-9110-y
23. Ghanei-Motlagh M et al (2016) A novel voltammetric sensor for sensitive detection of mercury(II) ions using glassy carbon electrode modified with graphene-based ion imprinted polymer. Mater Sci Eng: C 63:367–375. https://doi.org/10.1016/j.msec.2016.03.005
24. Giraldo JP, Kruss S (2023) Nanosensors for monitoring plant health. Nat Nanotechnol 18(2):107–108. https://doi.org/10.1038/s41565-022-01307-w
25. Gogotsi Y (2015) Transition metal carbides go 2D. Nat Mater 14(11):1079–1080. https://doi.org/10.1038/nmat4386

26. Gutiérrez-Ramírez JA et al (2021) Insecticidal effect of zinc oxide and titanium dioxide nanoparticles against bactericera cockerelli sulc. (Hemiptera: Triozidae) on tomato solanum lycopersicum. Agron 11(8):1460. https://doi.org/10.3390/agronomy11081460
27. Hackenberger DK et al (2018) Influence of soil temperature and moisture on biochemical biomarkers in earthworm and microbial activity after exposure to propiconazole and chlorantraniliprole. Ecotoxicol Environ Saf 148:480–489. https://doi.org/10.1016/j.ecoenv.2017.10.072
28. Hafez M, Khalil HF (2024) Nanoparticles in sustainable agriculture: recent advances, challenges, and future prospects. Commun Soil Sci Plant Anal 55(14):2181–2196. https://doi.org/10.1080/00103624.2024.2339950
29. Hatfield JL, Prueger JH (2015) Temperature extremes: effect on plant growth and development. Weather Clim Extrem 10:4–10. https://doi.org/10.1016/j.wace.2015.08.001
30. Henze M et al (2008) Biological wastewater treatment: principles, modelling and design. IWA Publishing. https://doi.org/10.2166/9781780401867
31. Huang X, Dong F, Xiang D, Zhang X, ZLU and selective nonenzymatic glucose detection based on A.P. core-shell nanoparticles decorated graphene. S. and A.B.C. 2019;279:227–233. (no date) 'No Title'
32. Huston M et al (2021) Green synthesis of nanomaterials. Nanomater 11(8):2130. https://doi.org/10.3390/nano11082130
33. Ikram M et al (2024) Flexible temperature and humidity sensors of plants for precision agriculture: current challenges and future roadmap. Comput Electron Agric 226:109449. https://doi.org/10.1016/j.compag.2024.109449
34. Jaberzadeh A et al (2013) Influence of bulk and nanoparticles titanium foliar application on some agronomic traits, seed gluten and starch contents of wheat subjected to water deficit stress. NotE Bot Horti Agrobot Cluj-Napoca 41(1):201. https://doi.org/10.15835/nbha4119093
35. Ji Q et al (2015) Chemical vapour deposition of group-VIB metal dichalcogenide monolayers: engineered substrates from amorphous to single crystalline. Chem Soc Rev 44(9):2587–2602. https://doi.org/10.1039/C4CS00258J
36. Johnson MS, Sajeev S, Nair RS (2021) Role of nanosensors in agriculture. In: 2021 international conference on computational intelligence and knowledge economy (ICCIKE). IEEE, pp 58–63. https://doi.org/10.1109/ICCIKE51210.2021.9410709
37. Jonckheere I et al (2004) Review of methods for in situ leaf area index determination. Agric For Meteorol 121(1–2):19–35. https://doi.org/10.1016/j.agrformet.2003.08.027
38. Kang S et al (2008) Antibacterial effects of carbon nanotubes: size does matter!. Langmuir 24(13):6409–6413. https://doi.org/10.1021/la800951v
39. Kashyap PL et al (2019) Nanosensors for plant disease diagnosis: current understanding and future perspectives. In: Nanoscience for sustainable agriculture. Springer International Publishing, Cham, pp 189–205. https://doi.org/10.1007/978-3-319-97852-9_9
40. Khan MR, Rizvi TF (2014) Nanotechnology: scope and application in plant disease management. Plant Pathol J 13(3):214–231. https://doi.org/10.3923/ppj.2014.214.231
41. Kumar S et al (2016) Mapping of spot blotch disease resistance using NDVI as a substitute to visual observation in wheat (Triticum aestivum L.). Mol Breed 36(7):95. https://doi.org/10.1007/s11032-016-0515-6
42. Kumar V, Guleria P (2020) Application of DNA-nanosensor for environmental monitoring: recent advances and perspectives. Curr Pollut Rep 10(4):765–785. https://doi.org/10.1007/s40726-020-00165-1
43. Kuo CW et al (2003) Fabrication of large-area periodic nanopillar arrays for nanoimprint lithography using polymer colloid masks. Adv Mater 15(13):1065–1068. https://doi.org/10.1002/adma.200304824
44. Lal R (2016) Soil health and carbon management. Food Energy Secur 5(4):212–222. https://doi.org/10.1002/fes3.96
45. Lan L et al (2020) One-step and large-scale fabrication of flexible and wearable humidity sensor based on laser-induced graphene for real-time tracking of plant transpiration at bio-interface. Biosens Bioelectron 165:112360. https://doi.org/10.1016/j.bios.2020.112360

46. Leaf area index (2020) Advanced remote sensing. Elsevier. https://doi.org/10.1016/B978-0-12-815826-5.00010-6
47. Li Q, Wu P, Ku Shang J (2009) Nanostructured visible-light photocatalysts for water purification. In: Nanotechnology applications for clean water. Elsevier, pp 17–37. https://doi.org/10.1016/B978-0-8155-1578-4.50011-1
48. Lichtfouse E et al (2009) Agronomy for sustainable agriculture. A review. Agron Sustain Dev 29(1):1–6. https://doi.org/10.1051/agro:2008054
49. Liu R, Zhang H, Lal R (2016) Effects of stabilized nanoparticles of copper, zinc, manganese, and iron oxides in low concentrations on lettuce (Lactuca sativa) seed germination: nanotoxicants or nanonutrients?. Water, Air, & Soil Pollut 227(1):42. https://doi.org/10.1007/s11270-015-2738-2
50. Liu X et al (2014) Black silicon: fabrication methods, properties and solar energy applications. Energy Environ Sci 7(10):3223–3263. https://doi.org/10.1039/C4EE01152J
51. Lui W et al (2015) Development of a texture/soil compaction sensor. pp 617–630. https://doi.org/10.2134/1996.precisionagproc3.c76
52. Malekzadeh M et al (2020) Laser pyrolysis synthesis of zinc-containing nanomaterials using low-cost ultrasonic spray delivery of precursors. Powder Technol 376:104–112. https://doi.org/10.1016/j.powtec.2020.08.011
53. Malik S et al (2023) Nanomaterials-based biosensor and their applications: a review. Heliyon 9(9):e19929. https://doi.org/10.1016/j.heliyon.2023.e19929
54. Martinez A, de la I, Labib SM (2023) Demystifying normalized difference vegetation index (NDVI) for greenness exposure assessments and policy interventions in urban greening. Environ Res 220:115155. https://doi.org/10.1016/j.envres.2022.115155
55. Mekuye B, Abera B (2023) Nanomaterials: an overview of synthesis, classification, characterization, and applications. Nano Select 4(8):486–501. https://doi.org/10.1002/nano.202300038
56. Mir AR, Pichtel J, Hayat S (2021) Copper: uptake, toxicity and tolerance in plants and management of Cu-contaminated soil. BioMetals 34(4):737–759. https://doi.org/10.1007/s10534-021-00306-z
57. Mishra V et al (2014) Interactions of nanoparticles with plants. In: Emerging technologies and management of crop stress tolerance. Elsevier, pp 159–180. https://doi.org/10.1016/B978-0-12-800876-8.00008-4
58. Muhammed M (2003) Engineering of nanostructured materials. In: Nanostructures: synthesis, functional properties and applications. Springer Netherlands, Dordrecht, pp 37–79. https://doi.org/10.1007/978-94-007-1019-1_2
59. Nadeem M et al (2020) Green synthesis of cerium oxide nanoparticles (CeO2 NPs) and their antimicrobial applications: a review. Int J Nanomedicine 15:5951–5961. https://doi.org/10.2147/IJN.S255784
60. Nair R, Kumar DS (2013) Plant diseases—control and remedy through nanotechnology. In: Crop improvement under adverse conditions. Springer New York, New York, NY, pp 231–243. https://doi.org/10.1007/978-1-4614-4633-0_10
61. Netto AT et al (2005) Photosynthetic pigments, nitrogen, chlorophyll a fluorescence and SPAD-502 readings in coffee leaves. Sci Hortic 104(2):199–209. https://doi.org/10.1016/j.scienta.2004.08.013
62. Nie M, Sun K, Meng DD (2009) Formation of metal nanoparticles by short-distance sputter deposition in a reactive ion etching chamber. J Appl Phys 106(5). https://doi.org/10.1063/1.3211326
63. Noah NM (2020) Design and synthesis of nanostructured materials for sensor applications. J Nanomater. In: Young SJ. pp 1–20. https://doi.org/10.1155/2020/8855321
64. Nongbet A et al (2022) Nanofertilizers: a smart and sustainable attribute to modern agriculture. Plants 11(19):2587. https://doi.org/10.3390/plants11192587
65. Parameswari P et al (2024) Nanotechnology-based sensors for real-time monitoring and assessment of soil health and quality: a review. Asian J Soil Sci Plant Nutr 10(2):157–173. https://doi.org/10.9734/ajsspn/2024/v10i2272

66. Pimpin A, Srituravanich W (2012) Review on micro-and nanolithography techniques and their applications. Eng J 16(1):37–56. https://doi.org/10.4186/ej.2012.16.1.37
67. Pottathara YB et al (2019) Synthesis and processing of emerging two-dimensional nanomaterials. In: Nanomaterials synthesis. Elsevier, pp 1–25. https://doi.org/10.1016/B978-0-12-815751-0.00001-8
68. Prasad TNVKV et al (2012) Effect of nanoscale zinc oxide particles on the germination, growth and yield of peanut. J Plant Nutr 35(6):905–927. https://doi.org/10.1080/01904167.2012.663443
69. Prasad Yadav T, Manohar Yadav R, Pratap Singh D (2012) Mechanical milling: a top down approach for the synthesis of nanomaterials and nanocomposites. Nanosci Nanotechnol 2(3):22–48. https://doi.org/10.5923/j.nn.20120203.01
70. Raliya R et al (2015) Mechanistic evaluation of translocation and physiological impact of titanium dioxide and zinc oxide nanoparticles on the tomato (Solanum lycopersicum L.) plant. Metallomics 7(12):1584–1594. https://doi.org/10.1039/C5MT00168D
71. Rashdan S et al (2013) Nanoparticles for biomedical applications: current status, trends and future challenges. In: Biomaterials and medical tribology. Elsevier, pp 1–132. https://doi.org/10.1533/9780857092205.1
72. Ravi-Kumar S et al (2019) Laser ablation of polymers: a review. Polym Int 68(8):1391–1401. https://doi.org/10.1002/pi.5834
73. Rawtani D et al (2018) Nanotechnology-based recent approaches for sensing and remediation of pesticides. J Environ Manag 206:749–762. https://doi.org/10.1016/j.jenvman.2017.11.037
74. Rinnan R, Rinnan Å (2007) Application of near infrared reflectance (NIR) and fluorescence spectroscopy to analysis of microbiological and chemical properties of arctic soil. Soil Biol Biochem 39(7):1664–1673. https://doi.org/10.1016/j.soilbio.2007.01.022
75. Romanovski V et al (2023) Nano metal oxide as nanosensors in agriculture and environment. In: Nanometal oxides in horticulture and agronomy. Elsevier, pp 321–352. https://doi.org/10.1016/B978-0-323-91809-1.00016-0
76. Rossi L et al (2016) The impact of cerium oxide nanoparticles on the salt stress responses of Brassica napus L. Environ Pollut 219:28–36. https://doi.org/10.1016/j.envpol.2016.09.060
77. Rui M et al (2016) Iron oxide nanoparticles as a potential iron fertilizer for peanut (Arachis hypogaea). Front Plant Sci 7. https://doi.org/10.3389/fpls.2016.00815
78. Rus A, Leordean VD, Berce P (2017) Silver nanoparticles (AgNP) impregnated filters in drinking water disinfection. In: Balc N (ed) MATEC web of conferences, vol 137. p 07007. https://doi.org/10.1051/matecconf/201713707007
79. Sabir S, Arshad M, Chaudhari SK (2014) Zinc oxide nanoparticles for revolutionizing agriculture: synthesis and applications. Sci World J 1–8. https://doi.org/10.1155/2014/925494
80. Sahraei M, Hassanpour H, Ziyadi H (2023) Iron oxide nanoparticle-induced changes in the growth, oxidative responses, and antioxidant capacity of physalis alkekengi seedlings. Russ J Plant Physiol 70(3):32. https://doi.org/10.1134/S1021443722602944
81. Schmidt MWI et al (2011) Persistence of soil organic matter as an ecosystem property. Nature 478(7367):49–56. https://doi.org/10.1038/nature10386
82. Schoumans OF et al (2014) Mitigation options to reduce phosphorus losses from the agricultural sector and improve surface water quality: a review. Sci Total Environ 468–469, 1255–1266. https://doi.org/10.1016/j.scitotenv.2013.08.061
83. Sharma P et al (2021) A review on biosensors and nanosensors application in agroecosystems. Nanoscale Res Lett 16(1):136. https://doi.org/10.1186/s11671-021-03593-0
84. Shi Y et al (2010) Synthesis of few-layer hexagonal boron nitride thin film by chemical vapor deposition. Nano Letters 10(10):4134–4139. https://doi.org/10.1021/nl1023707
85. Singh RP, Handa R, Manchanda G (2021) Nanoparticles in sustainable agriculture: an emerging opportunity. J Control Release 329:1234–1248. https://doi.org/10.1016/j.jconrel.2020.10.051
86. Somani PR, SomaniSP, Umeno M (2006) Planer nano-graphenes from camphor by CVD. Chem Phys Lett 430(1–3):56–59. https://doi.org/10.1016/j.cplett.2006.06.081

87. Sriramulu M, Balaji, Sumathi S (2021) Photo catalytic, antimicrobial and antifungal activity of biogenic iron oxide nanoparticles synthesised using aegle marmelos extracts. J Inorg Organomet Polym Mater 31(4):1738–1744. https://doi.org/10.1007/s10904-020-01812-2
88. Szabó Z et al (2013) Regular ZnO nanopillar arrays by nanosphere photolithography. Photonics Nanostructures-Fundam Appl 11(1):1–7. https://doi.org/10.1016/j.photonics.2012.06.009
89. Thorburn PJ, Wilkinson SN, Silburn DM (2013) Water quality in agricultural lands draining to the great barrier reef: a review of causes, management and priorities. Agric, Ecosyst & Environ 180:4–20. https://doi.org/10.1016/j.agee.2013.07.006
90. Tripathi RM, Sharma P (2021) Gold nanoparticles-based point-of-care colorimetric diagnostic for plant diseases. pp 191–204. https://doi.org/10.1007/978-3-030-66165-6_10
91. Uddling J et al (2007) Evaluating the relationship between leaf chlorophyll concentration and SPAD-502 chlorophyll meter readings. Photosynth Res 91(1):37–46. https://doi.org/10.1007/s11120-006-9077-5
92. Usman M et al (2020) Nanotechnology in agriculture: current status, challenges and future opportunities. Sci Total Environ 721:137778. https://doi.org/10.1016/j.scitotenv.2020.137778
93. Vigani G et al (2013) Cellular iron homeostasis and metabolism in plant. Front Plant Sci 4. https://doi.org/10.3389/fpls.2013.00490
94. Wang Y et al (2016) The application of nano-TiO2 photo semiconductors in agriculture. Nanoscale Res Lett 11(1):529. https://doi.org/10.1186/s11671-016-1721-1
95. Watson DJ (1947) Comparative physiological studies on the growth of field crops: I. Variation in net assimilation rate and leaf area between species and varieties, and within and between years. Ann Bot 11(1):41–76. https://doi.org/10.1093/oxfordjournals.aob.a083148
96. White PJ (2009) The use of nutrients in crop plants. In: Fageria NK (ed) Boca Raton. CRC Press, FL, USA, pp 430, US $129.95. ISBN 978–1–4200–7510–6. Exp Agric 45(3):380–380. https://doi.org/10.1017/S0014479709007789
97. Xia X et al (2023) Recent progress of Si-based anodes in the application of lithium-ion batteries. J Energy Storage 72:108715. https://doi.org/10.1016/j.est.2023.108715
98. Ghidan AY, Al Antary TM (2020) Applications of nanotechnology in agriculture. In: Applications of nanobiotechnology. IntechOpen. https://doi.org/10.5772/intechopen.88390
99. Yao C et al (2010) Development of a quartz crystal microbalance biosensor with aptamers as bio-recognition element. Sens 10(6):5859–5871. https://doi.org/10.3390/s100605859.
100. Yao KS et al (2007) Photocatalytic bactericidal effect of TiO2 thin film on plant pathogens. Surf Coat Technol 201(15):6886–6888. https://doi.org/10.1016/j.surfcoat.2006.09.068
101. Yin Y, Gates B, Xia Y (2000) A soft lithography approach to the fabrication of nanostructures of single crystalline silicon with well-defined dimensions and shapes. Adv Mater 12(19):1426–1430. https://doi.org/10.1002/1521-4095(200010)12:19<1426::AID-ADMA1426>3.0.CO;2-B
102. Ying S et al (2022) Green synthesis of nanoparticles: current developments and limitations. Environ Technol & Innov 26:102336. https://doi.org/10.1016/j.eti.2022.102336
103. Yu HD et al (2013) Chemical routes to top-down nanofabrication. Chem Soc Rev 42(14):6006. https://doi.org/10.1039/c3cs60113g
104. Yu H, Wang D, Han MY (2007) Top-down solid-phase fabrication of nanoporous cadmium oxide architectures. J Am Chem Soc 129(8):2333–2337. https://doi.org/10.1021/ja066884p
105. Zain M et al (2024) Nanotechnology in precision agriculture: advancing towards sustainable crop production. Plant Physiol Biochem 206:108244. https://doi.org/10.1016/j.plaphy.2023.108244
106. Zheng L et al (2005) Effect of nano-TiO2 on strength of naturally aged seeds and growth of spinach. Biol Trace Elem Res 104(1):083–092. https://doi.org/10.1385/BTER:104:1:083
107. Zubair MS et al (2022) First report of fruit rot of cherry and its control using Fe2O3 nanoparticles synthesized in calotropis procera. Molecules 27(14):4461. https://doi.org/10.3390/molecules27144461

Nanosensors in Plant Disease Diagnosis and Management

Sadia Haider, Pakeeza Afzal, Sabir Hussain, Faisal Mahmood, Tanvir Shahzad, Muhammad Sadiq, Hassan Nawaz, Awais Maqsood, and Muhammad Shahid

Abstract Nanobiotechnology has emerged as a promising field with diverse applications in the agriculture sector and it has significantly aided in plant disease diagnosis and management. Amongst the nanotechnology-based methods, nanosensors have quickly become essential tools to diagnose and treat plant diseases. Nanosensors have enhanced agricultural output by boosting crop yield. By influencing the targeted treatment, nanosensors have revolutionized phytodiagnostics with extreme sensitivity, rapid results and precise identification. Nanosensors can be employed to monitor soil quality and detect the presence of plant pathogens such as viruses, bacteria and fungi including their physiological indicators enabling early disease prognosis. These indicators often involve changes in the plant's metabolism, physiology, or biochemistry in response to infection. Early screening of pathogens prevents the spread of disease as well as safeguards leading to reduced economic losses. Accurate evaluation leads to effective treatments of crop diseases and helps to cut down the misuse of pesticides. Precise diagnosis ensures that the right pesticide is used at the right time and dosage, minimizing the risk of resistance. Optical, electrochemical, mechanical, transducable and resistive properties of nanosensors have led to the development of a large variety of these sensors which provides precision diagnostics and early intervention. This chapter hints on brief introduction of nanosensors, the mechanism on which they work and how they are classified according to their specific properties. Additionally, their applications have also been briefly explained. Moreover, some of the proposed drawbacks are appended to catch the attention of future research. Subsequently, the future prospects about transformation of plant disease diagnosis into new horizons of early plant disease detection for sustainable agriculture are also presented.

Keywords Plant diseases · Disease diagnosis · Sustainable agriculture · Efficient treatment · Nanosensors · Nanotechnology

---

S. Haider · P. Afzal · M. Sadiq · H. Nawaz · A. Maqsood · M. Shahid (✉)
Department of Bioinformatics and Biotechnology, Government College University, Faisalabad, Pakistan
e-mail: mshahid@gcuf.edu.pk

S. Hussain · F. Mahmood · T. Shahzad
Department of Environmental Sciences, Government College University, Faisalabad, Pakistan

A. Husen (ed.), *Nanobiosensors for Agricultural and Other Related Sectors*, Smart Nanomaterials Technology, https://doi.org/10.1007/978-981-96-6499-3_13

1 Introduction

The importance of plants in agriculture cannot be overstated. They are vital sources of food and sustenance for countless organisms, forming the backbone of life on Earth-In addition essentials in plants are vital for human life [23]. They are one of the principal forms of life on this planet and play a critical role in supporting the equilibrium of nature with the noteworthy capacity to produce their own food through photosynthesis [76]. Conserving and protecting plants is essential for ensuring the well-being and sustainability of future generations. Plant diseases can considerably affect the health and yield production of crops. Having acquaintance with the root causes of diseases and remedies for these plants is vital for keeping flourishing agriculture system healthy. Fungi, viruses, bacteria, oomycetes, nematodes, insect pests and climate change are the prime offenders in initiating countless plant disorders, which eventually leads to a prominent decrease in their yield, oscillating from 10 to 95% [52]. Researchers are developing different measures to control the widespread of pathogenic diseases. Biological control is one of those control methods in which living hosts are used to kill pathogens [27]. Amongst the phytopathogens, fungal and bacterial infections cause deleterious contagion of crops. The world's major crops, such as rice, wheat, potatoes, and soybeans, are most exposed to fungal infections in addition to almost all vegetables and fruit crops. On the other hand, the use of fungicides is an effective process, but it can also cause damage to human and animal health, our environment, and soil quality.

1.1 Conventional Diagnostic Approaches

Anciently, farmers used to detect pathogens by supernatural and superstitious methods without considering scientific approaches. The discipline of using conventional methods to investigate plant pathogens was founded in 1767 by Giovanni Targioni Tozzetti. He played an important role in investigating the cause of the famine in central Italy and discovered the agent of rust of wheat, *Puccinia graminis* [44]. Traditional methods for detecting and identifying plant pathogens predominantly rely on visual symptom examination, culturing techniques, serological detection, and isozyme analysis. While still under way, close observation of symptoms showed to be one of the oldest methods used to identify the disease-causing pathogens in plants [56]. Developmental progress made the identification of plant pathogens easier, quicker, and more accurate through new imaging-based approaches, among others [6]. Serology, the most widely recognized diagnostic technique, refers to the diagnoses that include specific antibodies that recognize the corresponding antigens within the chosen sample. Bacterial protein and nucleic acid amplification tests include conventional methodologies that include Polymerase Chain Reaction (PCR) [8], Recombinase Polymerase Amplification (RPA) [37], Enzyme Linked Immunosorbent Assay (ELISA) [59] and Loop Mediated Isothermal Amplification

(LAMP) [86], are among the traditional diagnostic methods. Though these present a very high degree of specificity and sensitivity, there are certain handicaps of many of the conventional molecular diagnostic methods that preclude their usage in day-to-day field testing. Among them are the specialized knowledge required at the laboratory stage, laborious sample preparation, long assay time, and complex and expensive equipment. These include laborious sample preparation, lengthy assay time and expensive equipment [39]. Furthermore, a trained individual performing the diagnosis influences the precision and trustworthiness of these traditional diagnostic technologies.

1.2 Importance of Early Disease Diagnosis

An accurate and timely diagnosis of the plant disease and its pathogen will assist in the development of practical disease management strategies. Bacterial infections will, in general, take a longer time to show symptoms since they usually attack the roots and the upper end of the stem. Usually, by the time the symptoms have appeared, it is often too late to take practical protection against the infecting agent [45]. This hassle in timely diagnosis and detection of the causative agent results in a decline in crop production and its quality. Moreover, annual agricultural losses of an approximation \$220 Billion are attributed to the failure of early detection emphasizing the critical need for efficient and accurate diagnostic methods [28]. Many of the serological, immunological and DNA based plant disease detection methods such as Fluorescence in situ hybridization FISH [88] and Next Generation Sequencing NGS [75] also ensure timely detection of plant diseases [44]. In all this chaos, a lot of plant pathogens disguise the farmers and researchers and hence many of the plant diseases either go undiagnosed or diagnosed later. In this scenario, nanotechnology has emerged as a new weapon to mount the challenge of plant disease detection and management. Nanosensors have quickly become important tools in early disease diagnosis of plant diseases by developing highly specific, selective and sensitive methods [18] (Fig. 2).

1.3 Nano-Sensors

Nano-sensors are mechanical sensors that indicate the presence of certain substances, and physical characteristics and play role in determining specific physiological characteristics of plants [84]. The adoption of nano-based agricultural techniques has accelerated as the overuse of fertilizers and pesticides steers to environmental pollution, soil degradation and increased production costs. Nanosensors developed for the examination of plant processes provide precise sensitivity and versatility, facilitating the presence of metabolic precursors and signaling ligands. The complex roles of these molecules in plants are explained paving way for more sustainable agricultural practices [66]. These State-of the art tools enable specific monitoring of soil conditions and encouraging sustainable farming practices [36].

Nano-biosensors development actually substitutes for the conventional methods of detection because of their cost effectiveness, effortless detection of pathogens and their phenomenal sensitivity [65]. Notable developments have been made which includes a large range of optical, electrochemical and wireless nanosensors [13, 77]. In addition, the enhancement of surface characteristics has paved the way for new discoveries focusing on achieving better interaction of nanosensors and target analytes [21].

2 Types of Nano-Sensors

In the era of precision agriculture, nanotechnology has proven to be a promising field in plant illness detection and management [26]. These fine-tuned nanosensors specifically chosen because of their unique nanomaterial's physical, biological and chemical properties have harnessed an early detection of plant pathogenic microbes [39]. Nanosensors have actually proved to be an extremely useful to monitor a wide range of parametric quantities like the presence of pathogens, nutrient level and a wide range of biomarkers in plants [30]. Growers and researchers are facilitated as the development of targeted fungicides reduce the spread of diseases [10]. Similarly, plant protection compounds like certain antimicrobial agents and agrochemicals have also been developed to ensure specific delivery through nanosensors [63]. A wide range of nanosensors are being employed in the agricultural sector to reduce crop loss and alleviate environmental hazards, offering innovative solutions for sustainable farming practices (Table 1).

2.1 Nanoparticle-Based Sensors

The distinctive properties of nanomaterials are significantly enhanced with integration into sensors that aim to detect plant pathogens [35]. A wide variety of analytes has been discovered with quantum dots and metallic nanoparticles, attributed to their extraordinary capability [39]. Quantum dots, being semiconductor nanocrystals, have been acknowledged for their excellent fluorescent properties, making them a significant area of interest in the nanosensing field [47]. It is a light-emitting nanoparticle designed to emit light at specific wavelengths and to allow for the detection of target analytes based on the fluorescent emitted [25, 46]. The properties of well-targeted performance with biomolecules and their nanoscale size make these semiconductor nanoclusters one of the very poise candidates for development in miniature, portable, and multiplex sensor platforms [74]. In addition, gold and silver nanoparticles are

Table 1 Different nanosensors used for detection of various diseases in different plants

Nanosensors	Disease	Pathogen	Plant	Key references
Gold nanoparticles	Wilt disease	*Ralstonia solanacearum*	Potato	[31]
Multi-welled carbon nanotubes	Stem & fruit rot, leaf blight	*Alternaria solani*	Tomatoes	[22]
Platinum nanoparticles	Potato late blight	*Phytophthora infestans*	Potato	[40]
FRET-based nanosensors	Tobacco mosaic disease	*Tobacco mosaic virus*	Tobacco	[87]
e-nose system	Stem pitting	*Citrus tristeza virus*	Grapefruit, & sweet orange	[38, 71]
Optical biosensors	Late blight	*Phytophthora infestans*	Potato	[16]
Photoelectric sensors (RGB)	Anthracnose	*Colletotrichum destructivum*	Tabaco	[42]
Infrared sensors	Powdery mildew	*Podosphaera xanthii, & Erysiphe cichoracearum*	Cucumber	[89]
Spectral sensors	Net blotch, & brown rust	*Pyrenophora teres, & Puccinia hordei*	Barley	[42]
Carbon nano-tubes (CNTs)	Bacterial wilt	*Ralstonia solanacearum*	Eggplant, & tomatoes	[53, 69]
Wearable sensors	Tomato spot wilt	*Tomato spotted wilt virus* (TSWV)	Tomato	[72]
Silver nanoparticles (AgNPs)	Sclerotium rolfsi	*Southern blight*	Groundnuts	[43]
Electrochemical biosensors	Walnut blight	*Xanthomonas arboricola*	Walnut plant	[51, 58]
DNA based biosensors	Crown galls	*Agrobacterium tumfaciens*	Rose plant	[79]
Thermal imaging sensors	Wheat leaf rust	*Puccinia triticina*	Wheat	[54, 89]
SPR based nanosensors	Black leaf streak disease/black sigatoka	*P. fijiensis*	Banana	[70, 78]
Quantum dot-based sensors (QD-based sensors)	Rhizomania	*Beet necrotic yellow vein virus*	Beetroot	(Enespa and [19]), [41]
Array based nanosensors	Sharka disease	*Plum pox virus*	Peaches	[61],
Plant nano bionic sensors	Stewart's wilt	*Pantoea stewarfii subsp. stewarfii*	Sweet corn	[81]
Ultrasonic sensors	Bacterial canker	*Xanthomonas citri pv.mangiferaeindicae*	Mango	[48, 60]

some of the metallic nanoparticles that have been analyzed in detail for their functionality as nanosensors [49]. Their biomarker evaluation ability is also supplemented by the ability to integrate with different varieties of biomolecules, for instance, DNA, aptamers, and antibodies [90].

2.2 Molecular Recognition Based Nanosensors

Recent advances in performing sensitive nanosensors have facilitated the management of intrinsic molecular recognition and transduction processes regarding specific biomolecules, with its applications in identifying and regulating plant diseases at the nanoscale [15]. The internal structure of the nanosensors includes a nanomaterial which has outstanding specificity and sensitivity toward biomolecules like proteins, nucleic acids, and metabolites that enable sensitive detection and analysis [11]. The nanosensors with these recognition elements detect proximity and efficacy of the target analytes [46]. The aforementioned molecular recognition and events are, subsequently, transduced into measurable mechanical, electrical and optical signals by various transducer components that are functional in the sensor [39, 64]. The identification of molecular recognition and transduction in nanosensors has enabled early diagnosis and targeted intervention for sensitive and specific detection of phytopathogens [32]. Delivered into the plant using nanomaterials acting as a tiny carrier, these elements enhance the innate defenses of the plants and increase their ability to survive and recover from disease.

2.3 Field Effect Transistors Based Nanosensors

FET-based nanosensors have promising potential for enhancing the diagnosis of plant diseases. Even minuscule nanoparticles have revealed that they can successfully identify and interact with harmful microorganisms that affect plant [29]. The FET-based nanosensors present alternative avenues for disease management in that the nanoparticles can take along specific biomolecules which enhance current methods for plant disease control [68]. Additionally, nano sensors can be designed to locate and respond to specific indicators of plant health. This is helpful in identifying and acknowledging problems early, before they normally appear in the plant [20]. It provides valuable information since it allows farmers to act early in reducing the impact of disease [17].

2.4 Electrochemiluminescence (ECL) Based Nano Sensors

Combining nanotechnology with electroluminescence has now achieved the development of nanoscale sensors that are strongly characterized by high sensitivity, selectivity, and their reputation as "Electrochemiluminescence Nano Sensors" Advanced technological instruments, because they finely utilize unique properties of nanomaterials, refine such ECL-based detection systems into effectively tackling a range of biomedical and environmental challenges [73]. One of the main advantages of the nanosensors based on electrochemiluminescence is that they increase significantly the surface area of the electrode and help in the identification of pathogens, making the sensor more selective and sensitive [73]. Better detection limits may be achieved by fine-tuning the sensor's performance to be more precise about those parameters. The advances related to micro, and nanotechnology considerably enhance electrochemiluminescence nanosensors, concerning the sophistication of the amplification of signals along with improved encapsulation elements of biological recognition processes. This is why current nanosensors are primarily sensitive, selective, and sophisticated [67].

2.5 Piezoelectric Transduction Based Nanosensors

This has led to a novel approach within biosensing, as the piezoresistive-based nanosensors provide the avenues for the design of platforms that highly specific and sensitive in recognizing and diagnosing plant disease through the exploitation of piezoresistive material distinctions [12]. The exceptional capability of piezoelectric materials to convert mechanical forces into electrical signals has facilitated the development of these advanced sensors [12]. The mechanical vibrations transmitted by piezoelectric substances can alter the cellular structure of plants by effectively inhibiting cellular growth and the multiplication of pathogens. This concept has been claimed to have inspired the development of nanosensors that exploit the use of piezoelectric technology for the detection of pathogens in plants [12].

2.6 Piezoresistive Transduction Based Nanosensors

A conceivable strategy may be looked for in the application of piezoresistive transduction, whereby there is a conversion of some deformation of the device by such specific biomarkers into an electric signal, which can facilitate point-of-care disease diagnosis. This technology has certain piezoresistive materials characteristics such as the change in electrical resistance caused by mechanical stress or strain [12]. They can be embedded with these sensors to allow for timely interventions strategies and real time noninvasive monitoring for disease progression. Metal nanoparticles can be used on biosensing systems to achieve improvement in their sensitivity and selectivity. These nanoparticles in piezoresistive nanosensors can not only integrate in terms of sensing disease but also prevent it, which is a dual functionality [82]. These nanosensors are able not only to recognize plant diseases, but also to monitor plant responses to external stress, demanded by application of precision agriculture [50]. Such a strategy reduces the probability of disease outbreaks and reduces the costs of both financial and environmental detrimental effects caused by overuse of agrochemicals [55]. Furthermore, piezoresistive, nano sensor based smart agricultural solution has been created with the means of wireless communication technologies with portable electronic devices. Using these solutions, producers can make proactive, data inferenced based disease management decisions on a real time basis with cloud based data analysis and decision assistance [15].

2.7 Optical Transduction Based Nanosensors

Plants have evolved many defense mechanisms to withstand both biotic and abiotic stressors [4]. Understanding the complex networks of signals involved in plant-pathogen interactions and metabolic processes is crucial to develop effective methods for monitoring and regulating plant health particularly for disease control [4]. Perceiving such complex networks of signals is essential to create methods that effectively and efficiently monitor and regulate plant health [4]. Optical transduction mechanisms that use nanomaterials that are sensitive to light generate signals that are proposed by [62, 83] to track plant health and to identify the beginning of diseases without any intervention. The nanomaterials like carbon nanotubes transform to produce early warning systems capable of detecting plant pathogens through interactions that provide an optical response that acts as the starting point [80].

3 Nano-Sensors' Applications

Nanosensors are highly appropriate to identify analytes in the plant tissues due to ease of integration into such tissues. Furthermore, they are ideal for carrying out in vivo studies of metabolic and cellular signaling. Some of the applications of nanosensors in the plants are given in the following subsections (Fig. 3).

3.1 Nanoparticle-Mediated Gene Delivery

Enhancing the accuracy of gene delivery and plant disease diagnoses for improved agricultural practices depends on nanosensors. The sensors use the distinct properties of nanomaterials, such as a huge surface area and increased reactivity, to interact with target analytes for the early diagnosis and treatment of diseases [2]. Nanosensors can pass through plant biofluids, thereby guaranteeing accurate gene delivery intended for the improvement of plant resistance against infections. By reducing the use of chemicals, this technology not only promotes sustainable farming methods but also presents exciting opportunities for disease prevention and real-time crop monitoring. It is anticipated that as research advances, the use of nanosensors in agriculture will transform plant health management, resulting in increased agricultural yields and resistance to environmental stressors. Nanotechnology based methods have also been used to deliver biomolecules to plant cell systems and the use of nanoparticles has proven to be more efficient in plant transformation due to the superior protection of genetic cargo from cellular enzymatic destruction through nucleases. Various nanomaterials are used for gene delivery e.g., carbon-based nanoparticles, silicon-based nanoparticles and polymer-based nanoparticles. Carbon nanotubes (CNTs) can transport both RNA and DNA, but metallic nanoparticles are limited to DNA delivery as genetic cargo. Furthermore, silicon based nanoparticles (such as PEG and polyethyleneimine) can deliver encapsulated RNA, DNA and proteins to cell [85].

3.2 *Nanomaterial-Based Nutrient Delivery Systems*

High nutrient value of nanofertilizers for crops are developed through nanotechnology by the integration of certain lucrative nutrients with the nanosensors for their targeted delivery to plants prolonged and uniform absorption has been proved suitable with the fertilizers made up of nano-formulative nutrients. The incorporation of nanosensors into plant disease control systems offers not only a novel solution but also a sustainable approach towards ensuring nutrition as the agricultural sector struggles with mounting demands from climate change and an expanding global population. By providing real-time monitoring and rapid detection of pathogens, these new instruments can greatly minimize dependency on chemical treatments, thus supporting ecologically friendly practices that correspond with modern sustainability ideals. Chemical, physical and biological based techniques are used to make nanomaterial for nanofertilizers. Nanofertilizers should be modified to have all the abilities and features that required for the modification, as such high solubility, stability, time-controlled release, efficiency, increased targeted action with effective concentration, lower eco-toxicity with a safe, and simple means of administration [33].

3.3 *Nanomaterial-Based Stress Tolerance Enhancement*

Nanoparticles (sized between 1 and 100 nm) compared to their larger counterparts have unique and varied physicochemical characteristics of strong adsorbate capability including high surface area-to-volume rate as well linking efficiency with excellent working performance. Because of these features, they serve for hosting disease management processes such as bactericides (antibiotics that target bacteria), fungicides or insecticides that result in increased plantation health. Additionally, by using them on plants as macro and micro-nano fertilizers Nanoparticles help relieve nutrient deficiency symptoms and restore essential nutrients. Recent studies have reported that the application of nanoparticles improves plant tolerance to abiotic stress by enhancing antioxidant activity, lowering hydrogen peroxide (H_2O_2) and malondialdehyde (MDA) buildup, and maintaining photosynthetic efficiency. Furthermore, they can reach the photo system-II (PS-II) reaction center and increase electron transmission and light absorption by accessing the chloroplast, resulting in improved photosynthetic efficiency and plant growth [91].

3.4 Nanomaterial-Based Quality Improvement

One of the most recent developments in nanotechnology is to ensure the quality of nutrients supply to crop plants at a controlled rate. Reducing particle size and wide area to promote improved interaction and effective nutrient absorption for crop fertilization is also qualitative improvement in nanomaterial-based technology. It also enhances considerable economic and environmental benefits. Silicon dioxide (SiO_2), iron oxide (Fe_3O_4), titanium oxide (TiO_2), copper (Cu-NPs), and selenium (Se-NPs) have recently caught attention to enhance the growth and development of plants. It improves soil quality releasing nutrient from nanofertilizer in slow and controlled condition for better absorption, assisting in the site-specific target delivery of agrochemicals and reducing the risk nanomaterial toxicity. It also improves the quality of gene editing by accurate targeting site, to increase crop productivity by using nanomaterial-based biomarkers which have been applied on PGPR rhizobacteria to produce phytohormones to improve the plant growth, a rise in photosynthesis rate, and a notable increase around leaves. Nanopesticides are the advanced form of nanotechnology which improves the qualitative and the most specific way to protect plants from different pathogens, it also prevents the disturbance of diversity of other microbes and soil fertility and are precisely targeted toward the desired area.

4 Mechanism of Nanosensors Function

The mechanism of nanosensors relies on the identification of target plant diseases through the unique chemical, optical, and thermal properties of nanomaterials. Early and precise diagnosis allows for timely intervention, providing valuable insight into the onset of illness and enabling better disease management. It is a viable approach to monitor the presence of disease-related biomarkers and environmental elements to monitor and control the spread of plant diseases by installing networks based on nanosensing in greenhouse or agricultural fields. The real-time data from these networks allows activation of early warning systems allowing farmers and other agricultural experts to lessen the effects of plant diseases. After the target disease and plant have been identified, the next step is to choose suitable nanosensors with the required sensing capabilities, such as carbon nanotubes, metal nanoparticles, or graphene quantum dots. The core of nanosensors is a biological recognition element (BRE) which is functionalized with certain biomolecules like aptamers or antibodies making it possible to identify the disease-related markers with preference. These functionalized nanomaterials and required transducer elements are included in the design as illustrated in Fig. 1. The specific target molecule associated with the disease selectively binds to a BRE, triggering a measurable signal and allowing for the detection and quantification of the target. These nanomaterials are then validated to be used for field- based plant disease diagnosis. Using known samples, analytical performance of sensors like sensitivity, selectivity and reproducibility, is evaluated. Thus, biologically, physically and chemically nanosensors are engineered to engage with certain specific analytes or biomolecules. The distinct characteristics of nanomaterials and their interactions with the target analytes are the main factors governing the mechanism of nanosensor function for the diagnosis of plant diseases (Fig. 1). These interactions facilitate the translation of biochemical signals into detectable responses, thereby enhancing the real-time monitoring of plant health. Furthermore, the nanosensors' ability to navigate plant biofluids ensures the targeted delivery of genetic material, improving plant resilience against pathogens [66]. Quantum dots, metal nanoparticles, and carbon nanotubes are examples of nanomaterials that have high surface-to-volume ratios and increased sensitivity to environmental changes. This results in a detectable change in the sensor's output. For instance, an optical nanosensor detects and coincide with the presence and concentration of the pathogen by measuring changes in the fluorescence intensity which is emitted by the binding of target analyte to a functionalized nanomaterial.

5 Challenges of Nano-Sensor Technology

Agricultural yield get more accurate and productive with nanosensors however they also create environmental risks [63]. The soil ecosystem, particularly in plant disease diagnosis and treatment are affected by unintentional release of nanoparticles due to incorrect handling, unintentional release, or wastewater discharge [5]. It could mess up the soil ecosystem intensely, harming plants and native soil creatures [10]. Nanomaterial behavior in the soil–plant system is unknown. This means that soil biota and plant growth may be at great risk since these defects do not heal and are permanent [77]. More importantly, nanoparticles can penetrate the food chain, which will cause severe damage to human health. And this is why scientists are calling for the implementation of mitigation strategies as well as a deeper understanding of how nanoparticles behave. It also constitutes risk and toxicity on the food side, controlling inputs of nanoparticles like nanofertilizers and nanopesticides resulting in genetically modified crops. This means that one must think about how perceptions, the complexity of regulations, and long-term societal liabilities are associated with nanotechnology.

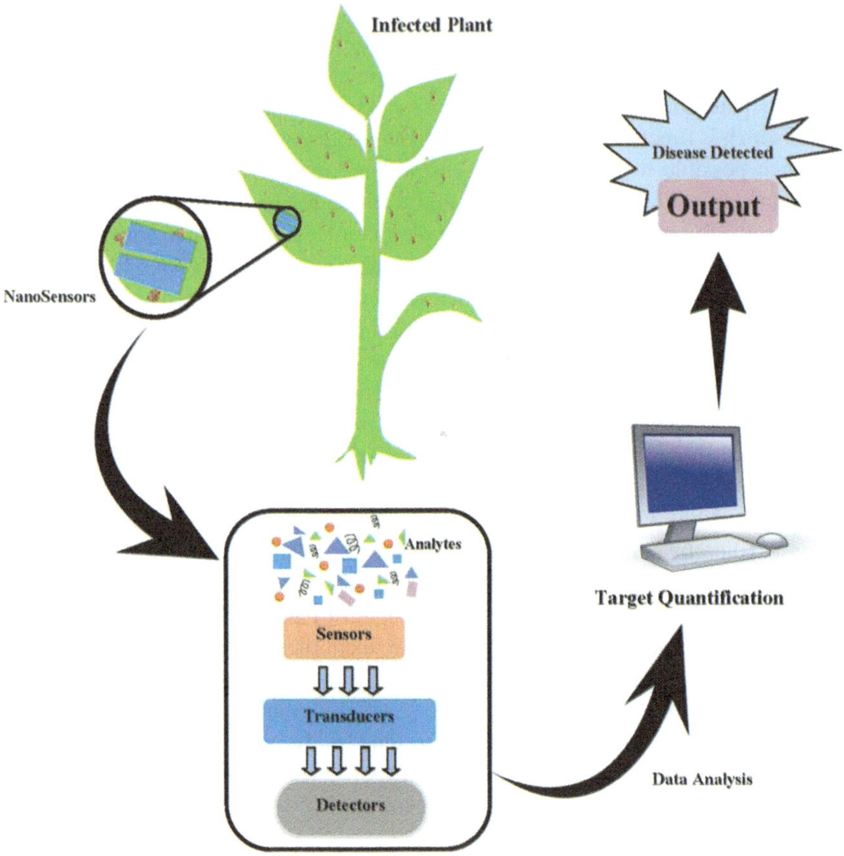

Fig. 1 Schematic representation of a nano-sensor-based disease detection system for plants. The nano-sensor detects analytes from the infected plant, with sensors, transducers, and detectors facilitating the detection process

The application of nanotechnology in agriculture and food science involves high ethical standards related to safety, risk assessment, transparency, consumer benefits, animal welfare, and environmental discretion [34]. All these concerns have led to the formulation of nanotechnological laws pertaining to safety, environmental protection, ethics, standardized practices, and confidence on public. The regulations of nanotechnology were first formulated concerning nanomaterial risks in food and agriculture, but there is a tremendous gap between consumer awareness and regulation as the current regulation does not oblige labelling agricultural products containing nanomaterials. This calls into question safety, openness, and the necessity of new regulations to guarantee appropriate labeling and risk control when using nanotechnology in agriculture. Instead of utilizing market set criteria researchers should use ethical standards to design nanotools [1]. Nanotechnology holds significant potential

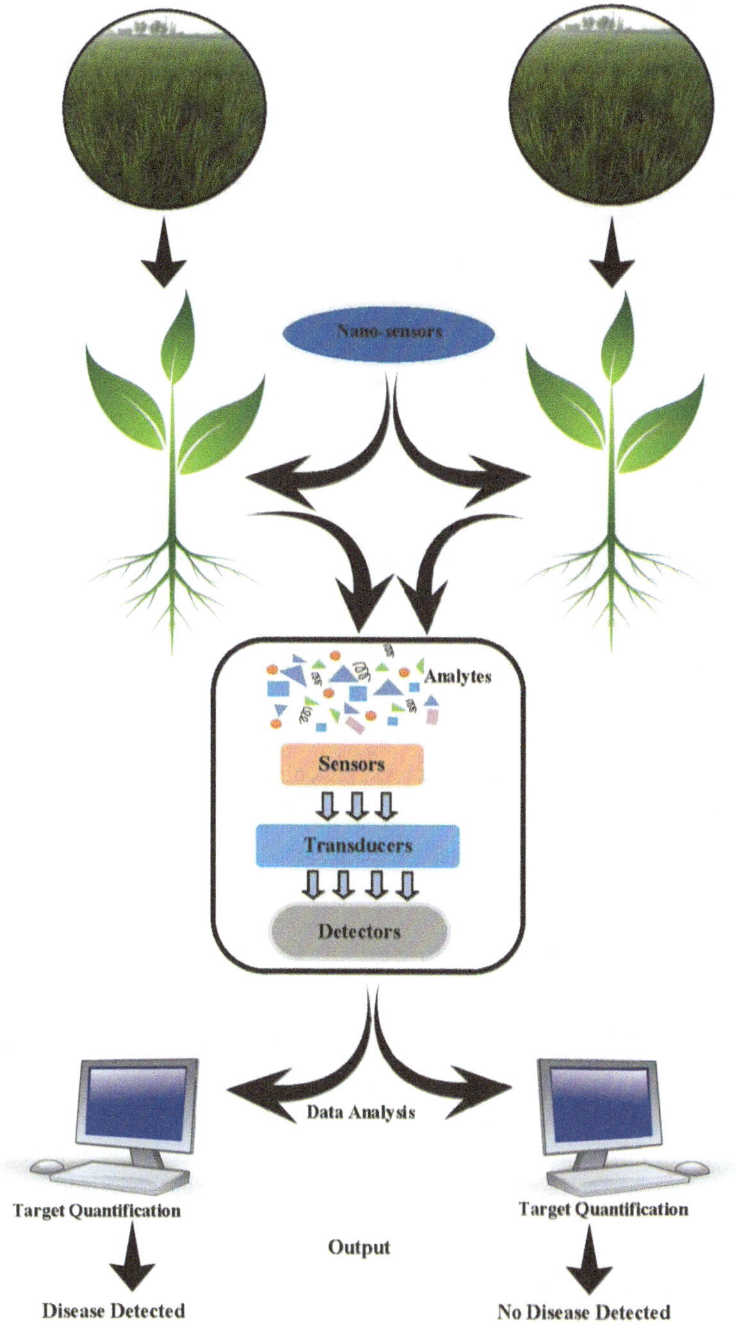

Fig. 2 Nano sensors are employed to plant samples collected from two distinct fields. Target quantification shows early disease identification of contaminated field through the corresponding sample plant after the formation of an analyte sensor complex aiding in early disease diagnostics

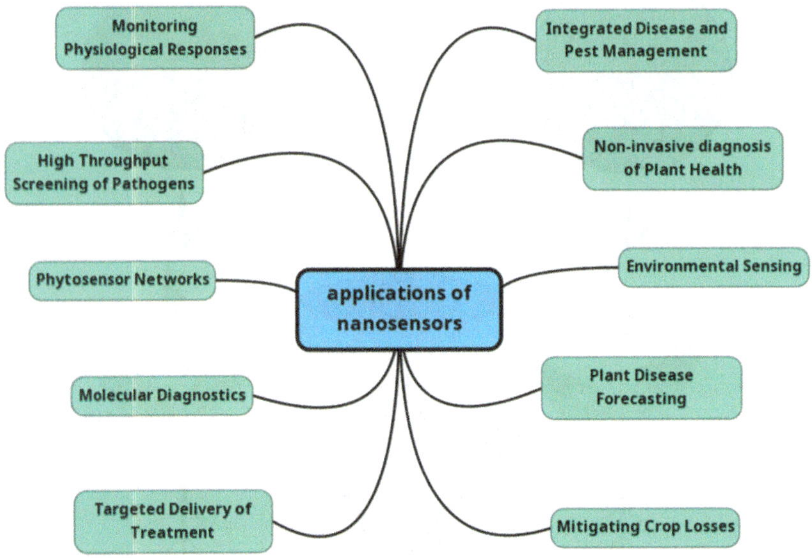

Fig. 3 The diverse use of nanosensors in monitoring physiological responses, pathogen screening, disease management, environmental sensing, and molecular diagnostics

for agricultural innovation, but technical challenges remain, especially controlling the synthesis and surface functionalization of nanoparticles, which are crucial for widespread industrial use [57].

6 Future Prospects of Nano-Sensor Technology

Nanosensors emerged as a key tool of nanotechnology for attaining a flourishing future of agricultural crops with increased enhancement in the disease detection methods. Introducing nanosensor-based methods to detect plant pathogens is setting a pace to make significant strides soon. In addition, these advancements will provide us with more precise monitoring methods for disease detection and management in plants. Developing biosensors techniques with higher sensitivity and specificity would be helpful in mapping the disease- prone crops while minimizing the environmental impact in parallel [39]. By encouraging multidisciplinary approaches to develop nanosensors with higher capacity of pathogen detection in crops, it would be helpful in early diagnosis of bacterial, fungal, nematode, phytoplankton, and viral diseases in plants [15]. These methods, involving nanosensors which collect information on environmental factors, soil characteristics, and prevalence of diseases, may now help in the earlier determination of onset stages and identify areas prone to these diseases; hence, this aids better control measures for diseases. Artificial intelligence-based analytics, used together with machine learning algorithms by nanotechnology, have tremendously enhanced accuracy and efficiencies with regard to disease-detection methods [9]. If these changes can be incorporated into the plant efficiently, researchers will be able to transform the health of plants by giving early warnings of specific diseases through monitoring real-time changes within the plants.

The nanosensors-based precision farming system promises to revolutionize the regenerative agricultural fields for the future. Sensors designed by using these advanced technologies will thus provide timely data regarding the prevalence of pathogens and the presence of plant nutrients, in turn aiding the farmer to quickly make optimal resource-use decisions. The applications are bound to get further propulsed from sophisticated nanosensors, including carbon nanotubes and quantum dots, to advanced precise nanosensors [3]. Sensors are a critical part of precision agriculture as they are used to measure the amounts of water and nutrients available in the crops, configure their nutrient requirements, and characterizing different plant pathogens, sometimes even coming from different ecological niches [7]. Amongst the most vital parts of nanotechnology, nanomaterials have a significant amount of potential in detecting disease. They can be developed in a very detailed fashion that could control the progression of the diseases by detecting them during early stages [24]. This indeed demands an integration of such methods and robust data analytics or decision support systems into the future developments in nanotechnology.

7 Conclusion

Nanotechnology has revealed promising outcomes with various nano-biosensors in identification of plant pathogens. The chapter discusses such instances that led to the discovery of plant pathogens using nanosensors, and laboratory-based developments along with prospective future developments to be done towards plant pathogen detection. The innovation of new and modified techniques using nanosensors offers timely detection of plant pathogens for the mitigation of plant diseases. Nanomaterials with unique characteristics of size and shape hold the immense potential to act as carriers for various important biomolecules in plants. In addition, remarkable advancements have been made to improve the selectivity, sensitivity, and stability of nanosensors. By developing cost-effective biosensors and expanding their use and availability to a lower level, we can improve crops' health and yield productivity with timely detection of major plant pathogens, i.e., bacteria, fungi, and viruses, which cause immense yearly yield loss worldwide.

Acknowledgements Authors are grateful to Department of Bioinformatic and Biotechnology, Government College University Faisalabad for providing resources in terms of library and literature subscriptions.

Competing Interests The authors declare that they have no competing interest.

Declaration of Generative AI and AI-Assisted Technologies in the Writing Process During the preparation of this work the book chapter author(s) have not taken any assistance from AI, and the book chapter author(s) reviewed and edited the content as needed and take(s) full responsibility for the content of the publication. All authors read and approved the final version of the manuscript.

References

1. Ansari MA (2023) Nanotechnology in food and plant science: challenges and future prospects. Plants 12(13):2565
2. Arjunan N, Thiruvengadam V, Sushil S (2024) Nanoparticle-mediated dsRNA delivery for precision insect pest control: a comprehensive review. Mol Biol Rep 51(1):355
3. Bajpai R, Puyam A, Kashyap PL (2023) Agro-nanodiagnostics for plant diseases. In: Nanotechnology in agriculture and agroecosystems. Elsevier, pp 169–188
4. Bal LM, Satpute G, Srivastava A (2017) Plant stress signaling through corresponding nanobiotechnology. Nanotechnol Appl Food 381–391. Elsevier
5. Banerjee K, Pramanik P, Maity A, Joshi D, Wani SH, Krishnan P (2019) Methods of using nanomaterials to plant systems and their delivery to plants (mode of entry, uptake, translocation, accumulation, biotransformation and barriers). Adv Phytonanotechnol 123–152. Elsevier
6. Barbedo JGA, Koenigkan LV, Santos TT (2016) Identifying multiple plant diseases using digital image processing. Biosys Eng 147:104–116
7. Beegum S, Das S (2022) Nanosensors in agriculture. In: Agricultural nanobiotechnology. Elsevier, pp 465–478
8. Bergeron M-J, Feau N, Stewart D, Tanguay P, Hamelin RC (2019) Genome-enhanced detection and identification of fungal pathogens responsible for pine and poplar rust diseases. PLoS ONE 14(2):e0210952
9. Bharti A, Jain U, Chauhan N (2024) From lab to field: nano-biosensors for real-time plant nutrient tracking. Plant Nano Biol 100079
10. Carrillo-Lopez LM, Villanueva-Verduzco C, Villanueva-Sánchez E, Fajardo-Franco ML, Aguilar-Tlatelpa M, Ventura-Aguilar RI et al (2024) Nanomaterials for plant disease diagnosis and treatment: a review. Plants 13(18):2634
11. Chelliah R, Wei S, Daliri EB-M, Rubab M, Elahi F, Yeon S-J et al (2021) Development of nanosensors based intelligent packaging systems: food quality and medicine. Nanomaterials 11(6):1515
12. Chorsi MT, Curry EJ, Chorsi HT, Das R, Baroody J, Purohit PK et al (2019) Piezoelectric biomaterials for sensors and actuators. Adv Mater 31(1):1802084
13. Dar FA, Qazi G, PirzadahTB (2020) Nano-biosensors: NextGen diagnostic tools in agriculture. In: Nanobiotechnology in agriculture: an approach towards sustainability. pp 129–144
14. Das S, Pattanayak S (2020) Nanotechnological approaches in sustainable agriculture and plant disease management. Org Agric 13–30
15. Das K, Jhan PK, Das SC, Aminuzzaman F, Ayim BY (2021) Nanotechnology: past, present and future prospects in crop protection. Technol Agric 1–22
16. Dyussembayev K, Sambasivam P, Bar I, Brownlie JC, Shiddiky MJ, Ford R (2021) Biosensor technologies for early detection and quantification of plant pathogens. Front Chem 9:636245
17. Elli G, Hamed S, Petrelli M, Ibba P, Ciocca M, Lugli P et al (2022) Field-effect transistor-based biosensors for environmental and agricultural monitoring. Sens 22(11):4178
18. Elmer W, White JC (2018) The future of nanotechnology in plant pathology. Annu Rev Phytopathol 56(1):111–133
19. Enespa, Chandra P (2024) Detection of phytopathogens in agricultural crops using nanodiagnostic techniques. IntechOpen
20. Feng X, Li P, Xiao M, Li T, Chen B, Wang X et al (2024) Recent advances in the detection of pathogenic microorganisms and toxins based on field-effect transistor biosensors. Crit Rev Food Sci Nutr 64(25):9161–9190

21. Forsan HF (2024) Applications of nanosensors in agriculture and food sectors. In: Handbook of nanosensors: materials and technological applications. Springer, pp 1331–1360
22. González-García Y, Cadenas-Pliego G, Alpuche-Solís ÁG, Cabrera RI, Juárez-Maldonado A (2021) Carbon nanotubes decrease the negative impact of Alternaria solani in tomato crop. Nanomater 11(5):1080
23. Hermawan (2021) The transfer of productive land for industrial interest. Jurnal Daulat Hukum 4(3):231
24. Ibrahim AA, Abd-Ellatif S, Abdel Razik E-SS, Hamedo HA, Salem KF (2024) Nanosensors for enhancing plant growth and productivity innovative methods in horticultural crop improvement. In: Biosensors and nanosensors. Springer, pp 25–50
25. Jia Z, Shi C, Yang X, Zhang J, Sun X, Guo Y et al (2023) QD-based fluorescent nanosensors: production methods, optoelectronic properties, and recent food applications. Compr Rev Food Sci Food Saf 22(6):4644–4669
26. Johnson MS, Sajeev S, Nair RS (2021) Role of nanosensors in agriculture. In: Paper presented at the 2021 international conference on computational intelligence and knowledge economy (ICCIKE)
27. Kamel Madbouly A (2018) Efficacy of pseudomonads as biocontrol agents of phytopathogens. Nov Res Microbiol J 2(3):48–52
28. Khakimov A, Salakhutdinov I, Omolikov A, Utaganov S (2022) Traditional and current-prospective methods of agricultural plant diseases detection: a review. IOP Conf Ser: Earth Environ Sci 951(1):012002
29. Khan MR, Ahamad F, Rizvi TF (2019) Effect of nanoparticles on plant pathogens. In: Advances in phytonanotechnology. Elsevier, pp 215–240
30. Kim DY, Kadam A, Shinde S, Saratale RG, Patra J, Ghodake G (2018) Recent developments in nanotechnology transforming the agricultural sector: a transition replete with opportunities. J Sci Food Agric 98(3):849–864
31. Kulabhusan PK, Tripathi A, Kant K (2022) Gold nanoparticles and plant pathogens: an overview and prospective for biosensing in forestry. Sens 22(3):1259
32. Kumar V, Arora K (2020) Trends in nano-inspired biosensors for plants. Mater Sci Energy Technol 3:255–273
33. Kumar P, Raut A (2024) Microbes-assisted bioaugmentation process in the reduction of emerging industrial pollutants from soil. In: Bioremediation of emerging contaminants from soils. Elsevier, pp 519–540
34. Kumari A, Rana V, Yadav SK, Kumar V (2023) Nanotechnology as a powerful tool in plant sciences: recent developments, challenges and perspectives. Plant Nano Biol 100046
35. Kurochkina M, Konshina E, Oseev A, Hirsch S (2018) Hybrid structures based on gold nanoparticles and semiconductor quantum dots for biosensor applications. Nanotechnol, Sci Appl 15–21
36. Kwak SY, Wong MH, Lew TTS, Bisker G, Lee MA, Kaplan A et al (2017) Nanosensor technology applied to living plant systems. Annu Rev Anal Chem (Palo Alto Calif) 10(1):113–140
37. Lau HY, Wang Y, Wee EJ, Botella JR, Trau M (2016) Field demonstration of a multiplexed point-of-care diagnostic platform for plant pathogens. Anal Chem 88(16):8074–8081
38. Lee RF (2015) Control of virus diseases of citrus. Adv Virus Res 91:143–173
39. Li Z, Yu T, Paul R, Fan J, Yang Y, Wei Q (2020) Agricultural nanodiagnostics for plant diseases: recent advances and challenges. Nanoscale Adv 2(8):3083–3094
40. Liang M, Yan X (2019) Nanozymes: from new concepts, mechanisms, and standards to applications. Acc Chem Res 52(8):2190–2200
41. Liebe S, Maiss E, Varrelmann M (2023) The arms race between beet necrotic yellow vein virus and host resistance in sugar beet. Front Plant Sci 14
42. Mahlein A-K (2016) Plant disease detection by imaging sensors–parallels and specific demands for precision agriculture and plant phenotyping. Plant Dis 100(2):241–251
43. Mansoor S, Zahoor I, Baba TR, Padder SA, Bhat ZA, Koul AM et al (2021) Fabrication of silver nanoparticles against fungal pathogens. Front Nanotechnol 3

44. Martinelli F, Scalenghe R, Davino S, Panno S, Scuderi G, Ruisi P et al (2015) Advanced methods of plant disease detection. A review. Agron Sustain Dev 35:1–25
45. Miller SA, Beed FD, Harmon CL (2009) Plant disease diagnostic capabilities and networks. Annu Rev Phytopathol 47(1):15–38
46. Mondal R, Dam P, Chakraborty J, Paret ML, Katı A, Altuntas S et al (2022) Potential of nanobiosensor in sustainable agriculture: the state-of-art. Heliyon 8(12)
47. Munir N, Gulzar W, Abideen Z, Hancock JT, El-Keblawy A, Radicetti E (2023) Nanotechnology improves disease resistance in plants for food security: applications and challenges. Biocatal Agric Biotechnol 51:102781
48. Naidu GG, Ramesh G (2022) Mango leaf disease detection using ultrasonic sensor. In: Paper presented at the 2022 IEEE international conference on data science and information system (ICDSIS)
49. Nandi NB, Das N, Ghanta S, Puzari KR, Dutta P, Kłak J et al (2023) Atropisomers and a copper (ii) complex derived from 1, 3-dimethyl-5-(8′-quinolinylazo)-6-aminouracil: structures, magnetism and biological properties. New J Chem 47(47):21633–21647
50. Naresh V, Lee N (2021) A review on biosensors and recent development of nanostructured materials-enabled biosensors. Sens 21(4):1109
51. Nguyen HA, Thi NPA, Trang NPT, Ho TT, Trinh TND, Tran NKS (2024) Recent advances in biosensors for screening plant pathogens. Anal Methods
52. Orchi H, Sadik M, Khaldoun M (2021) On using artificial intelligence and the internet of things for crop disease detection: a contemporary survey. Agriculture 12(1):9
53. Peeters N, Guidot A, Vailleau F, Valls M (2013) R alstonia solanacearum, a widespread bacterial plant pathogen in the post-genomic era. Mol Plant Pathol 14(7):651–662
54. Prasad P, Savadi S, Bhardwaj S, Gupta P (2020) The progress of leaf rust research in wheat. Fungal Biol 124(6):537–550
55. Presti DL, Di Tocco J, Massaroni C, Cimini S, De Gara L, Singh S et al (2023) Current understanding, challenges and perspective on portable systems applied to plant monitoring and precision agriculture. Biosens Bioelectron 222:115005
56. Ray M, Ray A, Dash S, Mishra A, Achary KG, Nayak S et al (2017) Fungal disease detection in plants: traditional assays, novel diagnostic techniques and biosensors. Biosens Bioelectron 87:708–723
57. Ray MK, Mishra AK, Mohanta YK, Mahanta S, Chakrabartty I, Kungwani NA et al (2023) Nanotechnology as a promising tool against phytopathogens: a futuristic approach to agriculture. Agriculture 13(9):1856
58. Retamales J, Núñez P, Alvarado R, Campan EDM, Otto T, Segovia C et al (2022) Characterization of xanthomonas arboricola pv. juglandis bacteriophages against bacterial walnut blight and field evaluation. Viruses 14(7):1380
59. Sakamoto S, Putalun W, Vimolmangkang S, Phoolcharoen W, Shoyama Y, Tanaka H et al (2018) Enzyme-linked immunosorbent assay for the quantitative/qualitative analysis of plant secondary metabolites. J Nat Med 72:32–42
60. Sanahuja G, Ploetz R, Lopez P, Konkol J, Palmateer A, Pruvost O (2016) Bacterial canker of mango, mangifera indica, caused by xanthomonas citri pv. mangiferaeindicae, confirmed for the first time in the Americas. Plant Disease 100(12):2520
61. Santonocito R, Parlascino R, Cavallaro A, Puglisi R, Pappalardo A, Aloi F et al (2023) Detection of plant pathogenic fungi by a fluorescent sensor array. Sens Actuators, B Chem 393:134305
62. Sewelam N, Kazan K, Schenk PM (2016) Global plant stress signaling: reactive oxygen species at the cross-road. Front Plant Sci 7:187
63. Shang Y, Hasan MK, Ahammed GJ, Li M, Yin H, Zhou J (2019) Applications of nanotechnology in plant growth and crop protection: a review. Molecules 24(14):2558
64. Sharma P, Kumar S, Patel A, Datta B, DeLong RK (2021) Nanomaterials for agricultural and ecological defense applications: active agents and sensors. Wiley Interdiscip Rev: Nanomedicine Nanobiotechnology 13(5):e1713
65. Sharma P, Sharma S (2016) Paradigm shift in plant disease diagnostics: a journey from conventional diagnostics to nano-diagnostics. Curr Trends Plant Dis Diagn Manag Pract 237–264

66. Shaw DS, Honeychurch KC (2022) Nanosensor applications in plant science. Biosens 12(9):675
67. Shelash Al-Hawary SI, Malviya J, Althomali RH, Almalki SG, Kim K, Romero-Parra RM, et al (2023) Emerging insights into the use of advanced nanomaterials for the electrochemiluminescence biosensor of pesticide residues in plant-derived foodstuff. Crital Rev Anal Chem 1–18
68. Shivashakarappa K, Reddy V, Tupakula VK, Farnian A, Vuppula A, Gunnaiah R (2022) Nanotechnology for the detection of plant pathogens. Plant Nano Biol 2:100018
69. Shoala T (2020) Carbon nanostructures: detection, controlling plant diseases and mycotoxins. Carbon Nanomater Agri-Food Environ Appl. Elsevier, pp 261–277
70. Shpacovitch V, Hergenröder R (2020) Surface plasmon resonance (SPR)-based biosensors as instruments with high versatility and sensitivity. Sens 20(11):3010
71. Taha Roodbar Shojaei TRS, Mohamad Amran Mohd Salleh MAMS, Kamaruzaman Sijam KS, Raha Abdul Rahim RAR, Mohsenifar A, Safarnejad R et al (2016) Detection of citrus tristeza virus by using fluorescence resonance energy transfer-based biosensor
72. Teixeira SC, Gomes NO, Calegaro ML, Machado SA, de Oliveira TV, Soares NDFF et al (2023) Sustainable plant-wearable sensors for on-site, rapid decentralized detection of pesticides toward precision agriculture and food safety. Biomater Adv 155:213676
73. Tovar-Lopez FJ (2023) Recent progress in micro-and nanotechnology-enabled sensors for biomedical and environmental challenges. Sens 23(12):5406
74. Tripathi A, Gupta P, Mittal AK, Mistri TK (2024) Outlooks of nanosensors in medical and smart devices. Agric Food Technol
75. Trippa D, Scalenghe R, Basso MF, Panno S, Davino S, Morone C et al (2024) Next-generation methods for early disease detection in crops. Pest Manag Sci 80(2):245–261
76. Usman AB, Abubakar S, Alaku C, Nnadi O (2014) Plant: a necessity of life. Int Lett Nat Sci 20:151–159
77. Usman M, Farooq M, Wakeel A, Nawaz A, Cheema SA, ur Rehman H et al (2020) Nanotechnology in agriculture: current status, challenges and future opportunities. Sci Total Environ 721:137778
78. Vally V, Jouen E, Pecheur B, Carval D, Bellaire DLD, Teycheney PY et al (2024) First report of black sigatoka disease in bananas caused by Pseudocercospora fijiensis on Mauritius Island. Plant Dis 108(3):802
79. Vatankhah A, Reezi S, Izadi Z, Ghasemi-Varnamkhasti M, Motamedi A (2022) Development of an ultrasensitive electrochemical biosensor for detection of Agrobacterium tumefaciens in Rosa hybrida L. Meas 187:110320
80. Verma G, Mishra KK, Raigond B, Jeevan B, Singh AK, Paschapur AU et al (2023) Nanodiagnostics: tool for diagnosis of plant pathogens. In: Biotic stress management of crop plants using nanomaterials. CRC Press, pp 113–123
81. Voke E, Pinals RL, Goh NS, Landry MP (2021) In planta nanosensors: understanding biocorona formation for functional design. ACS sens 6(8):2802–2814
82. Worrall EA, Hamid A, Mody KT, Mitter N, Pappu HR (2018) Nanotechnology for plant disease management. Agron 8(12):285
83. Xia X-J, Zhou Y-H, Shi K, Zhou J, Foyer CH, Yu J-Q (2015) Interplay between reactive oxygen species and hormones in the control of plant development and stress tolerance. J Exp Bot 66(10):2839–2856
84. Yusof NA, Isha A (2020) Nanosensors for early detection of plant diseases. In: Nanomaterials for agriculture and forestry applications. Elsevier, pp 407–419
85. Zarrabian M, Sherif SM (2024) Silence is not always golden: a closer look at potential environmental and ecotoxicological impacts of large-scale dsRNA application. Sci Total Environ 175311
86. Zhang L, Gleason C (2019) Loop-mediated isothermal amplification for the diagnostic detection of Meloidogyne chitwoodi and M. fallax. Plant Disease 103(1):12–18
87. Zhang Y, Zhang H, Wang J, Zhang J (2019) Fluorescence resonance energy transfer-based detection of Tobacco Mosaic Virus using aptamer probes. Biosens Bioelectron 130:103–110

88. Zhao H, Wang D, Li H, Li S, Wang Y, XuA, et al (2024) Fluorescence in situ hybridization analysis of Oligonucleotide 5S rDNA, 45S rDNA, and (TTTAGGG) 3 locations in Gloriosa superba L. Cytogenet Genome Res 1–14
89. Zhu W, Chen H, Ciechanowska I, Spaner D (2018) Application of infrared thermal imaging for the rapid diagnosis of crop disease. IFAC-PapersOnLine 51(17):424–430
90. Zuhrotun A, Oktaviani DJ, Hasanah AN (2023) Biosynthesis of gold and silver nanoparticles using phytochemical compounds. Molecules 28(7):3240
91. Zulfiqar HF, Afroze B, Shakoor S, Bhutta MS, Ahmed M, Hassan S, et al (2024) Nanoparticles in agriculture: enhancing crop resilience and productivity against abiotic stresses

Nanosensors for Wastewater Treatment and Disinfection

R. Ridhi

Abstract Present chapter veritable on indispensable need to avert water deficiency amongst existing accelerating living population to maintain ecological balance via utilizing waste water with ensued effective treatment and disinfection. There is an imperative need to work in the domain of modern wastewater treatment technologies with proliferating efficacies and low capital requirement. Therefore, water contaminants need to be shackled at appropriate time to ensure its effective treatment and disinfection. Nanosensors on account of their high aspect ratio, enhanced surface, optical, magnetic, electrical and spectroscopic properties, portability and flexibility in comparison to bulk materials may prove as promising early warning alarms with ensued effective waste water treatments. Moreover, high adsorption capacity of nanosensors ensures effective detection and obviation of microorganisms and other metal contaminants in water. The role of chemresistors and other nanosensors techniques for the detection and treatment of waste water are demonstrated and analyzed in this chapter.

Keywords Nanotechnology · Nanosensors · Chemresistors · Microorganisms · Water contaminations

1 Introduction

Current world of emerging Artificial Intelligence with substantive growth in modern technologies has resulted in various impediments in quality water supplies with integration of water scarcity [113]. Therefore, there is an imperative need for purifying and disinfecting waste water for its effective utilization by living organisms [69]. Nanotechnology has been reported as promulgating approach for disinfecting waste water on account of high adsorption capacity of nanomaterials due to high surface area, application in nanocatalysis, amenability to obviate microorganisms and other impeding contaminations in waste water [13, 39, 69]. However, the waste

R. Ridhi (✉)
Department of Physics, Mehr Chand Mahajan DAV College for Women, Chandigarh, India
e-mail: r.ridhi@mcmdavcwchd.in

A. Husen (ed.), *Nanobiosensors for Agricultural and Other Related Sectors*, Smart Nanomaterials Technology, https://doi.org/10.1007/978-981-96-6499-3_14

water contaminations to be obviated need to be shackled in a paradigm manner to use the target specific nanotechnology or other related concomitant technology for water disinfection [115]. This is because impure or contaminated water can result in harmful diseases like typhoid fever, hepatitis A and Cholera. Therefore, suitable detection techniques articulating as early warning alarms need to be adopted for proliferating waste water treatment and disinfection [5, 115].

Figure 1 demonstrates varied contaminations affluent in waste water as reported by [104]. Recent reviews by [86, 103] has highlighted imperative consortium of micro and nano manufacturing for waste water treatment due to amenable nanofiltration, degradation using suitable photocatalyst and effective nano absorbents as depicted in Fig. 2.

Nanosensors integrate on the principles of Fluorescence, electrochemistry and Surface Plasmon Resonance for the detection of analytes in water [33]. The interaction between target analytes present as water contaminants and nanomaterials such as nanoparticles, nanotubes and nanowires that have higher surface area, sensitivity and selectivity act as an asset for waste water disinfection [74, 100]. A recent review by [106] has identified five technical bottlenecks for practical implication of nanosensors in waste water treatment and has also highlighted embracing advances to overcome those barriers. Therefore, depending on the event and application, suitable nanosensors type can be chosen for the target analyte sensing. Oukhatar et al.

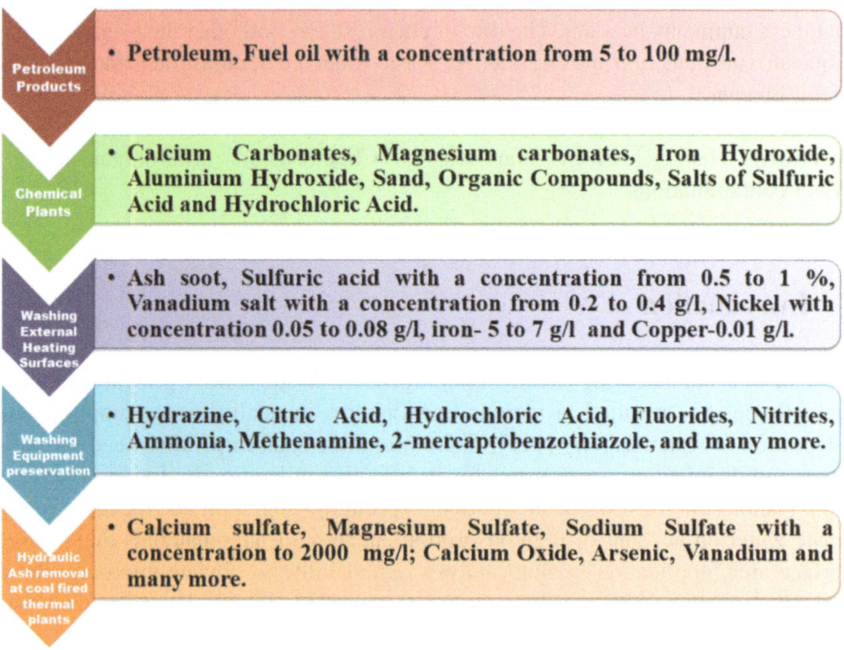

Fig. 1 Sources and types of water contaminations

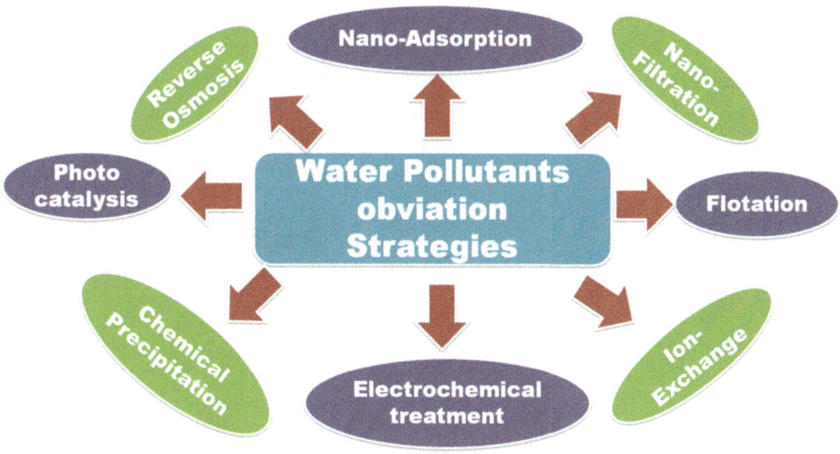

Fig. 2 Different strategies for effective obviation of water impurities

[68] has demonstrated the paradigm approach for the classifications of nanosensors as shown in Fig. 3.

Ridhi et al. [80] has demonstrated the various principles of nanosensors imperative to articulate the interaction mechanism between target analytes and nanoparticles to enhance sensitivity and selectivity of nanosensors towards environmental and water pollutants. Figures 4, 5, 6 and 7 demonstrate the sensing principles of varied nanosensors types with their veritable role in the detection of varied target analytes on account of their seizure behaviors due to interaction with nanomaterials incorporating nanosensors [1, 22]. The principle of optical nanosensors based on interaction between light and nanomaterials leading to variation in optical properties like transmission, reflection, refractive index or phase shift. These seizure behaviors in these properties are detected with amended detection limit via interaction with

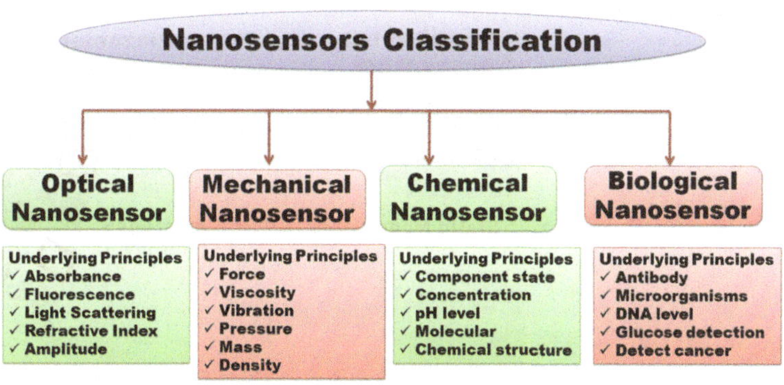

Fig. 3 Classification of nanosensors on the basis of distinct properties of materials

target analytes (Fig. 4). For instance, in fluorescence, nanosensors can incorporate fluorescent molecule that emit light when excited by an external stimulus.

Optical detection of specific analytes is indicated by variation in intensity or shift in the wavelength of the emitted light. Similarly, in Surface Plasmon Resonance (SPR), light interacts with free electrons on metal surfaces, leading to enhanced

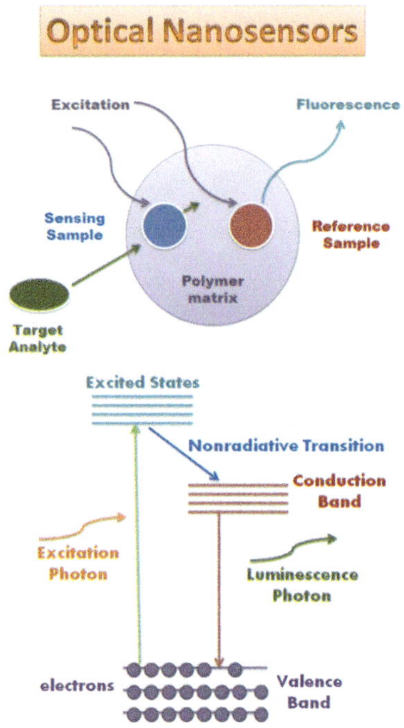

Fig. 4 Schematic representation of detection principle of optical nanosensors

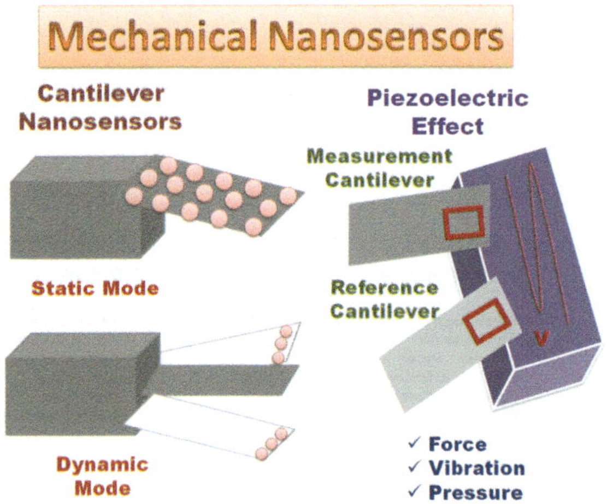

Fig. 5 Schematic representation of detection principle of mechanical nanosensors

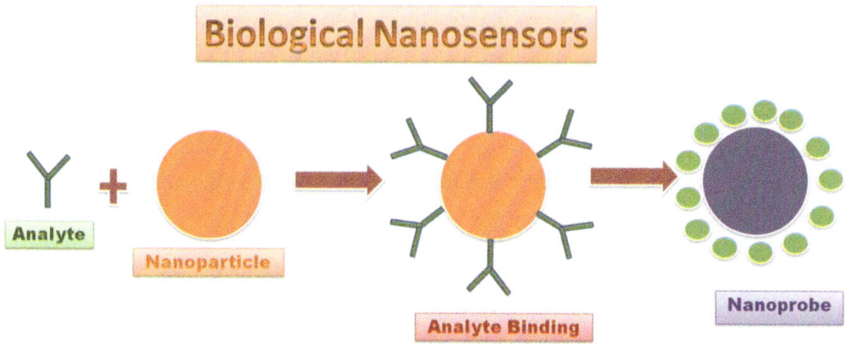

Fig. 6 Schematic representation of detection principle of biological nanosensors

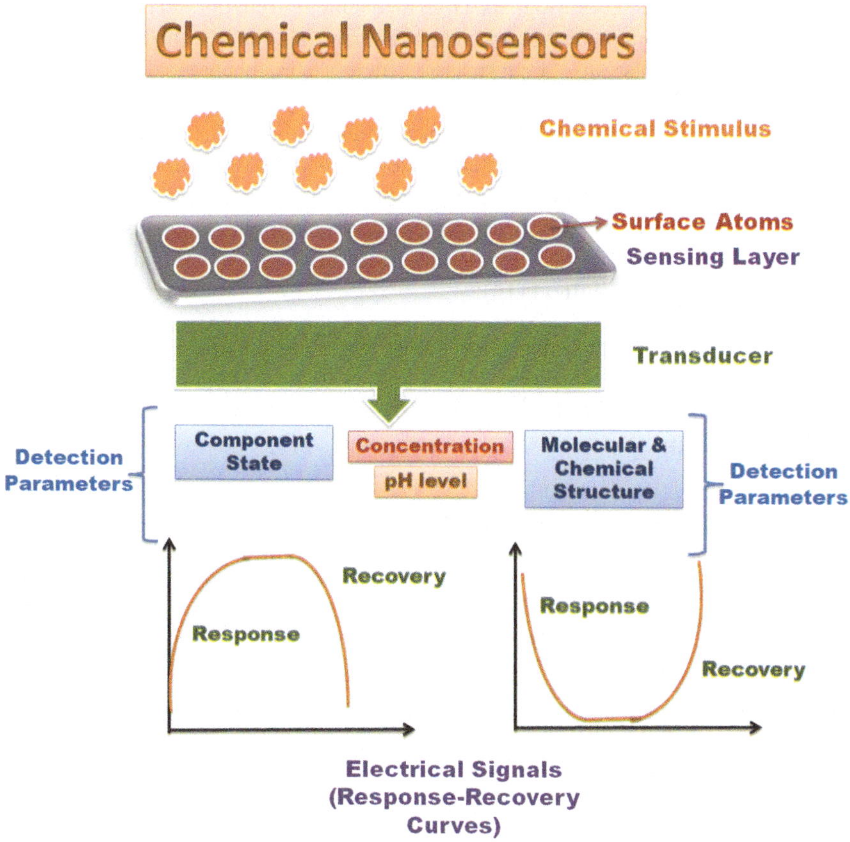

Fig. 7 Schematic representation of detection principle of chemical nanosensors

electromagnetic fields. Variations in refractive index on account of molecular binding events are utilized as detection parameter. Güler et al. [32] has veritable the role of plasmonic nanosensors for efficient detection of heavy metals and other water contaminants and environmental pollutants. The significant advantages of Plasmonic nanosensors inculcate rapid response, higher sensitivity-selectivity with amenability to operate in real time obviating the requirement of complex sample preparation [32, 48]. Some nanosensors form consortium of optical and mechanical properties of nanomaterials allowing for ultra-sensitive measurements.

Nanomechanical chemresistors (Fig. 5) relies on the principle of variation in mass, stiffness or strain of a nanostructure with the interaction of target analyte. The various modes of detection such as resonance frequency shift, deflection or variations in oscillation modes can be detected using various nanomechanical sensors inculcating piezoelectric sensors, capacitance variations, Laser Doppler vibrometry or other mechanical detection signals. High mechanical strengths and light weights of Graphene and Carbon nanotubes and other 2D materials can be utilized for removal of water contaminants and for fabricating efficient nanomechanical sensors using nanomembranes [14, 92].

Bionanosensors (Fig. 6) inculcate transduction mechanism to analyze the effect of target analyte binding to the recognition element. This transduction mechanism converts this analyzed effect into a detectable signal via various modes such as Optical (Fluorescence, Absorbance or SPR), Electrochemical (Changes in electrical properties (Fig. 7) on account of biochemical reactions) or Mechanical (mass, strain or frequency resonance shifts) [115].

2 Parameters of Waste Water and Their Detection Mechanism

Waste water refers to storm water and water utilized by living population for industrial or residential purposes [97]. Waste water parameters forms consortium of analytical tools for determining the degree of water pollution [97]. These parameters inculcate Total Suspended Solids (TSS), Heavy Metals, Chemical Oxygen Demand (COD), Biological Oxygen Demand $(BOD)_5$, Total Phosphorous and Total Nitrogen [3, 97]. Nitrate and Phosphorous utilized in agriculture are common chemical pollutant present in surface and ground waters [84]. TSS corresponds to dry weight of the un-dissolved particles or non-filterable residue. COD refers to the active organic compounds in water that have amenability to react with oxygen to form CO_2. BOD_5 measures the amount of organic contaminants that can be oxidized by biological organisms. A recent study by [3] has concluded that COD and BOD_5 are strongly affected by TSS with 7.92% and 7.94%, amendments respectively with 10% TSS enhancement. An indispensable need for the detection of viruses and bacteria in waste water veritable the role of nanobiosensors for regular surveillance and infection screening of Severe Acute Respiratory Syndrome Coronavirus 2(SARS-CoV-2) that has caused millions of deaths [42]. Nanobiosensors with concomitant use with point-of-care-testing (POCT) for real samples testing enables feasibility of practical implication of nanobiosensors for solving real problems associated with waste water enabling effective solution [47]. Micro-pollutants and TSS have the tendency to accrue rapidly when entered in waste water and it becomes impeding to obviate them completely [34]. Therefore, it becomes imperative to use suitable nanosensors as early warning alarms to avert their accruing and disinfect water at appropriate time [34, 66].

The sources of metal contaminations in water inculcate domestic waste, agriculture practices, industries like coal mining and leather, urban development and unpredictable climate risks [87, 96]. A recent review by [87] has identified the various health risks associated with varied metal contaminations in the water. For instance, excessive consumption of Cadmium hampers bones and kidneys of human population [70, 87]. Similarly, Mercury and Lead are neuro toxic metals that have substantive effect on blood levels with amendment in their dosage taken [52, 70]. Fakayode et al. [27] has elaborated the ill effects of chronic and acute levels of heavy metal ions exposure inculcating immune system deterioration, hypertension, dwindling children's intelligence quotients, impediments in development stages, oxidative cellular damage and cytotoxicity. A recent review by [30] has revitalized the role of nanostructured chemical sensors for trace metals or toxic metal ions such as Lead (Pd), Cadmium (Cd), Uranium (Ur), Arsenic (As) and Chromium (Cr) detection in waste water samples. Significance of electrochemical sensors using functionalized nanomaterials for the detection of heavy metals with Limit of detection lower than that reported by World Health Organization (WHO) has been reviewed by Mahmoud et al. [61]. Therefore, these waste water parameters and their concentration levels hampering water or required to disinfect water need to be detected using low detection limit devices to enable effective disinfection and purification of waste water.

Moufid et al. [64] has elicited electronic tongue and electronic nose as portable economical and simple analytical tools for waste water contaminations detection with better selectivity and discrimination among various concomitant existing waste water contaminations. Similarly, Bhaskar et al. [11] while highlighting impeding issues of various nanosensors and nanotechnology for waste water treatment in developing countries has emphasized utilization of sustainable and low cost nanotechnology approaches for monitoring water quality. They have trickled this track using hybrid techniques inculcating photonic crystal-coupled emission (PCCE), surface plasmon-coupled emission (SPCE) and cost-effective Smartphone based techniques. Hejabi et al. [36] has also emphasized significance of artificial intelligence (AI) models to understand and evaluate the dynamic and complex systems of Waste Water Treatment Plants (WWTP). Two AI models namely, artificial neural network (ANN) and Support Vector Machine (SVM) were utilized to assess the efficacy and quality of WWTP. It was concluded that ensemble approaches of the two models demonstrated better efficacy as compared to their individual counterparts [26, 36]. Epelle et al. [26] has demonstrated limitations of primary, secondary and tertiary treatment of waste water for large scale applications in WWTP and veritable role of nanaomaterials like CNTs, Graphene based Fullerenes, silver, Cooper, Iron and Gold nanoparticles for obviation of organic and inorganic water contaminations. Lesoinne et al. [53] has reviewed the significance of three promising electrodes Bi, Co and Cu for electrochemical detection of three categories of pollutants namely, Heavy metals, organic matter, phenolic compounds and nutrients. Environment friendly nature of Bismuth has made ease replace of Hg electrodes with Bi electrodes having reproducible and high-resolution sensing properties [53, 85].

3 Nanosensors for the Detection of Waste Water Contaminations

Tarannum [98] has highlighted distinct properties of nanosensors that make them as preferable portable detection devices as compared to existing labor-intensive, costly, and skill-required tools. These properties inculcate: amended specificity-selectivity, precise, accurate, reproducible sensing responses, stable under ambient conditions, portable, non-toxic, biocompatible and faster response-recovery times. A recent review by [50] has identified the significance of DNA nanosensors for the concomitant detection of affluent parameters in waste water and pond water samples. The technological advancements, safety issues, and choice of specific nanomaterial for nano-engineered material applied for the detection of suitable pollutants in waste water need to be shackled to ensure better detection limits to avert water scarcity with amenable water quality for the living population [10, 109].

3.1 Graphene Based Nanomaterials for Waste Water Treatment

Graphene is a 2-D material with significant, electrical, optical, chemical mechanical and thermal properties that make them amenable to environmental applications [71, 72]. The clientele of Graphene hybrid membranes should be concomitant with water permeability to obviate and detect the water pollutant with suitable LOD [6, 12, 62]. A recent review by [31] has reviewed imperative role of graphene nanostructures in biosensors coupled with significant influence of synthesis method in affecting the sensing phenomena. With suitable optimization of pore size of graphene nanoporous membranes and applied pressure, graphene efficacy to filter and desalinate water can enhance from 33 to 100 % [31, 110]. Moreover, high optical transparency of Graphene with amenability towards flexible optical sensing techniques inculcating Raman Scattering, Plasmon Resonance and Fluorescence makes it a promising nanosensor using different detection tools [89].

Cheng et al. [18] has demonstrated the significance of Graphene nanocomposites and its related derivatives for the detection of biological materials inculcating amino acids, drug molecules, saccharides, and many more, inorganic and metal ions.

Ridhi et al. [77] has also revitalized the significance of nanocomposite materials in amending sensing responses towards target analytes. A recent review by [9] has spurred numerous reports on amenable efficacy of Graphene based nanomaterials for sensing pathogens, virus, bacteria and other biosensing allowing for rapid diagnosis of various medical issues and diseases. He et al. [35] has highlighted different forms of graphene namely,pristine Graphene, Graphene Oxide (GO) and reduced Graphene Oxide (r-GO) having spurring electrochemical properties to be utilized for sensing prototypes. With the advancement of 3D- printing techniques, fabrication of graphene bases sensors has been easy, portable, biocompatible and more flexible in efficacy as compared to other materials [101].

Table 1 list the various graphene based nanosensors for the detection of varied target analytes with their nanosensor devices, sensing responses and detection limit. From Table 1, we observe that Grpahene Oxide (GO) and its functionalized forms in terms of quantum dots, nanocomposites and other additives have been reported as significant sensors using fluorescence, electrochemical and optical sensors. A recent review by [65] has elaborated an articulate paradigm methodological approach of fabricating GO composites using various forms with its applications. The trend of GO as new avenues in material science for nanosensors and other applications are attributed to high electron mobility, mechanical strength, large surface area with amended electrical conductivity and amenability of ease flexibility to incorporate easily with other materials [65, 75].

3.2 Carbon Nanotubes (CNTs) as Substantive Asset for Waste Water Treatment

Revitalizing photophysical properties of Single Walled Carbon Nanotubes (SWCNTs) makes them promising optical nanosensors for the detection of organic and biological molecules in wastewater [23]. A recent study by Khan et al. [43, 44] has also reported piezoelectric effect utilization for the fabrication of mechanical sensor based on Multi-walled carbon nanotubes (MWCNTs) dispersed in polyvinyl acetate (PVA) using ethylene glycol methacrylate (PVAc) matrix as a substrate material. The composite materials showed decrease in resistance with increase in pressure after being subjected to compression in the pressure range of 9.3–2348.8 kPa. Generally CNTs lack selectivity; therefore, they are functionalized with suitable additives to amend their sensing amenabilities [77, 88]. For instance, Ridhi et al. [78, 79] has used Metal Phthalocyanines (MPcs) as spurring materials for amending sensitivity and selectivity of SWCNTs towards ammonia gas. The amenability of band gap tenability of Metal Oxides with their high electrical conductivity and porosity makes them suitable integrating materials for CNTs for outpaced nanosensors [43, 44, 59].

Shooshtari et al. [95] has reported three fold amendment in sensitivity and dwindling of response time (by 30 s) of CNT with TiO_2 hybridization. Physi-sorption and Chemisorption ability of CNTs makes them efficient chemresistor devices with

Table 1 Graphene based nanosensors for detection of water contaminants

S. No	Material used	Target analyte	Sensing device	Sensing principle	Detection limit	Sensing response	Reference
1	Alizarin red aluminum (III) complex covalently binding to graphene oxide (GO)	Lysine (25 to 250 mg/L) (Amino acid)	Fluorescence	Photo-induced electron transfer (PET)	2.0 mg/L	pH value was between 6.5 and 7.2	[19]
2	CdSe/ZnS quantum dots-GO using imprinting technique	Tyramine (0.07 and 12 mg/L)(Amino aicd)	Fluorescence	Concentration-dependent fluorescence signal	0.021 mg L^{-1}	Selectively detected around 0.08 or 2 mg L^{-1} of tyramine in spiked rice wine samples	[107]
3	GQD as donor and MoS_2 nanosheets as acceptor	Epithelial cell adhesion molecule (EpCAM) detection (3 nM to 54 nM)	Fluorescence	Change (quenching) of fluorescence signal with fluorescence resonance energy transfer (FRET) mechanism	450 pM	–	[93]
4	Europium-functionalized graphene quantum dots (GQDs) (Eu-GQDs)	Cu^{2+} (0.31 μM) and l-cysteine (0.5–50 μM) in serum samples	Fluorescence		0.31 μM for l-cysteine and 0.056 μM for Cu^{2+}		[55]
5	Nitrogen and sulfur doped graphene quantum dots	2,4,6-trinitrophenol	Fluorescence	Photo-induced electron transfer (PET) mechanism	19.05 ppb	42% at 100 ppm and 385% at 1000 ppm,	[63]
6	Polydopamine (PDA) polymer, thiacloprid, and N-GQDs	Thiacloprid (0.1 and 10 mg/L)	Fluorescence		0.03 mg/L		[56]

(continued)

Table 1 (continued)

S. No	Material used	Target analyte	Sensing device	Sensing principle	Detection limit	Sensing response	Reference
7	Oxygen-rich nitrogen-doped GQDs	Hg (II) ion (0.04 to 6 μM)	pH sensitive sensor	Nonradiative electron transfer	8.6 nM	Quantum yield was 18%	[94]
8	Complex of graphene nanoplatelets (GNPs) and molecularly imprinted pyrrole MIPs	Metronidazole (MNZ)	Electrochemical sensor	Electrocatalytic property	0.52 nmol/L	Displayed two linear ranges within 0.005–0.75 μmol/L and 0.75–10.0 μmol/L	[25]
9	GQDs coated on hollow nickel nanospheres (hNiNS) modified with glass carbon electrode (GCE)	Bisphenol S (BPS) (0.1 and 50 μM)	Molecularly imprinted electrochemical sensor (MIECS)	Electrochemical variations	0.03 μM	–	[76]
10	Pd nanoparticles coated on N-GQDs	H_2O_2	Electrochemical sensor	Electrochemical activity	–	95%	[112]
11	GO, T-rGO, erGO, and IC-rGO	Isoniazid (INZ) (antibiotic drug)	Electrochemical sensing	Differential pulse voltammetry	1.2 μM	1.02 μA μM^{-1} cm^{-2}	[73]

significant portability, sensitivity and selectivity for their utilization as practical nanosensor devices [21]. The versatility of Metal Oxides with varied binding affinity and other chemical properties of different metal ions allow utilization of different Metal Oxides nanosensors for the detection of different target analytes [15]. A recent work by Ridhi et al. [81] has reported the effect of central metal ions in affecting sensing responses of composite materials towards organic vapors. Flexibility in accommodating lone pairs or donated charge carriers in vacant orbitals of different metal ions allows varied sensing mechanism on account of charge transfer mechanism between target gases and sensing device [78, 82, 83]. Therefore, functionalization of CNTs with varied metal ions aid in fabrication of promising nanosensors as early warning alarms for averting contaminations in water utilizing low detection limit nanosensors. An imperative stride of integration of nanotechnology and sensing has been reviewed recently by [60]. Low power consumption, ease of miniaturization and portability are significant revitalizing parameters that make CNTs as promulgating nanosensors for waste water treatment [60]. Moreover, higher electrical conductivity ($\sim 10^8$ S·m^{-1}), thermal conductivity ($\sim$3000 W·m·K^{-1} at room temperature) and spurring mechanical properties (Tensile strength $\approx$ 50 GPa and Young's modulus $\approx$ 1.0 TPa) of CNTs makes their fabrication into different types of nanosensors [2, 67]. Table 2 lists various CNTs functionalized nanosensors for the detection of various contaminations present in waste water. From Table 2, we observe that functionalization of CNTs have shown spurring sensing properties with amended properties in comparison to pristine CNTs. Microscopic dimensional effects integrated with fractal geometric theory of CNTs have also shown significant impact in enhancement of sensing responses towards nitric oxide from traditional inter-digital electrode [114].

3.3 Significance of Metal Oxides (MO) as Effective Nanosensors for Waste Water Treatment

Metal oxides such as MgO and TiO_2 have outpaced antimicrobial activity on account of their distinct physical-chemical characteristics [8, 69]. Metal oxides are also useful for obviation of contaminants from waste water [69]. For instance, $Fe(OH)_3$ assists in the removal of Cr (VI) as an effective flocculent [37]. An imperative role of electro-spinning synthesis method, shapes and nature of integrated nanomaterials in semiconductor metal oxide chemresistive gas sensors has been reviewed by [17] for the detection of various gases inculcating Ammonia, NO, methane, ethanol, acetone, H_2S, Toluene, etc. Galvani et al. [29] has categorized MOs on the basis of electronic structure for their effective utilization as chemresistors. These inculcate pre-transition MOs (example: MgO), transition MOs (Fe_2O_3, TiO_2, etc.) and post-transition MOs (examples: ZnO, SiO_2, etc.).

A recent review by [45] demonstrates the impediments and promising potential of MOs gas sensors for the detection of NO_x for fabricating low detection limit

Table 2 Carbonanotubes based nanosensors for detection of water contaminants

S. No	Material used	Target analyte	Sensing device	Sensing Principle	Detection limit	Sensing response	Reference
1	ZnO-CNT networks	Ammonia	Chemresistor	Rise and decay times	0.4 ppm	4 to 10 ppm (room temperature)	[59]
2	SWCNTs/Au nanoparticles (NPs)	NO_2	Chemresistor	Change in resistance and shift of fermi level after NO_2 exposure	40 ppm	–	[51]
3	Pd@VO_2 composite-modified CNTs-based Thin Film Transistor (TFT) sensor	Methane gas	Thin film transistor	Charge transfer intercation between methane and sensing film	50 ppm	4.63%	[40]
4	Semiconducting carbon nanotube (CNT) thin-film	H_2 gas	Gate assistant technology and modified concentration calculation method	Amending desorption amenability and dwindling recovery times	10 ppm	–	[57]
5	Single-walled carbon nanotubes (SWCNTs) functionalized with 2,2,6,6-tetramethylpiperidine-1-oxyl (TEMPO) as a catalyst	H_2 gas		Redox reactions	5 ppm	420% at 60% relative humidity (RH)	[41]
6	F_{16}ZnPc/MWCNTs-COOH	Cl_2	Chemresistive	Raman, electrochemical impedance spectroscopy (EIS) and X-ray photoelectron spectroscopy (XPS) studies	0.06 ppb	21.28% for 2 ppm of Cl_2	[90]

(continued)

Table 2 (continued)

S. No	Material used	Target analyte	Sensing device	Sensing Principle	Detection limit	Sensing response	Reference
7	$F_{16}CuPc$/CNTs hybrids	Cl_2	Chemresistors	Response-recovery with adsorption–desorption of Cl_2 gas	0.27 ppb	35.82%	[91]
8	Nitrogen doped SWCNT	NO_2	Chemresistor	Variation in resistance with exposure of NO_2	500 ppb	27.7% towards 10 ppm of NO_2	[99]
9	CNTs	Noble gases (Ar, Ne and He)	Nanomechanical resonators	Strengthening and inertia effects of gases on impulse wave propagation of CNTs	–	Decrease in wave velocity from 3507.81 to 3394.64 m/s	[7]
10	CNT on SiO_2 substrate	H_2	Chemresistor	Variation in electrical current with exposure of H_2 gas	20 ppm	315% at 140 ppm H_2	[24]
11	Pd doped SWCNT	H_2	Chemresistor	Response-recovery curves	–	33.79	[98]

chemresistors using electrospun nanofibres. Noble metals tend to amend the chemical and electronic sensitization of Semiconductor MOs enabling highly efficient low operating temperature, faster response-recovery speeds, ultra low detection limit and higher sensing responses MOs gas sensors [117]. Zhu et al. [117] has reviewed Ag, Au, Ru, Rh, Pt, Pd and other metals decorated semiconducting MOS inculcating CuO, ZnO, SnO_2, WO_3, Fe_2O_3, In_2O_3, etc as amenable gas sensors as compared to pristine MOs sensors. Similarly, [46] has highlighted the influence of Bimetallic organic framework (BMOFs) for amending sensing properties of Semiconducting MOs gas sensors. In addition to the integrating materials incorporated in MOs gas sensors, the role of synthesis techniques, structures and optimized concentrations is also emphasized to understand the various sensing parameters that need to be tapped to amend their sensing efficacy [46].

Printable MOs chemresistors for the detection of non-biological analytes such as micro-organisms, cells and tissues promulgates an incentivizing paradigm approach for the detection of micro-contaminants in waste water for effective waste water treatment [49]. Consortium of room temperature conducting polymers and high operating temperature MOs gas sensors is a promising paradigm for fabricating efficient chemresistor as it can obviate the limitation of low sensitivity of conducting polymers while integrating highly sensitive, selective and low detection limit MOs chemresistors [111]. Innovative sensor designs using pattern recognize algorithms have been reviewed by Wawrzyniak et al. [108] to obviate the impediment of low selectivity of MOs gas sensors and demonstrated their application for the detection of volatile organic compounds and other toxic substances. Table 3 displays various MOs gas sensors operating at room temperatures and higher temperatures. From Table 3, we observe that TiO_2, NiO and Fe_3O_4 can undergo significant sensing responses at room temperature. ZnO and other MOs require higher operating temperatures for yielding appreciable sensing responses towards target gases. However, their significant doping with organic materials like Carbon nanotubes, graphene and conducting polymer nanocomposites can lead to outpacing sensing responses of MOs gas sensors under ambient conditions as well.

4 Conclusions

Present chapter veritable on nanosensors acting as early warning alarms to avert water contamination enabling effective waste water treatment at appropriate time and dosage. It is indispensable in current increasing population with rise in water scarcity impediments to gave pragmatism effective waste water treatment methodologies. Nanosensors are substantive assets with proliferating role to avert water contamination issues while acting as early warning alarms utilizing low detection limit devices. Organic–Inorganic materials with their composites can integrate to form suitable nanosensors accentuating advantages of both consorting materials. The role of synthesis techniques, selection of suitable material, suitable sensing device depending on optical, spectroscopic, mechanical, electrical, biological and

Table 3 Metal oxides based nanosensors for detection of water contaminants

S. No	Material Used	Target analyte	Sensing device	Sensing Principle	Detection limit	Sensing response	Reference
1	Pd functionalized $CuCrO_2$	H_2S (50 ppm)	Chemresistor	Charge transfer interaction	0.5 ppm	72.3% at 150 °C	[50]
2	Al_2O_3 stabilized ZrO_2	NO_2	Chemresistor	Charge transfer interaction	–	–	[54]
3	TiO_2	NO_2 (100 ppm)	TiO_2 nanowires	Charge transfer interaction		3.10 at room temperature	[116]
4	Al doped NiO	CH_4 (100 ppm)	Chemresistor	Charge transfer interaction	–	58% at room temperature	[28]
5	Fe_3O_4	NH_3 (200 ppm)	Chemresistor	Transient current variations with NH_3 exposure		6% at room temperature	[102]
6	SnO_2	CO_2 (2000 ppm)	Chemresistor	Density functional theory (DFT) and charge transfer mechansim	–	71.1% at 240 °C	[105]
7	Pt functionalized ZnO	Triethylamine (100 ppm)	Kelvin probe	DFT simulation	–	242 at 200 °C	[58]
8	Pd/Au doped ZnO nanorod	Acetone (1 ppm)	Chemresistor	Charge transfer interaction	5 ppb	17 at 150 °C	[38]
9	$BiFeO_3$ (BFO) nanoparticles	n-Butanol (100 ppm)	Chemresistor	Variation in resistance with exposure of vapors	0.28 ppm	19.37 at 280 °C	[20]

(continued)

Table 3 (continued)

S. No	Material Used	Target analyte	Sensing device	Sensing Principle	Detection limit	Sensing response	Reference
10	Co_3O_4/ZnO	CO (100 ppm)	Chemresistor	Surface interface interaction	–	38% at 210 °C	[16]
11	CuO nanostructures	Ethanol (50–800 nM)	Cyclic Voltammeter	Varition in current	10 ppm	36% at 250 °C	[4]

other surface interface interaction mechanisms are discussed and analyzed in detail in this chapter.

Acknowledgements Dr. R. Ridhi is thankful to Mehr Chand Mahajan DAV College for Women, Sector-36 A, Chandigarh for providing facilities for this work.

Competing Interests The authors declare that they have no competing interest.

Declaration of generative AI and AI-assisted technologies in the writing process During the preparation of this work the book chapter author(s) have not taken any assistance from AI, and the book chapter author(s) reviewed and edited the content as needed and take(s) full responsibility for the content of the publication. All authors read and approved the final version of the manuscript.

References

1. Adam T, Gopinath SCB (2022) Nanosensors: recent perspectives on attainments and future promise of downstream applications. Process Biochem 117:153–173
2. Aftab S, Iqbal MZ, Hussain S, Hegazy HH, Kabir F, Jaffery SHA, Koyyada G (2023) New developments in gas sensing using various two-dimensional architectural designs. Chem Eng J 469:144039
3. Aghdam E, Mohandes SR, Manu P, Cheung C, Kaltungo AY, Zayed T (2023) Predicting quality parameters of wastewater treatment plants using artificial intelligence techniques. J Clean Prod 405:137019
4. Ahmad R, Yewale MA, Khan M, Nakate UT, Ahmad A, Alshammari MB, Bhalerao KD, Bhat KS, Lee B (2024) Bifunctional CuO nanostructured materials preparation for ethanol gas and riboflavin sensing applications. Sens Actuators, B Chem 415:135979
5. Ahmed SF, Mofijur M, Ahmed B, Mehnaz T, Mehejabin F, Maliat D, Hoang A, Shafiullah GM (2022) Nanomaterials as a sustainable choice for treating wastewater. Environ Res 214:113807
6. Anegbe B, Ifijen IH, Maliki M, Uwidia IE, Aigbodion AI (2024) Graphene oxide synthesis and applications in emerging contaminant removal: a comprehensive review. Environ Sci Eur 36:15
7. Arash B, Wang Q (2013) Detection of gas atoms with carbon nanotubes. Sci Rep 3:1782
8. Atacan K, Guy N, Ozacar M (2022) Recent advances in photocatalytic coatings for antimicrobial coatings. Curr Opin Chem Eng 36:100777
9. Baruah A, Newar R, Das S, Kalita N, Nath M, Ghosh P, Chinnam S, Sarma H, Narayan M (2024) Biomedical applications of graphene-based nanomaterials: recent progress, challenges, and prospects in highly sensitive biosensors. Discover Nano 19:103
10. Bashir O, Bhat SA, Basharat A, Qamar M, Qamar SA, Bilal M, Iqbal HMN (2022) Nano-engineered materials for sensing food pollutants: technological advancements and safety issues. Chemosphere 292:133320
11. Bhaskar S, Ramamurthy SS (2024) Low-cost sustainable nanotechnology approaches for water quality monitoring in developing countries. Water Resour Dev Manag 155–188
12. Bhol P, Yadav S, Altaee A, Saxena M, Misra PK, Samal AK (2021) Graphene-based membranes for water and wastewater treatment: a review. ACS Appl Nano Mater 4(4). https://doi.org/10.1021/acsanm.0c03439
13. Bodzeka M, Konieczny K, Mydlak AK (2023) Recent advances in water and wastewater disinfection by nano-photocatalysis. Desalin Water Treat 305:2–16
14. Chahar M, Khaturia S, Singh HL, Solanki VS, Agarwal N, Sahoo DK, Yadav VK, Patel A (2023) Recent advances in the effective removal of hazardous pollutants from wastewater by using nanomaterials—a review. Front Environ Sci 11:1–24

15. Chavali MS, Nikolova MP (2019) Metal oxide nanoparticles and their applications in nanotechnology. SN Applied Sciences 1:607
16. Chen M, Li X, Li Y, Li Y, Qin Z, Wang Q (2024) MOF-derived Co_3O_4 nanoparticles over direct grown ZnO nanoflower on ceramic for CO sensor with high selectivity. Sens Actuators, B Chem 401:134951
17. Chen L, Yu Q, Pan C, Song Y, Dong H, Xie X, Li Y, Liu J, Wang D, Chen X (2022) Chemiresistive gas sensors based on electrospun semiconductor metal oxides: a review. Talanta 246:123527
18. Cheng R, Ou S (2019) Nanocomposite-based graphene for nanosensor applications. Nanorods Nanocomposites. In: Ghamsari MS, Dhara S. https://doi.org/10.5772/intechopen.85136
19. Cheng R, Peng Y, Ge C, Bu Y, Liu H, Huang H, Ou S, Xue Y, Dai L (2017) A turn-on fluorescent lysine nanoprobe based on the use of the Alizarin red aluminum(III) complex conjugated to graphene oxide, and its application to cellular imaging of lysine. Microchim Acta 9:3521–3528
20. Cheng Z, Zhu H, Ji H, Yuan Z, Meng F (2023) Perovskite-structured BiFeO3 nanoparticles with abundant oxygen vacancies for selective n-butanol gas sensing. ACS Appl Nano Mater 24:23370–23380
21. Dariyal P, Sharma S, Chauhan GS, Singh BP, Dhakate SR (2021) Recent trends in gas sensing via carbon nanomaterials: outlook and challenges. Nanoscale Adv 3:6514–6544
22. Darwish MA, Elaziem WA, Elsheikh A, Zayed AA (2024) Advancement in nanomaterials for nanosensors: a comprehensive review. Nanoscale Adv 6:4015–4046
23. Dewey HM, Lamb A, Uprety JB (2024) Recent advances on applications of single-walled carbon nanotubes as cutting-edge optical nanosensors for biosensing technologies. Nanoscale 16:16344
24. Diabat AMA, Algadri NA, Ahmed NM, Abuelsamen A, Bidier SA (2021) A high-sensitivity hydrogen gas sensor based on carbon nanotubes fabricated on SiO_2 substrate. Nanocomposites 7:172–183
25. Ensafi AA, Esfahani PN, Rezaei M (2018) Metronidazole determination with an extremely sensitive and selective electrochemical sensor based on graphene nanoplatelets and molecularly imprinted polymers on graphene quantum dots. Sens Actuators B 270:192–199
26. Epelle EI, Okoye PU, Roddy S, Gunes B, Okolie JA (2022) Advances in the applications of nanomaterials for wastewater treatment. Environ 9(11):141
27. Fakayode SO, Walgama C, Narcisse VEF, Grant C (2023) Electrochemical and colorimetric nanosensors for detection of heavy metal ions: a review. Sens 23:9080
28. Gagaoudakis E, Michail G, Katerinopoulou D, Moschovis K, Iliopoulos E, Kiriakidis G, Binas V, Aperathitis E (2020) Transparent p-type NiO: Al thin films as room temperature hydrogen and methane gas sensors. Mater Sci Semicond Process 109:104922
29. Galvani M, Freddi S, Sangaletti L (2024) Disclosing fast detection opportunities with nanostructured chemiresistor gas sensors based on metal oxides, carbon, and transition metal dichalcogenides. Sens 24:584
30. Gezahegn TF, Ambaye AD, Mekoyete TM, Malefane ME, Oyedotun KO, Mokrani T (2024) Breakthroughs in nanostructured-based chemical sensors for the detection of toxic metals. Talanta Open 10:100354
31. Goodrum R, Weldekidan H, Li H, Mohanty AK, Misra M (2024) Graphene-based nanostructures from green processes and their applications in biomedical sensors. Adv Ind Eng Polym Res 7:37–53
32. Güler KÇ, Göktürk I, Yılmaz F, Araz A, Denizli A (2024) Plasmonic nanosensors for environmental pollutants sensing: recent advancesand perspectives. Essential Chem 1:1–19
33. Gupte T, Pradeep T (2022) Chapter 3-nanosensors for water quality monitoring. Sep Sci Technol 15:37–53
34. Hairom NHH, Soon CF, Mohamed RMSR, Morsin M, Zainal N, Nayan N, Zulkifli CZ, Harun NH (2021) A review of nanotechnological applications to detect and control surface water pollution. Environ Technol Innov 24:102032

35. He S, Ma Y, Maulik G, Jellicoe M, Nag A, Powell W, Deng S, Fang J, Wu Y (2024) A review on graphene-based sensors for tactile applications. Sens Actuators, A 372:115363
36. Hejabi N, Saghebian SM, Aalami MT, Nourani V (2021) Evaluation of the effluent quality parameters of wastewater treatment plant based on uncertainty analysis and post-processing approaches (case study). Water Sci Technol 83:1633–1648
37. Hoag GE, Collins JB, Holcomb JL, Hoag JR, Nadagouda MN, Varma RS (2009) Degradation of bromothymol blue by greener nanoscale zero valent iron synthesized using polyphenols. J Mater Chem 19:8671–8688
38. Huang J, Zhou J, Liu Z, Li X, Geng Y, Xiaoqing T, Du Y, Qian Z (2020) Enhanced acetone-sensing properties to ppb detection level using Au/Pd-doped ZnO nanorod. Sens Actuators, B Chem 310:127129
39. Jain K, Patel AS, Pardhi VP, Flora SJS (2021) Nanotechnology in wastewater management: a new paradigm towards wastewater treatment. Molecules 26:1797
40. Ji F, Hu J, Zhang Y (2023) Functionalized carbon-nanotubes-based thin-film transistor sensor for highly selective detection of methane at room temperature. Chemosens 11:365
41. Jung HY, Park KYL, Datar A, Lee HJ, Huang J, Somu S, Busnaina A, Jungab YJ, Kwon YK (2013) High-performance H2S detection by redox reactions in semiconducting carbon nanotube-based devices. Analyst 138:7206–7211
42. Kadadou D, Tizani L, Wadi VS, Banat F, Alsafar H, Yousef AF, Barceló D, Hasan SW (2021) Recent advances in the biosensors application for the detection of bacteria and viruses in wastewater. J Environ Chem Eng 10:107070
43. Khan S, Misra V, Singh A, Singh V (2023) Metal oxide decorated CNT nanosensors: a brief overview. Met Oxide-Based Carbon Nanocomposites Environ Remediat Saf 9781003323464:15
44. Khan F, Mubashir T, Ahmed K, Mateen A, Lee S, Ahmed T (2023) Fabrication and characterization of carbon nanotubes-based pressure nanosensors: a study on piezoresistive behavior. Trans Electr Electron Mater 24:518–527
45. Khomarloo N, Mohsenzadeh E, Gidik H, Bagherzadeh R, Latifi M (2024) Overall perspective of electrospun semiconductor metal oxides as high-performance gas sensor materials for NO_x detection. RSC Adv 14:7806–7824
46. Kong L, Yuan Z, Gao H, Meng F (2023) Recent progress of gas sensors based on metal oxide composites derived from bimetallic metal-organic frameworks. TrAC, Trends Anal Chem 166:117199
47. Kulkarni MB, Ayachit NH, Aminabhavi TM (2022) Recent advancements in nanobiosensors: current trends, challenges, applications, and future scope. Biosens 12:892
48. Kumar V, Guleria P (2024) Application of DNA-nanosensor for environmental monitoring: recent advances and perspectives. Biol Pollut 10:765–785
49. Kumar A, Kim SS, Kim HW, Kumar M (2023) Printable metal oxide nanostructures based chemiresistive nonbiological analyte sensors. Nano Express 4:042003
50. Kumar A, Kumar A, Kwoka M, Gang M, Kumar M (2024) IoT-enabled surface-active Pd-anchored metal oxide chemiresistor for H2S gas detection. Sens Actuators, B Chem 402:135065
51. Kumar S, Pavelyev V, Mishra P, Tripathi N (2017) Sensitive detection of nitrogen dioxide using gold nanoparticles decorated single walled carbon nanotubes. In: Proceedings of the international conference ITNT
52. Ledezma CZ, Bolagay DN, Figueroa F, Ledezma EZ, Ni M, Alexis F, Guerrero VH (2021) Heavy metal water pollution: a fresh look about hazards, novel and conventional remediation methods. Environ Technol Innov 22:101504
53. Lesoinne HAP, Alvarez FR, Gonzalez VHP, Chapa SOM, Villanueva RCG (2022) Nanomaterials for electrochemical detection of pollutants in water: a review. Electrophor 43:249–262
54. Li X, Guan M, Liu Y, Dong H, Li X, Shao C, Lu D, Liu Y (2024) Wearable inorganic oxide chemiresistor based on flexible Al_2O_3-stabilized ZrO_2 ceramic sponge substrate for NO_2 sensing. ACS Sens 9:4841–4850

55. Lin L, Song X, Chen Y, Rong M, Wang Y, Zhao L, Zhao T (2015) Chen X (2015) Europium-decorated graphene quantum dots as a fluorescent probe for label-free, rapid and sensitive detection of $Cu^{(2+)}$ and L-cysteine. Anal Chim Acta 3(891):261–268. https://doi.org/10.1016/j.aca.2015.08.011
56. Liu Y, Cao N, Gui W (2018) Nitrogen-doped graphene quantum dots-based fluorescence molecularly imprinted sensor for thiacloprid detection. Talanta 183:339–344
57. Liu F, Xiao M, Ning Y, Zhou S, He J, Lin Y, Zhang Z (2022) Toward practical gas sensing with rapid recovery semiconducting carbon nanotube film sensors. Science China Inf Sci 65:162402
58. Liu J, Zhang L, Fan J, Zhu B, Yu J (2021) Triethylamine gas sensor based on Pt-functionalized hierarchical ZnO microspheres. Sens Actuators, B Chem 331:129425
59. Lupan O, Schütt F, Postica V, Smazna D, Mishra YK, Adelung R (2017) Sensing performances of pure and hybridized carbon nanotubes-ZnO nanowire networks: a detailed study. Sci Rep 7:14715
60. Luo K, Peng H, Zhang B, Chen L, Zhang P, Peng Z, Fu X (2024) Advances in carbon nanotube-based gas sensors: exploring the path to the future. Coord Chem Rev 518:216049
61. Mahmoud R, Hady EEA, Mohamed HFM, Ibrahim M, Fatah GAE, Zaher A, Gadelhak Y (2024) Nanomaterials for electrochemical sensing of heavy metals in wastewater streams. Handb Nanosensors 1219–1242. Springer, Cham. https://doi.org/10.1007/978-3-031-47180-3_48
62. Mohammed MN, Aljibori HSS, Jweeg MJ, Oqaili FA, Abdullah TA, Abdullah OI, Meharban F, Rashed RT, Aldulaimi M, Azawi KA (2024) A comprehensive review on graphene oxide based nanocomposites for wastewater treatment. Pol J Chem Technol 26:64–79
63. Mondal TK, Dinda D, Saha SK (2018) Nitrogen, sulphur co-doped graphene quantum dot: an excellent sensor for nitroexplosives. Sens Actuators B 257:586–593
64. Moufid M, Tiebe C, Bari NE, Fakra DAH, Bartholmai M, Bouchikhi B (2022) Pollution parameters evaluation of wastewater collected at different treatment stages from wastewater treatment plant based on E-nose and E-tongue systems combined with chemometric techniques. Chemom Intell Lab Syst 227:104593
65. Mushahary N, Sarkar A, Basumatary F, Brahma S, Das B, Basumatary S (2024) Recent developments on graphene oxide and its composite materials: from fundamentals to applications in biodiesel synthesis, adsorption, photocatalysis, supercapacitors, sensors and antimicrobial activity. Results Surf Interfaces 15:100225
66. Nishu KS (2023) Smart and innovative nanotechnology applications for water purification. Hybrid Advances 3:100044
67. Onyancha RB, Ukhurebor KE, Aigbe UO, Osibote OA, Kusuma HS, Darmokoesoemo H, Balogun VA (2021) A systematic review on the detection and monitoring of toxic gases using carbon nanotube-based biosensors. Sens Bio-Sens Res 34:100463
68. Oukhatar A, Bakhouya M, Ouadghiri DE (2021) Electromagnetic based wireless nano-sensors network: architectures and applications. J Commun 16:8–19
69. Pérez H, Garcíaa OJQ, Allierib MAA, Vázqueza RR (2023) Nanotechnology as an efficient and effective alternative for wastewater treatment: an overview. Water Sci Technol 87:2971
70. Potharaju R, Aruna M (2024) Chapter 13-a study on heavy metal pollution in water. pp 111–120
71. Rana K, Kaur H, Singh N, Sithole T, Siwal SS (2024) Graphene-based materials: unravelling its impact in wastewater treatment for sustainable environments. Next Materials 3:100107
72. Priyadharshini SD, Manikandan S, Kiruthiga R, Rednam U, Babu PS, Subbaiya R, Karmegam N, Kim W, Govarthanan M (2022) Graphene oxide-based nanomaterials for the treatment of pollutants in the aquatic environment: recent trends and perspectives–a review. Environ Pollut 306:119377
73. Qian L, Thiruppathi AR, Zalm JVD, Chen A (2021) Graphene oxide-based nanomaterials for the electrochemical sensing of Isoniazid. ACS Appl Nano Mater 4:3696
74. Ramezani M, Taghdisi SM, Robati RY, Oroojalian F, Abnous K, Alibolandi M (2020) Chapter 17-nanosensors for water safety. Nanosensors Smart Cities: Micro Nano Technol 285–301

75. Ramirez CA, Thiruppathi AR, Ozoemena O, Chen A (2024) Recent advances in graphene oxide-based electrochemical sensors. Can J Chem 102. https://doi.org/10.1139/cjc-2024-0010
76. Rao H, Zhao X, Liu X (2018) A novel molecularly imprinted electrochemical sensor based on graphene quantum dots coated on hollow nickel nanospheres with high sensitivity and selectivity for the rapid determination of bisphenol S. Biosens Bioelectron 100:341–347. https://doi.org/10.1016/j.bios.2017.09.016
77. Ridhi R, Chouksey A, Gautam S, Rawat JS, Jha P (2021) Study of the effect of flow rate and decomposition temperature on sensing of ammonium nitrate by carbon nanotubes. Sens Actuators, B Chem 334:129658
78. Ridhi R, Neeru GS, Saini GSS, Tripathi SK, Rawat JS, Jha P (2021) Study of the effect of orbital on interaction behaviour of SWCNT-metal phthalocyanines composites with ammonia gas. Sens Actuators, B Chem 337:129767
79. Ridhi R, Neeru GS, Saini GSS, Tripathi SK, Rawat JS, Jha P (2020) Amendment in sensing response of Single Walled Carbon nanotube (SWCNT) towards ammonia gas with copper phthalocyanine functionalization. Mater Today: Proc 28:1759–1763
80. Ridhi R, Saini GSS, Tripathi SK (2024) Nanotechnology as a sustainable solution for proliferating agriculture sector. Mater Sci Eng, B 304:117383
81. Ridhi R, Saini GSS, Tripathi SK (2024) Effect of central metal ions on interaction of metal phthalocyanines composites with chemical analytes. Appl Phys A 130:359
82. Ridhi R, Saini GSS, Tripathi SK (2023) Study of sensing mechanism of heterocyclic hazardous vapors with Metal Phthalocyanines. Dyes Pigm 216:111328
83. Ridhi R, Singh S, Saini GSS, Tripathi SK (2018) Comparison of interaction mechanisms of copper phthalocyanine and nickel phthalocyanine thin films with chemical vapours. J Phys Chem Solids 115:119–126
84. Sagasta JM, Zadeh SM, Turral H (2018) More people, more food, worse water?: a global review of water pollution from agriculture (2018). In: Food and agriculture organization of the United Nations Rome and the international water management institute on behalf of the water land and ecosystems research program of the CGIAR Colombo 2018
85. Sahoo PK, Panigrahy B, Sahoo S, Satpati AK, Li D, Bahadur D (2013) In situ synthesis and properties of reduced graphene oxide/Bi nanocomposites: as an electroactive material for analysis of heavy metals. Biosens Bioelectron 43:293–299
86. Saleh TA (2024) Materials, nanomaterials, nanocomposites and methods used for the treatment and removal of hazardous pollutants from wastewater: treatment technologies for water recycling and sustainability. Nano-Struct Nano-Objects 39:101231
87. Saravanan P, Saravanan V, Rajeshkannan R, Arnica G, Rajasimman M, Baskar G, Pugazhendhi A (2024) Comprehensive review on toxic heavy metals in the aquatic system: sources, identification, treatment strategies, and health risk assessment. Environ Res 258:119440
88. Schroeder V, Savagatrup S, He M, Lin S, Swage TM (2019) Carbon nanotube chemical sensors. Chem Rev 119:599–663
89. Sengupta J, Hussain CM (2023) Graphene-based electrochemical nano-biosensors for detection of SARS-CoV-2. Inorg 11:197
90. Sharma AK, Mahajan A, Bedi RK, Kumar S, Debnath AK, Aswal DK (2018) Non-covalently anchored multi-walled carbon nanotubes with hexa-decafluorinated zinc phthalocyanine as ppb level chemiresistive chlorine sensor. Appl Surf Sci 427:202–209
91. Sharma AK, Mahajan A, Saini R, Bedi RK, Kumar S, Debnath AK, Aswal DK (2018) Reversible and fast responding ppb level Cl_2 sensor based on noncovalent modified carbon nanotubes with Hexadecafluorinated copper phthalocyanine. Sens Actuators, B Chem 255:87–99
92. Shehata N, Egirani D, Olabi AG, Inayat A, Abdelkareem MA, Chae KJ (2023) Membrane-based water and wastewater treatment technologies: Issues, current trends, challenges, and role in achieving sustainable development goals, and circular economy. Chemosphere 320:137993. https://doi.org/10.1016/j.chemosphere.2023.137993
93. Shi J, Lyu J, Tian F, Yang M (2017) A fluorescence turn-on biosensor based on graphene quantum dots (GQDs) and molybdenum disulfide (MoS_2) nanosheets for epithelial cell adhesion molecule (EpCAM) detection. Biosens Bioelectron 93:182–188

94. Shi B, Zhang L, Lan C (2015) One-pot green synthesis of oxygen-rich nitrogen-doped graphene quantum dots and their potential application in pH-sensitive photo-luminescence and detection of mercury (II) ions. Talanta 142:131–139. https://doi.org/10.1016/j.talanta.2015.04.059
95. Shooshtari M, Salehi A (2022) An electronic nose based on carbon nanotube -titanium dioxide hybrid nanostructures for detection and discrimination of volatile organic compounds. Sens Actuators, B Chem 357:131418
96. Singh V, Ahmed G, Vedika S, Kumar P, Chaturvedi SK, Rai SN, Vamanu E, Kumar A (2024) Toxic heavy metal ions contamination in water and their sustainable reduction by eco-friendly methods: isotherms, thermodynamics and kinetics study. Sci Rep 14:7595
97. Straathof J, Hull NM (2024) Evaluating wastewater quality parameters as an alternative or complement to molecular indicators for normalization during SARS-CoV-2 wastewater-based epidemiology. Environ 11:80
98. Tarannum N, Gautam A, Chauhan T, Kumar D (2024) Nanomaterial based sensors for detection of food contaminants: a prospect. Sens Technol 2:2373196
99. Tian XH, Zhou TY, Meng Y, Zhao YM, Shi C, Hou PX, Zhang LL, Liu C, Cheng HM (2022) A flexible NO_2 gas sensor based on single-wall carbon nanotube films doped with a high level of nitrogen. Molecules 27:6523
100. Tripathy J, Mishra A, Pandey M, Thakur RR, Chand S, Rout PR, Shahid MK (2024) Advances in nanoparticles and nanocomposites for water and wastewater treatment: a review. Water 16:1481
101. Tu H, Wei Z, Bahrami A, Kahla NB, Ahmad A, Özkılıç YO (2023) Recent advancements and future trends in 3D concrete printing using waste materials. Dev Built Environ 16:100187
102. Varpe AS, Deshpande MD (2022) Effect of annealing temperature on optical, dielectric and NH3 gas sensing properties of $Zn_xFe_{3-x}O_4$ (x = 0 and 0.5) nanoparticles synthesized by sol-gel auto-combustion method. J Alloy Compd 892:162124
103. Verma G, Mondal K, Islam M, Gupta A (2024) Recent advances in advanced micro and nanomanufacturing for wastewater purification. ACS Appl Eng Mater 2:262–285
104. Vysokomornaya OV, Kurilenko EY, Shcherbinina AA (2015) Major contaminants in industrial and domestic wastewater. MATEC Web Conf 23:01041
105. Wang D, Chen Y, Liu Z, Li L, Shi C, Qin H, Hu J (2016) CO_2-sensing properties and mechanism of nano-SnO2 thick-film sensor. Sens Actuators, B Chem 227:73–84
106. Wang W, Srivastava S, Vikesland PJ (2024) Overcoming barriers and embracing advances: nanosensor implementation for practical water contaminant surveillance. One Earth 7:1351–1361
107. Wang Q, Zhang D (2018) A novel fluorescence sensing method based on quantum dot-graphene and a molecular imprinting technique for the detection of tyramine in rice wine. Anal Methods 10:3884–3889
108. Wawrzyniak J (2023) Advancements in improving selectivity of metal oxide semiconductor gas sensors opening new perspectives for their application in food industry. Sens 23:9548
109. Willner MR, Vikesland PJ (2018) Nanomaterial enabled sensors for environmental contaminants. J Nanobiotechnol 16:95
110. Worku AK, Ayele DW (2023) Recent advances of graphene-based materials for emerging technologies. Results Chem 5:100971
111. Wuloh J, Agorku ES, Boadi NO (2023) Modification of metal oxide semiconductor gas sensors using conducting polymer materials. J Sens 427986:1–11
112. Xi J, Xie C, Zhang Y (2016) Pd nanoparticles decorated N-doped graphene quantum dots@Ndoped carbon hollow nanospheres with high electrochemical sensing performance in cancer detection. ACS Appl Mater Interfaces 8:22563–22573
113. Yadav KK, Pinto MMSC, Gacem A, Fallatah AM, Ravindran B, Rezania S, Algethami JS, Eltayeb LB, Abbas M, Al-shareef TH, Vinayak V, Son CT, Alreshidi MA, Díaz JMR, Homod RZ (2024) Recent advances in the application of nanoparticle-based strategies for water remediation as a novel clean technology–a comprehensive review. Mater Today Chem 40:102226

114. Yang T (2021) Optimizing electrode structure of carbon nanotube gas sensors for sensitivity improvement based on electric field enhancement effect of fractal geometry. Sci Rep 11:16675
115. Zahmatkesh S, Keshteli MH, Bokhari A, Sundaramurthy S, Panneerselvam B, Rezakhani Y (2023) Wastewater treatment with nanomaterials for the future: a state-of-the-art review. Environ Res 216:114652
116. Zhu Z, Lin SJ, Wu CH, Wu RJ (2018) Synthesis of TiO_2 nanowires for rapid NO_2 detection. Sens Actuators, A 272:288–294
117. Zhu LY, Ou LX, Mao LW, Wu XY, Liu YP, Lu HL (2023) Advances in noble metal-decorated metal oxide nanomaterials for chemiresistive gas sensors: overview. Nano-Micro Lett 15. https://doi.org/10.1007/s40820-023-01047-z

Nanosensors and Their Role in Recycling of Agricultural Wastes

Mohammad Areeb Siddiqui, Reena Vishvakarma, and Archana Vimal

Abstract A huge of agriculutural waste is generated every year throughout the world leading to environmental burden due to pollutant and greenhouse gas emitted by them. The advancement in nanotechnology and uses of nanosensors helps in recycling the agricultural waste and in turn promoting sustainable agriculture practices. Nanosensors are the analytical tools that are highly specific and sensitive and helps tracking and surveillance of important agricultural parameters. In this chapter the various types of nanosensors like optical, chemical, biosensors and their utility in monitoring soil quality, seed quality, plant diseases, plant nutrition, irrigation etc. are discussed. Nanosensors also aids in various process of recycling agriculture wastes like composting, biogas production, mulching. They assists in real time monitoring of biological process and contaminants detection that improves product quality along with cost-effectiveness. They help out in monitoring and regulating various inhibitory factors. They also participates in biomass conversion process like fermentation, pyrolysis, combustion and gasification for better yields and waste valorization. Another important role that biosensors plays is that evaluation of environmental emissions and managing carbon footprint emissions. However, there are still some challenges to be addressed that are scalability, economic viability, specificity, farmers acceptance. Nanosensors are future technology designed to revolutionized agriculture practices by optimizing resources, waste recycling and management in a eco-friendly manner. Scientists are also integrating IoT systems to improve their potential. This chapter deals with the role of nanosensors in sustainable agriculture practices and recycling agricultural products to generate valuable products in a economical way to reduce environmental burden.

Keywords Composting · Chemical nanosensors · Nanotechnology · Precision agriculture · Pesticides

---

M. A. Siddiqui · R. Vishvakarma · A. Vimal (✉)
Department of Bioengineering, Integral University, Lucknow, Uttar Pradesh, India
e-mail: vimal.archana@gmail.com

A. Husen (ed.), *Nanobiosensors for Agricultural and Other Related Sectors*, Smart Nanomaterials Technology, https://doi.org/10.1007/978-981-96-6499-3_15

1 Introduction

Agro-waste is the leftover material from farming that isn't useful anymore [39]. It can be solid or liquid, and it comes from things like pesticide use, leftover crops, or waste from animals. Natural plant fibers, especially those from agricultural waste, are being looked at more as a way to create eco-friendly products. Scientists are studying how we can use these fibers, which could be turned into renewable and earth-friendly materials [10].

Recycling and reusing waste from nature helps to protect the environment. If managed well, agro-waste can even be profitable. Some common agro-wastes, like those that contain cellulose (a natural fiber), are already being used to make valuable products like biocomposites, car parts, and medical supplies. However, in the present time only about 10% of agro-waste is used for such applications [10].

Cellulose is one of the most common natural materials on Earth, found in plants, some aquatic species, algae, fungi, and even some small organisms [28]. It's strong, water-resistant, and helps plants keep their structure in intact form. Due to the property like lightweight, sustainability, and easy to work with, cellulose has potential as a key material in making eco-friendly products. The main source of cellulose comes from wood, but it's also found in things like crop leftovers, algae, and grasses. Cellulose has a unique and important structure among polysaccharides. Right from their natural formation, cellulose strands follow a basic pattern. They are made up of tiny fibers, called nanofibers, that range in diameter from 2 to 20 nm and can be longer than a few micrometers [4, 36].

Scientists and researchers use special methods to extract cellulose from plants and plant waste like wheat straw, rice husks, and empty palm oil fruit. These methods include using chemicals like acids or alkaline substances, as well as chlorine-free techniques. In plant waste, cellulose fibers are mixed with materials like lignin, hemicellulose, and pectin, making them composites. Depending on the type and structure of the plant, these materials vary [6].

To separate cellulose fibers from agricultural waste, a lot of care is needed. There are different methods for this, such as chemical, mechanical, and a mix of both treatments. Chemicals like potassium hydroxide and sodium hydroxide are often used, but overuse can harm the environment if the chemicals are not disposed of properly [31].

2 Role of Nanotechnology in Recycling Agriculture Waste

Eco-friendly methods, such as using superheated steam, hot compressed water, or steam explosion, have been proven to work well. The amount of acid and the time used for treating the fibers play a key role in how effectively the cellulose fibers are separated and their final quality. Nanotechnology can help to reduce the waste in agricultural industries like cotton, beverage production, and rice milling. In the cotton industry, when cotton is turned into fibers, some by-products like cellulose or fibers are thrown away or used in low-value products. By using a technique called electrospinning and new types of solvents, researchers are now able to produce very thin fibers (100 nm). These nanofibers can be used as fertilizers or to absorb pesticides. In the rice milling industry, rice husk is a by-product that can be a valuable source of renewable energy. When rice husk is burned for thermal energy or biofuel, it produces high-quality nanosilica. This nanosilica can be used to make useful materials like concrete and glass [14]. This way, rice husk can be reused effectively instead of being wasted [10].

Chemical treatments are often used to remove non-cellulosic materials from fibers, but this can weaken their structure. Many researchers have studied how to extract cellulose from agricultural waste using processes like chlorite bleaching, alkali treatment, and acid hydrolysis. Scientists had explored sulfuric acid (H_2SO_4) hydrolysis to obtain purified cellulose. These treatments, such as using sodium chlorite ($NaClO_2$) and alkali, help remove lignin and hemicellulose from plant fibers. A previous study showed that acid hydrolysis of maize husks increased the crystallinity of cellulose, making it purer and better structured for nanocellulose production. Overall, chemical treatments improve cellulose composition and crystallinity [8, 19, 21, 25, 47].

In mechanical treatment we can use homogenization, ultrasonics, and electrospinning to isolate cellulose fibers. Other methods, such as refining, cryocrushing, and microfluidization, also help in this process. Electrospinning, which uses electric force on a polymer solution, is common in making microfibrillated cellulose (MFC). Studies done by Zhang et al. [48] explored techniques like high-pressure microfluidization and steam explosion to extract cellulose from different sources like flax, banana trees, and kenaf fibers. However, extreme pressures can damage the cellulose structure. Chen [7] used ultrasonication to remove lignin from poplar wood and isolate cellulose nanofibers [15, 48].

In chemo-mechanical treatment, a mix of chemical and mechanical methods is used to extract cellulose fibers from plants. It was observed that steam explosion, combined with xylanase pretreatment and bleaching, to extract cellulose from sugarcane bagasse. This process removed most of the hemicellulose and lignin, leaving more pure form of cellulose. The xylanase treatment also reduced fading by 23%. Similar methods, like steam explosion with acid hydrolysis, have been successfully applied to other plant fibers like banana, pineapple leaves, jute, and coir, producing nanoscale cellulose fibers [1, 18].

3 Types of Nanosensors Used in Agriculture

Nanotechnology has made big advancements in sensors by introducing nanosensors. Since nanosensors were developed, science and technology have changed a lot. These tiny sensors are even found in our phones and other electronic devices. Because they are small with a large surface area, they can detect large amounts of material [2].

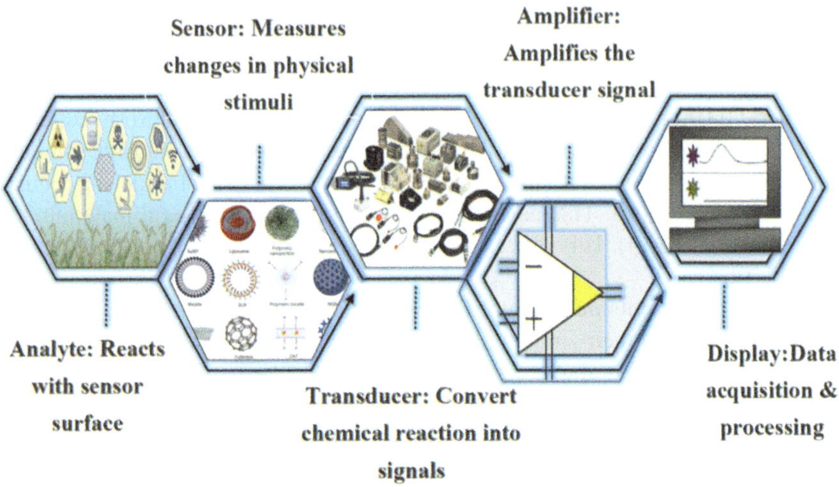

Fig. 1 Different components of a nanosensors

Nanosensors are made from materials like graphene, metal oxide, and carbon nanotubes. They are highly efficient, sensitive, specific, and work quickly. They are used in many areas, such as agriculture, medicine, and industry. In farming, nanosensors help to solve many problems by monitoring and detecting issues at early stage, allowing farmers to take the right actions before problems get worse [2].

The main parts of a nanosensor are the probe, transducer, and detector. In the first step, the substance being measured (the analyte) attaches to the sensor. This causes a reaction, like a change in vibration frequency. In the second step, the sensor sends signals based on this reaction. Finally, these signals are passed to the transducers, completing the process. The transducer, which converts the reaction into signals, is made up of components like thermistors, piezoelectric crystals, optrodes, and electrodes. The signal is amplified with help of a amplifier and send to the display the represent the signal in a readable form [2]. The important components of the nanosensors are shown in Fig. 1.

There are various types of nanosensors used in the agriculure sector for various application. Each sensors have their unique features, utility and limitations. The various kind of naosensors are summarized in Fig. 2.

3.1 Chemical Nanosensors

These sensors are used to detect and monitor certain chemicals, like carbon monoxide and ozone found in pollution, as well as drugs. They can also measure things like

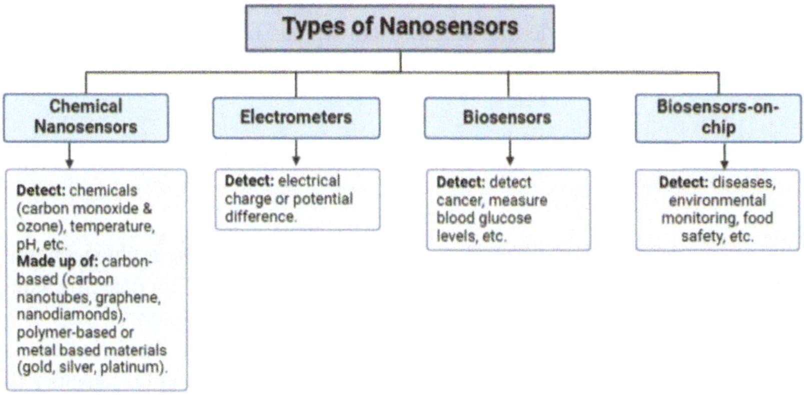

Fig. 2 Different types of nanosensors based on type of transducers and their applications

temperature and pH levels. Chemical nanosensors can be made using different materials, such as carbon-based ones (like carbon nanotubes, graphene, and nanodiamonds), polymer-based materials, or metal-based ones (such as gold, silver, and platinum). These nanosensors are mainly used to detect various chemicals, which is helpful in checking pollution, testing drugs, and analyzing environmental samples. They are also used to measure specific harmful compounds like organophosphorus compounds. Besides that, chemical nanosensors are also used for medical diagnostics. To work, they are combined with devices called transducers, which can detect changes in electricity, light, heat, or pressure. This combination helps in analyzing pharmaceuticals more effectively [2, 24].

3.2 Electrometers

A device called a charge sensor or charge electrometer is used to measure tiny electric charges or currents, including single electrons. There are two ways to measure electrostatic charge. The first one is contacting, where there is direct physical or electrical contact. The second one is non-contacting, where no direct contact is made. Charge sensors are commonly used to directly measure electric charge or very small currents.

Non-contacting devices, like electrostatic voltmeters and electric field meters, have been used for a long time to measure the surface charge of objects without touching them. Highly sensitive charge electrometers are used in many areas, such as mass spectrometry, surface charge analysis, measuring ionization currents in ionization chambers, radiation or particle detection for nuclear studies, and in powder and aerosol technology. In short, electrometers convert the presence of charge or a potential difference on an electrode into an electrical signal. They are mostly used in ion chambers in radiochemistry [2, 17].

3.3 Biosensors

Biosensors work similarly to chemical sensors because both use a physiochemical transducer. Biosensors are commonly used in healthcare, such as for detecting cancer or measuring blood glucose levels.

Nanotechnology is being used more and more in agriculture. Different types of tiny capsules and devices help detect and treat plant diseases. Special sensors called nanobiosensors can be aimed at specific parts of a plant to make sure the treatment goes exactly where it's needed. Nanotechnology is also helpful for cleaning industrial wastewater, filtering water, and helping plants absorb nutrients better. One big benefit of using these tiny technologies is that they reduce the harm to other parts of the plant and cut down the negative impact that chemical fertilizers and pesticides have on the environment. Because nanomaterials have special chemical and physical properties, they work well in many areas, including breeding animals and plants, and even in plant growing processes [2, 12, 35, 38, 44].

3.4 Biosensors-On-Chip

Microfluidic chips with built-in biosensors, also called biosensors-on-chip, are designed for point-of-care (POC) diagnostics, which means they can be used for quick and easy testing. Microfluidics allows these chips to work with small sample sizes and handle multiple tests at the same time. Multiple sensors can also be added to a single chip to make it more useful. Ideally, a biosensor-on-chip should be affordable, small, fast, and highly sensitive.

In short, this device combines microfluidic technology with biosensors, bringing together chemical and biosensor components. It makes the device more compact and improves its ability to detect substances on a single platform [7, 35].

4 Role of Nanosensors in Agriculture

Nanosensors play a major role in agriculture sector. They help in the detection of various agriculture parameter like nutrients present in soil, soil quality, moisture & pH of soil, presence of useful & harmful microorganisms, heavy metals as contaminants or any other pollutants etc. The detailed working is shown in Fig. 3. The detailed application is discussed in the upcoming sections.

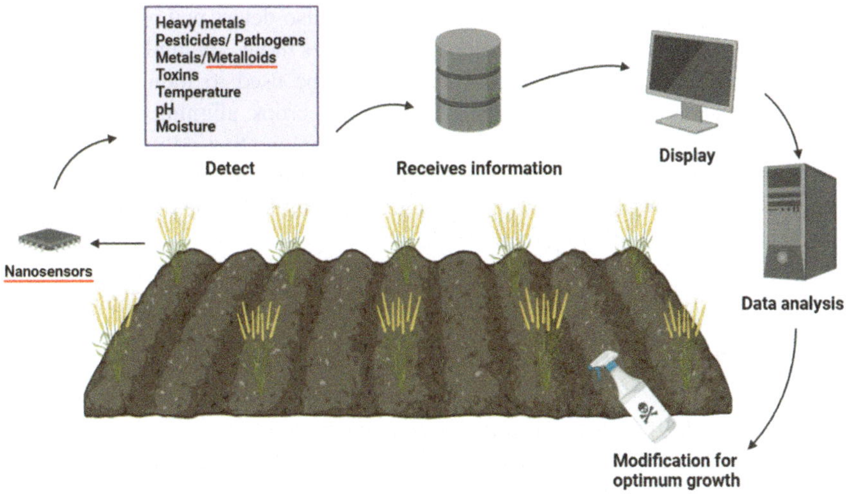

Fig. 3 Detection of different agricultural parameters using nanosensors

4.1 Real-Time Monitoring

One of the key uses of nanotechnology devices is the development of autonomous nanosensors placed throughout farmlands. These sensors work together with GPS to provide real-time and detailed monitoring of crop growth [34]. This technology helps gather important data, such as the best times to plant and harvest crops, leading to improved farming practices. Intel has created chips with nanoscale features and installed wireless sensor nodes, called "motes," throughout a vineyard in Oregon, USA. These sensors take temperature readings every minute, which is a step toward fully automating the vineyard. Crossbow Technologies has developed similar motes for agriculture that can help estimate irrigation needs, detect frost, alert farmers when pesticides are needed, and determine the best time to harvest. They can also be used for bioremediation and measuring water quality. Nanobarcodes and nanoprocessing are other innovations that could be used to track agricultural production. For example, the concept of barcoding used in grocery stores has been adapted to help identify crop diseases. Nanobarcodes have been designed to label different pathogens in a field, and these can be monitored using tools that rely on fluorescence [11].

4.2 Precision Agriculture

Nanotechnology is helping farmers monitor crops and farm conditions in real-time by tracking things like moisture, soil fertility, temperature, crop nutrients, pathogens, and plant diseases. Nanosensors can accurately measure soil conditions, such as pH, nutrients, pesticides, and moisture levels, and also detect pathogens and track nitrogen uptake [34]. This helps farmers use resources more efficiently and supports sustainable farming. These nanosensors can also be used to monitor when and how much water or chemicals should be applied to crops, aligning with the goals of precision agriculture. Researchers are working on creating advanced irrigation systems that deliver water precisely to the roots based on crop needs. This includes improving water storage, controlling how much water is held in the soil, and releasing it according to the plant's requirements [29].

4.3 Smart Nanofertilizers

Monreal et al. [29] suggested that the communication between plant roots and microorganisms in the soil (especially in crops like wheat) is part of a chemical signaling network. This helps release nutrients into the soil. This idea has been used to create smart nanofertilizers that use nanobiosensors to release nutrients like zinc and iron when needed. These smart fertilizers are coated in a special biopolymer film, which controls when the nutrients are released based on signals from the plant roots. Testing these nanofertilizers in different crops and conditions will help improve how efficiently nutrients are used, reducing waste and environmental harm. Nanotechnology experts are also studying how plants control their growth hormones, which affect root growth and seed development. This will help scientists understand how plants adapt to their environment.

4.4 Nano-Chips

Additionally, nanosensors based on microelectromechanical systems (MEMS) have been developed to detect and respond to environmental changes. For example, researchers have created wireless nanosensors with MEMS cantilever beams coated in water-sensitive nanopolymers that help optimize moisture levels. Nano-chips are tiny microarrays made of fluorescent probes that are used to detect genetic changes in bacteria and viruses. These chips are very specific and sensitive, allowing them to identify even small changes in the DNA of these microorganisms. An electrochemical sensor has been created using a screen-printed electrode (SPE) coated with a special agarose/nano-gold membrane and an antibody to detect *Vibrio parahaemolyticus*, a type of bacteria. This antibody is marked with horseradish peroxidase (HRP), which helps in identifying the bacteria. Scientists had designed a biochip sensor with titanium contact pads and a tiny titanium well on a special $LiNbO_3$ surface. When the bacteria were not infected by viruses, small voltage changes were noticed in the well, showing some electrical activity. A nanogold-based sensor using surface plasmon resonance (SPR) technology was used to detect Karnal bunt disease in wheat. In another study, scientists had used a gold electrode modified with copper nanoparticles to detect a fungus called *Sclerotonia sclerotiorum* by monitoring the levels of salicylic acid in oil seeds [20].

4.5 Detection of Pesticides

Nanosensors for detecting pesticide residues have many benefits, including being compact, highly sensitive, having a low detection range, being very selective, and providing quick results. Enzyme biosensors, which measure how pesticides affect enzymes, are one way to check for pesticides. Optical biosensors use techniques like resonant mirrors, surface plasmon resonance (SPR), and waveguides to study how biomolecules interact. Many biosensors have been developed to detect different pesticides. For example, they have been used for pesticides like methyl parathion, parathion, fenitrothion, and more. Some of these sensors work by detecting how certain enzymes, like acetylcholinesterase (AChE), are inhibited by pesticides. A sensor was designed to detect organophosphorus pesticides at very low levels (0.5 nM) using AChE on a modified electrode, which could identify pesticides like paraoxon [45].

Another sensor was made to detect pesticides like methyl parathion and chlorpyrifos by observing how enzyme reactions change the pH near the electrode. A more advanced AChE-based sensor with hollow gold nanospheres was created to detect very small amounts of chlorpyrifos and carbofuran. There are also sensors that use carbon nanotubes (CNTs) to better anchor the AChE enzyme, making it more effective at detecting pesticides in vegetables. Other biosensors use liposomes-tiny fluid-filled pouches-to improve detection. These liposomes, combined with multi-walled carbon nanotubes (MWCNTs), were used to develop an AChE sensor to detect organophosphates. Additionally, a sensor using quantum dots (tiny particles that emit light) was made to detect organophosphates like methyl parathion, with very sensitive detection limits. Quantum dots (QDs) are tiny particles usuaLLY 2–8 nm in size that light up when exposed to a certain light source. They have an inorganic core that determines the color of the light they emit. These QDs are useful for detecting different biomarkers because they emit sharp, bright colors and are more sensitive than regular dyes. Their high resolution makes them perfect for detailed biological studies. For example, an enzyme biosensor using CdS quantum dots was designed to detect the pesticide trichlorfon. Similarly, an electrochemical biosensor was developed using gold nanoparticles to detect pesticides like monocrotophos, carbaryl, and methyl parathion. The use of gold nanoparticles (AuNPs) combined with a sol–gel silicate material made it easier for electrons to move, improving the hydrolysis reaction and increasing the sensitivity of the sensor. A powerful biosensor was created to detect the pesticide monocrotophos by combining the special properties of AuNPs and quantum dots (QDs). A newly developed electrochemical sensor that used both cadmium telluride quantum dots (CdTe QDs) and AuNPs on the same electrode showed greater sensitivity than using either QDs or AuNPs alone. CdTe QDs were also used in a fluorescence-based method to detect 2,4-D (a common herbicide) in very small amounts (250 pg/l). An electrochemical sensor was designed to detect diuron, a type of herbicide. This sensor used a Prussian blue (PB)-AuNP coating, which increased electron movement and improved sensitivity [26]. Another sensor created by Kaushik et al. (2009) used a chitosan nanobiocomposite with iron oxide nanoparticles to detect pesticides like cypermethrin and permethrin quickly, with detection happening in under 50 s. A sensor using the enzyme organophosphate hydrolase (OPH) was developed to detect organophosphate pesticides. The OPH enzyme was attached to tiny silica nanoparticles, which made the sensor stable for up to 60 days and able to detect paraoxon pesticide. Another nanobiosensor was created to detect metsulfuron-methyl, herbicide, by using a specialized atomic force microscopy tip. In 1990, a system called SELEX (Selection Evolution of Ligands by Exponential Enrichment) was developed to create aptamers-synthetic single-stranded nucleic acids that fold into specific shapes, allowing them to bind tightly to their target molecules. These nano-aptamers were used to detect herbicides and pesticides and check for toxicity in food. One aptamer sensor was designed to detect acetamiprid (a pesticide) by combining the enzyme-like activity of AuNPs with the aptamer's ability to bind strongly to acetamiprid. This sensor could detect acetamiprid at very low levels (0.1 ppm) in just 10 min [20]. Another advanced aptamer-based sensor was developed using gold nanoparticle-decorated carbon nanotubes and graphene nanoribbon composites. This sensor was extremely sensitive, with a detection limit as low as 1.7×10^-14 mol/dm^3. It worked by measuring changes in electron resistance when acetamiprid and the aptamer bound to the electrode [20].

5 Role of Nanosensors in Composting

The composting process, which determines the quality of the final product, is mainly driven by the activity of microbes. Understanding these microbes is important for predicting the quality of compost. Since all biological systems, including compost microbes, are made up of nanoscale components like DNA and proteins, compost microbiology has always been studied at the nanoscale. Recently, there's been more interest in using nanotechnology (like nanodevices and nanoparticles) to solve biological problems. By applying nanotechnology during composting, we can better understand, detect, and control microbial processes [5].

The advances in nanotechnology and photolithography, which is used to create tiny patterns on materials, have led to the development of nanochips. These nanochips can be used to study many biological targets at once and can help control the physical and chemical environment during processes like composting. Because they are so small, these systems need less sample material, speed up testing, and use fewer expensive chemicals [42].

One example is Lab-on-a-Chip (LOC), which allows fluids to be controlled and analyzed in tiny amounts. LOC can perform tasks like separating and detecting biological molecules quickly and with high accuracy. This type of system is ideal for studying complex environments like soil because it solves challenges like uneven soil composition and the difficulty of observing soil systems. Since compost is similar to soil, LOC could be useful for studying the microbes in compost and how they interact with plants. Another technology, BioMEMS, works similarly to LOC but has more control and decision-making capabilities. It can help study how cells react to different environments and could be used to improve compost quality, though more research is needed. Other advanced technologies, like Oxford Nanopore (which can sequence genes quickly) and IChip (which can culture hard-to-grow bacteria), may also help with composting research in the future. While these technologies are promising, they are still expensive and not widely available. However, ongoing developments in microfluidic devices for studying genes, proteins, and other cell functions should make them more accessible soon [42].

Using compost without testing its quality can be dangerous for people and the environment because it might contain harmful germs, chemicals, or other hazardous materials. Traditionally, detecting germs takes time with culture-based methods. However, nanobiosensors made from tiny materials like nanoparticles and nanostructures (such as nanocapsules, nanotubes, and quantum dots) can quickly find and measure these microbes. These sensors can detect harmful bacteria like *E. coli*, *Salmonella*, and others in water sources. Nanobiosensors are highly sensitive and can detect even very small amounts of contamination. They have also been used to rapidly find DNA of pathogens through special nanobarcodes. These sensors are becoming more important, especially since bacteria like *Legionella* have been found in compost and linked to illnesses like Legionnaire's disease. This has led to warnings on compost products. Similarly, nanobiosensors can detect plant diseases in compost, such as bacterial spot disease, witches' broom disease in lime, and Rhizomania. This helps make composting safer for plants and people [9, 16, 33, 40].

Nanobiosensors can also detect harmful chemicals like pesticides in compost. For instance, researchers Dasary and Dong developed nanobiosensors that can find organophosphate pesticides in food and water. These sensors are very sensitive, with the ability to detect pesticide levels as low as 0.2 $nmol/dm^3$, which is enough to check compost and take any necessary safety measures. There is very little information about the economic or financial benefits of using nanotechnology on farms or in composting operations. Many researchers have pointed out that the costs of nanotechnology are very high, which could lead to increased production or operation costs. Agrawal and Rathore [3] noted that the full potential of nanotechnology in agriculture and the food industry has not yet been fully realized, it is slowly moving from theory to practical use on farms [3]. Most research in agricultural nanotechnology is still happening in labs and hasn't yet turned into real products that can be used in composting systems [30, 41, 43, 46].

Parisi et al. [32] mentioned that new nanotech products for agriculture are having a hard time reaching the market, keeping agriculture as a minor player in the nanotechnology field. This is mainly because the production costs for these nanotech products are high, and agriculture needs them in large amounts. Additionally, the practical benefits of using nanotechnology are not yet proven [32]. Another report also confirmed that the high costs are partly due to the newness of the technology and the complexity of the equipment and skilled workers needed to create nanoparticles and nanomaterials. Despite the high production costs, it's important to consider how nanotechnology might affect the selling price of compost products at farms [42].

6 Evaluation of Environmental Emissions During Agricultural Waste Recycling

Environmental nanosensors and nanobiosensors are small devices that can monitor important things like moisture, temperature, nutrients, and water levels in real-time and on-site. Developing these real-time monitoring sensors makes them more effective and reliable for use in different fields. There are various types of nanosensors, including chemical sensors, electrical sensors, and biosensors, which are used in many industries. Generally, a nanosensor consists of three main parts; Nanomaterials, these can be single or mixed materials like metals, magnets, quantum dots, or carbon-based materials; Analytes, these are the substances the sensor is designed to detect; Transducer, this part converts the interaction between the analyte and the sensor into a signal that can be easily read. Precision farming is a useful method for increasing crop yields by monitoring environmental conditions and taking specific actions. It involves using nanosensors to manage the use of fertilizers and pesticides effectively, ensuring that they are applied at the right times and in the right amounts. Farmers are particularly interested in detecting pesticides and fertilizers because these substances help protect crops from pests and provide essential nutrients. For example, fertilizers account for about 37.5% of crop productivity. One type of nanosensor is an electrochemical sensor made from zinc oxide (ZnO) and multiwalled carbon nanotubes (MWCNTs), which was used to detect a pesticide called dinosulfon in water. Researchers found that this combination is effective for sensors and can improve their performance. The sensor works well at a pH of 8.0. Another type is a fluorescence nanosensor made from cadmium selenide/zinc sulfide (CdSe/ZnS) quantum dots. This sensor successfully detected harmful compounds like anthracene, phenanthrene, and pyrene in water, with a very low detection limit of 5 nmol [23].

A study by Sahoo et al. tested a zinc oxide quantum dot (ZnO_2 QD) nanosensor for detecting and breaking down four commonly used pesticides: aldrin, tetradifon, glyphosate, and atrazine. They found that the quantum dots formed strong bonds with aldrin and tetradifon because of a special group in the quantum dots that can replace certain parts of these pesticides. For glyphosate, the binding happened through ionic interactions [37]. Imazapyr is herbicide often used to control grass growth in corn, sunflower, and sugarcane fields. Kumar et al. created a nanosensor coated with APTES and Yb_2O_3 to detect imazapyr. They confirmed that the sensor was sensitive and selective using simple fluorescence tests. The sensor works through interactions between the carboxylic group of imazapyr and the amine group of APTES. There's also a type called APTA-nanosensors, which are used for detecting various pesticides.

Verdian summarized the use of these sensors for detecting acetamiprid [22]. Environmental pollution from heavy metals is a major global problem because it affects our natural resources, such as biodiversity and agricultural productivity, as well as our health. To address this issue, it's important to test water quality for drinking and various industrial uses. Nanosensors can help monitor several key factors, including; physical factors such as temperature, turbidity, and conductivity; chemical factors such as pH levels, heavy metals, organic substances, dissolved oxygen, and nitrates; Biological factors such as biological oxygen demand and bacterial content [23]. One example is a carbon dot-based nanosensor that was developed to detect chromium (VI) in water. It has a very low detection limit of 2.3 nmol, making it effective. This nanosensor is advantageous because it is not toxic and remains stable when exposed to light [23].

Another study by Maric et al. introduced platinum-halloysite nanoclay nanojets as a nanosensor for detecting heavy metals like cadmium (Cd^{2+}), zinc (Zn^{2+}), lead (Pb^{2+}), and mercury (Hg^{2+}). The nanosensor is more effective at detecting certain metals, with the order of effectiveness being mercury > lead > zinc > cadmium. This is because mercury and lead have higher electronegativity, which means they stick strongly to the surface of the platinum nanojets [27].

7 Limitations of Nanosensors

Nanotechnology is very useful, but it also comes with some concerns and limitations. Some nanoparticles, which are tiny particles used in nanotechnology, can be harmful to humans, animals, plants, and the environment because they might not dissolve and could be toxic. For example, Gaurab and Dattatrya [13] mentioned that people are still worried about the safety of using nanotechnology in medicine, especially for diagnosing and treating diseases. One big concern is that these tiny particles might harm living systems by causing oxidative stress, inflammation, immune system damage, genetic damage, or even cancer. This is because many engineered nanoparticles are not compatible with biological systems [13].

As nanotechnology grows, there's also worry about these tiny materials getting into the environment and food chain. Engineered nanomaterials, which are often made of heavy metals, can damage the soil, plants, and water. Agrawal [3] categorized common nanoparticles into five types: carbon-based particles, metal oxides, quantum dots, zero-valent metals, and nanopolymers. Because these particles are so small, they can stick to plant roots, be absorbed, and build up inside plants. This can harm plants chemically and physically. Humans are at risk too because these nanoparticles can stay in the environment, enter the food chain, and be eaten by people and animals. Studies also show that toxic nanoparticles can float in the air and enter the body through the skin or by breathing them in [4].

8 Conclusion and Future Prospects

To address the global problem of environmental pollution and promote sustainable agriculture practice recycling of agricultural waste is necessary. The incorporation of nanotechnology and nanosensors in particular promotes effective resource management and waste valorization. A wide variety of nanosensors are available that are highly precise, sensitive and offers real-time monitoring that helps in effective agro-waste recycling. They improvise the methods of farming leading to waste reduction and high productivity by detecting soil quality, plant health, soil moisture, pollutant and contaminants, presence of useful and harmful microbes, soil fertility etc. This helps optimal use of pesticides and fertilizers are possible that increases crop production and waste reduction. Additionally, they supports the conversion of agro-waste into valuable products like biogas, compost, bio fuels. However, in present time the real world application of nanosensors are limited due to certain challenges associated with them. Scientist are working to address these limitations that are economic viability, scalability, durability, sensitivity in complex agricultural environments so that small scale farmers can also take their benefit. Researchers are also embedding Internet of Things (IoT) in nanosensors that will help in real-time monitoring, continuous feedback, remote monitoring, high precision that increases efficient waste recycling. The present situation also demands for development of environment friendly and biodegradable nanosensors to reduce nanotoxicity to promote sustainability. Standard Government policy and regulations needs to established. This will ensure safety and accessibility to all the farmers working in different scale. To conclude, nanosensors has improved sustainable agricultural practice and play pivotal role in recycling agro-waste. By working on current challenges and formulating strong policies pave path for sustainable and productive farming.

Competing Interests The authors declare that they have no competing interest.

Declaration of generative AI and AI-assisted technologies in the writing process During the preparation of this work the book chapter author(s) have not taken any assistance from AI, and the book chapter author(s) reviewed and edited the content as needed and take(s) full responsibility for the content of the publication. All authors read and approved the final version of the manuscript.

References

1. Abdullah NA, Rani MSA, Mohammad M, Sainorudin MH, Asim N, Yaakob Z, Razali H, Emdadi Z (2021) Nanocellulose from agricultural waste as an emerging nanotechnology material for nanotechnology applications-an overview. Polim/Polym 66(3):157–168
2. Agrawal S, Prajapati R (2012) Nanosensors and their pharmaceutical applications: a review. Int J Pharm Sci Technol 4:1528–1535
3. Agrawal S, Rathore P (2014) Nanotechnology pros and cons to agriculture: a review. Int J Curr Microbiol App Sci 3(3):43–55
4. Ani FN (2016) Utilization of bioresources as fuels and energy generation. In: Electric renewable energy systems. Elsevier Inc, pp 140–155

5. Baalousha M, Cornelis G, Kuhlbusch TAJ, Lynch I, Nickel C, Peijnenburg W, Van Den Brink NW (2016) Modeling nanomaterial fate and uptake in the environment: current knowledge and future trends. Environ Sci Nano 3(2):323–345
6. Börjesson M, Westman G (2015) Crystalline nanocellulose—preparation, modification, and properties. In: Cellulose: fundamental aspects and current trends. pp 7
7. Chen S, Shamsi MH (2017) Biosensors-on-chip: a topical review. J Micromech Microeng 27(8):083001
8. Dos Santos RM, Neto WPF, Silvério HA, Martins DF, Dantas NO, Pasquini D (2013) Cellulose nanocrystals from pineapple leaf: a new approach for the reuse of this agro-waste. Ind Crops Prod 50:707–714
9. Duhan JS, Kumar R, Kumar N, Kaur P, Nehra K, Duhan S (2017) Nanotechnology: the new perspective in precision agriculture. Biotechnology Reports 15:11–23
10. Dungani R, Karina M, Subyakto S, Sulaeman A, Hermawan D, Hadiyane A (2016) Agricultural waste fibers towards sustainability and advanced utilization: a review
11. El Beyrouthya M, El Azzi D (2014) Nanotechnologies: novel solutions for sustainable agriculture. Adv Crop Sci Technol 2(3):8863
12. Fan X, White IM, Shopova SI, Zhu H, Suter JD, Sun Y (2008) Sensitive optical biosensors for unlabeled targets: a review. Anal Chim Acta 620(1–2):8–26
13. Gaurab R, Dattatrya S, Amit Y, Gopal CK (2015) Nanomedicine. In: Handbook of research on diverse applications of nanotechnology in biomedicine, chemistry, and engineering. IGI Global, pp 64–89
14. Hamdy AS, Gomaa M, Ali AM (2021) Waste recycling technologies for nanomaterials manufacturing. In: Topics in mining, metallurgy and materials engineering. Springer
15. Hornsby P, Mohamad M, McMullan C, Sharma S, Carmichael E (2012) Nanocellulose-reinforced polymers derived from banana trees and other fiber sources. In: ECCM15–15th European conference on composite materials, Venice, Italy. pp 24–28
16. Jahanban L, Davari M (2014) Organic agriculture and nanotechnology. Build Org Bridg 3:679–682
17. Jalil J, Zhu Y, Ekanayake C, Ruan Y (2017) Sensing of single electrons using micro and nanotechnologies: A review. Nanotechnol 28(14)
18. Jonoobi M, Oladi R, Davoudpour Y, Oksman K, Dufresne A, Hamzeh Y, Davoodi R (2015) Different preparation methods and properties of nanostructured cellulose from various natural resources and residues: a review. Cellulose 22:935–969
19. Kampeerapappun P (2015) Extraction and characterization of cellulose nanocrystals produced by acid hydrolysis from corn husk. J Metals, Mater Miner 25(1):19–26
20. Kaushal M, Wani SP (2017) Nanosensors: frontiers in precision agriculture. In: Prasad R, Kumar M, Kumar V (eds) Nanotechnology: an agricultural paradigm. Springer Singapore, pp 279–291
21. Kumar A, Negi YS, Choudhary V, Bhardwaj NK (2014) Characterization of cellulose nanocrystals produced by acid hydrolysis from sugarcane bagasse as agro-waste. J Mater Phys Chem 2(1):1–8
22. Kumar V, Guleria P, Mehta SK (2017) Nanosensors for food quality and safety assessment. Environ Chem Lett 15:165–177
23. Kumar V, Guleria P (2020) Application of DNA-nanosensor for environmental monitoring: recent advances and perspectives. Curr Pollut Rep 1–21
24. Kurbanoglu S, Bakirhan NK, Shah A, Ozkan SA (2019) Chemical nanosensors in pharmaceutical analysis. In: Ozkan SA, Shah A (eds) New developments in nanosensors for pharmaceutical analysis. Academic Press, pp 141–170
25. Lani NS, Ngadi N, Johari A, Jusoh M (2014) Isolation, characterization, and application of nanocellulose from oil palm empty fruit bunch fiber as nanocomposites. J Nanomater 1:702538 (2014)
26. Lisa M, Chouhan RS, Vinayaka AC, Manonmani HK, Thakur MS (2009) Gold nanoparticles-based dipstick immunoassay for the rapid detection of dichlorodiphenyltrichloroethane: an organochlorine pesticide. Biosens Bioelectron 25(1):224–227

27. Maric T, Mayorga-Martinez CC, Nasir MZM, Pumera M (2019) Platinum–halloysite nanoclay nanojets as sensitive and selective mobile nanosensors for mercury detection. Adv Mater Technol 4(2):1800502
28. Mishra RK, Sabu A, Tiwari SK (2018) Materials chemistry and the futurist eco-friendly applications of nanocellulose: status and prospect. J Saudi Chem Soc 22(8):949–978
29. Monreal CM, DeRosa M, Mallubhotla SC, Bindraban PS, Dimkpa C (2016) Nanotechnologies for increasing the crop use efficiency of fertilizer-micronutrients. Biol Fertil Soils 52:423–437
30. Nichols D, Cahoon N, Trakhtenberg EM, Pham L, Mehta A, Belanger A, Kanigan T, Lewis K, Epstein S (2010) Use of ichip for high-throughput in situ cultivation of "uncultivable" microbial species. Appl Environ Microbiol 76(8):2445–2450
31. Norrahim MNF, Ariffin H, Yasim-Anuar TAT, Ghaemi F, Hassan MA, Ibrahim NA, Ngee JL, Yunus WMZW (2018) Superheated steam pretreatment of cellulose affects its electrospinnability for microfibrillated cellulose production. Cellulose 25:3853–3859
32. Parisi C, Vigani M, Rodríguez-Cerezo E (2015) Agricultural nanotechnologies: what are the current possibilities? Nano Today 10(2):124–127
33. Priyanka S, Shashank PS, Prashant SP, Singh KP (2014) Nanotechnology and its role in pathogen detection: a short review
34. Prasad R, Bhattacharyya A, Nguyen QD (2017) Nanotechnology in sustainable agriculture: Recent developments, challenges, and perspectives. Front Microbiol 8:1014
35. Rai V, Acharya S, Dey N (2012) Implications of nanobiosensors in agriculture
36. Rani MSA, Rudhziah S, Ahmad A, Mohamed NS (2014) Biopolymer electrolyte based on derivatives of cellulose from kenaf bast fiber. Polymers 6(9):2371–2385
37. Sahoo M, Vishwakarma S, Panigrahi C, Kumar J (2021) Nanotechnology: current applications and future scope in food. Food Frontiers 2(1):3–22
38. Shawon ZBZ, Hoque ME, Chowdhury SR (2020) Nanosensors and nanobiosensors: Agricultural and food technology aspects. In: Nanofabrication for smart nanosensor applications. Elsevier, pp 135–161
39. Shehrawat PS, Sindhu N (2015) Agricultural waste utilization for healthy environment and sustainable lifestyle. Ann Agric Biol Res 20:110–114
40. Shyla KK, Natarajan N, Nakkeeran S (2014) Antifungal activity of zinc oxide, silver, and titanium dioxide nanoparticles against. Macrophomina phaseolina
41. Simpson JT, Workman R, Zuzarte PC, David M, Dursi LJ, Timp W (2016) Detecting DNA methylation using the Oxford Nanopore Technologies MinION sequencer. BioRxiv 047142
42. St Martin CCG, Rouse-Miller J, Vilpigue P, Rampersaud R (2020) Application of nanotechnology to research on the microbiology of composting. In: Meghvansi MK, Varma A (eds) Biology of composts. Springer International Publishing, pp 203–214
43. Timm AC, Shankles PG, Foster CM, Doktycz MJ, Retterer ST (2016) Toward microfluidic reactors for cell-free protein synthesis at the point-of-care. Small 12(6):810–817
44. Velasco-Garcia MN (2009) Optical biosensors for probing at the cellular level: a review of recent progress and future prospects. Semin Cell Dev Biol 20(1):27–33
45. Weerathunge P, Ramanathan R, Shukla R, Sharma TK, Bansal V (2014) Aptamer-controlled reversible inhibition of gold nanozyme activity for pesticide sensing. Anal Chem 86(24):11937–11941
46. Woyke T, Rubin EM (2014) Searching for new branches on the tree of life. Science 346(6210):698–699
47. Wulandari WT, Rochliadi A, Arcana IM (2016) Nanocellulose prepared by acid hydrolysis of isolated cellulose from sugarcane bagasse. IOP Conf Ser: Mater Sci Eng 107(1):012045
48. Zhang X, Han G, Jiang W, Zhang Y, Li X, Li M (2016) Effect of steam pressure on chemical and structural properties of kenaf fibers during the steam explosion process. BioResources 11(3)

The manufacturer's authorised representative in the EU is Springer Nature Customer Service Centre GmbH, Europaplatz 3, 69115 Heidelberg, Germany. If you have any concerns regarding our products, please contact ProductSafety@springernature.com

Printed and bound by CPI Group (UK) Ltd, Croydon, CR0 4YY
30/09/2026
02225303-0014